Dying and Dead Seas
Climatic Versus Anthropic Causes

NATO Science Series

A Series presenting the results of scientific meetings supported under the NATO Science Programme.

The Series is published by IOS Press, Amsterdam, and Kluwer Academic Publishers in conjunction with the NATO Scientific Affairs Division

Sub-Series

I. Life and Behavioural Sciences	IOS Press
II. Mathematics, Physics and Chemistry	Kluwer Academic Publishers
III. Computer and Systems Science	IOS Press
IV. Earth and Environmental Sciences	Kluwer Academic Publishers
V. Science and Technology Policy	IOS Press

The NATO Science Series continues the series of books published formerly as the NATO ASI Series.

The NATO Science Programme offers support for collaboration in civil science between scientists of countries of the Euro-Atlantic Partnership Council. The types of scientific meeting generally supported are "Advanced Study Institutes" and "Advanced Research Workshops", although other types of meeting are supported from time to time. The NATO Science Series collects together the results of these meetings. The meetings are co-organized bij scientists from NATO countries and scientists from NATO's Partner countries – countries of the CIS and Central and Eastern Europe.

Advanced Study Institutes are high-level tutorial courses offering in-depth study of latest advances in a field.
Advanced Research Workshops are expert meetings aimed at critical assessment of a field, and identification of directions for future action.

As a consequence of the restructuring of the NATO Science Programme in 1999, the NATO Science Series has been re-organised and there are currently five sub-series as noted above. Please consult the following web sites for information on previous volumes published in the Series, as well as details of earlier sub-series.

http://www.nato.int/science
http://www.wkap.nl
http://www.iospress.nl
http://www.wtv-books.de/nato-pco.htm

Series IV: Earth and Environmental Sciences – Vol. 36

Dying and Dead Seas
Climatic Versus Anthropic Causes

edited by

Jacques C.J. Nihoul

University of Liège,
Belgium

Peter O. Zavialov

P.P. Shirshov Institute of Oceanology,
Moscow, Russia

and

Philip P. Micklin

Western Michigan University,
Kalamazoo, MI, U.S.A.

Kluwer Academic Publishers

Dordrecht / Boston / London

Published in cooperation with NATO Scientific Affairs Division

Proceedings of the NATO Advanced Research Workshop on
Dying and Dead Seas – Climatic Versus Anthropic Causes
Liège, Belgium
7–10 May 2003

A C.I.P. Catalogue record for this book is available from the Library of Congress.

ISBN 1-4020-1901-7

Published by Kluwer Academic Publishers,
P.O. Box 17, 3300 AA Dordrecht, The Netherlands.

Sold and distributed in North, Central and South America
by Kluwer Academic Publishers,
101 Philip Drive, Norwell, MA 02061, U.S.A.

In all other countries, sold and distributed
by Kluwer Academic Publishers,
P.O. Box 322, 3300 AH Dordrecht, The Netherlands.

Printed on acid-free paper

Printed in the Netherlands.

TABLE OF CONTENTS

PREFACE

There are incentive indications that the growth of human population, the increasing use and abuse of natural resources combined with climate changes (probably due to anthropic pollution, to some extent) exert a considerable stress on closed (or semi-enclosed) seas and lakes.

In many regions of the world, marine and lacustrine hydrosystems are (or have been) the object of severe or fatal alterations, from changes in regional hydrological regimes and/or modifications of the quantity or the quality of water resources associated with (*natural or man-made*) land *reclamation*, deterioration of geochemical balances (increased salinity, oxygen's depletion ...), mutations of ecosystems (eutrophication, dramatic decrease in biological diversity ...) to geological disturbances and to the socio-economic perturbations which have been – or may be in the near future – the consequences of them.

Seas and lakes are dying all over the world and some may be regarded as already dead and there is an urgent need to try to understand how this is happening and identify the causes of the observed mutations, weighing the relative effects of climatic evolution and anthropic interferences.

This book is the outcome of the NATO Advanced Research Workshop, held in Liège in May 2003. The Workshop was organized at the University of Liège as a follow on meeting to the 35th International Liège Colloquium on Ocean Dynamics, dedicated in 2003 to *Dying and Dead Seas*. The book contains the synthesis of the lectures given by 16 main speakers during the ARW.

The major objective of the Workshop was to assemble a dedicated group of *invited* and active prominent scientists from various countries and all disciplines (mathematical modelling, physics, chemistry, biology, geology, socio-economy and human sciences, including historical studies of past events and ancient civilizations) to foster a mutually beneficial exchange of information, opening on to a survey of major recent discoveries, essential mechanisms, impelling question marks and valuable recommendations for future research and management decisions.

The case studies presented range from the seas and lakes of Central Asia to Australian, African and American enclosed seas or lakes and to the Dead Sea. These studies provide a time-space synoptic view of dying and endangered seas and lakes, all over the world, contrasting what is happening now, assessing the respective roles of natural and anthropic causes, with what happened in the past and what one might expect to happen in the future.

The main support for the Advanced Research Workshop was provided by NATO under the auspices of the Scientific and Environmental Affairs Division. We gratefully acknowledge the NATO panel and especially the Program Director, Environmental and Earth Science and Technology, Dr. Alain M. Jubier for their support and constant assistance in the funding and organization of the ARW.

Professor Jacques C.J. Nihoul

Chapter 1

WHAT DO WE KNOW ABOUT DEAD, DYING AND ENDANGERED LAKES AND SEAS?

Andrey G. Kostianoy and Peter O. Zavialov

P.P. Shirshov Institute of Oceanology, Russian Academy of Sciences, 36 Nakhimovsky Ave., Moscow, 117997, Russia

Sergey A. Lebedev

Geophysical Center, Russian Academy of Sciences, 3 Molodezhnaya Str., Moscow, 117236, Russia

1. INTRODUCTION

Shallowing and degradation of some freshwater and salt lakes and inland seas are among major environmental problems at the beginning of the century. 96.5% of the planet's water is salty water in oceans and seas. Nearly 70% of the world's fresh water is frozen in the Antarctic and Greenland ice sheets, glaciers, and permanent snow cover and ice. About 30% of all fresh water is groundwater. Lakes and rivers contain only 0.25% of all fresh water. All over the globe, water is being diverted for industrial, agricultural and household uses. The United Nations announced that about 3 billion people would face severe water shortages by 2025 if the present consumption rates persist. More than five million people die every year from water-related diseases such as cholera and dysentery. Significant desiccation, salinization and water quality degradation have taken place in many inland water bodies over the past century. The most striking examples are: (1) the Lobnor Lake in China which completely dried up by 1972; (2) the Dead Sea whose level has dropped by 14 m since 1977 and whose present salinity is about 340 g/l;

J.C.J. Nihoul et al. (eds.), Dying and Dead Seas, 1–48.

(3) the Aral Sea whose level has dropped by about 23 m and the salinity increased eightfold since 1960; and (4) Lake Chad which has shrunk to about a twentieth of its size of 1963.

The examples above are the most well-known. However, the complete list of endangered lakes and seas is much longer. Even the world's largest freshwater system, namely the Great Lakes, is shrinking. The 2002 aggregate level of the five Great Lakes was the lowest in more than 30 years. The Superior and Huron Lakes are near their record lows. In each specific case, the water level drop may be a result of either natural climate variability or anthropogenic impact, or a combination of the two factors. Thus, an important goal of research is to assess relative contributions from the two mechanisms. The former mechanism is connected with global warming, lower precipitation and smaller snow cover in the basins of lake feeding rivers, higher air temperatures resulting in increased rates of evaporation and transpiration from plants. The latter one is manifested through anthropogenic diversions of water from inflowing rivers. In many cases, most of the water withdrawn for agriculture and industrial uses eventually returns to the lake system in the recharged groundwater flow. However, this is not always so. For example, Chicago sends its 8 billion liter a day draw on taken from the Great Lakes system to the Mississippi River and not back to the lakes (Mitchell, 2002). Uncontrolled fresh water consumption and irrigation pose a serious threat to systems of lakes, inland seas and even oceanic shelves adjacent to river mouths. The waters of the Amu Darya River in Uzbekistan and the Yellow River in China, withdrawn by farmers, industries and cities, have practically failed to reach the Aral and Yellow Seas, respectively, for many years during the past decade. In North America, not only does the Colorado River barely reach the Gulf of California, but last year even the Rio Grande River has dried before it could merge with the Gulf of Mexico. Similar situation often takes place with Nile and Ganges, for example (Montaigne, 2002).

The consequences of desiccation of inland seas and lakes can be grouped in four interrelated categories:

1). **Climatic consequences:** local climate change (air and water temperature, humidity, precipitation regime, wind systems), increase of climate continentality, desertification, salt and dust storms, etc;

2). **Ecological consequences:** decay of the sea/lake and river delta ecosystems, loss of biodiversity, pollution and salinization of soils;

3). **Economical and social consequences**: decay or collapse of fisheries, fishery-related industries and sea transport, tourism, decrease of agricultural productivity, growth of unemployment;

4). **Health consequences:** poor potable water quality, increase of related disease incidence (e.g., cholera, typhus, gastritis, blood cancer), increase of respiratory system disease incidence, congenital defects and high infant mortality.

In this descriptive article, we address several examples of critical inland water bodies all over the world. Here, the term "critical" means that the lake or inland sea is facing either a severe anthropogenic pressure in any form or a rapid change of its physical conditions due to climate change. In particular, we present data on the sea/lake level variability derived from the TOPEX/Poseidon satellite altimetry analysis for 1992-2002 for the Aral and Caspian Seas, Kara Bogaz-Gol Bay, and lakes Sarykamysh, Balkhash, Issyk-Kul, Chad, Victoria, Tanganyika, Rukwa, Tana, and others. For many of these water bodies, direct level measurements have been very sparse or absent. A part of the altimetry data presented in this paper, as well as some hydrographic data on the Aral and Caspian Seas, are original results by the authors. However, this article is intended as a review rather than a research paper. The main part of this report is therefore a compilation made of information published in literature and on the Internet as cited in the reference list.

This review was presented as a report at the 35th International Liege Colloquium on Ocean Dynamics and the NATO Advanced Research Workshop *"Dying and Dead Seas"* held on 5-10 May 2003 in Liege, Belgium. This meeting was a unique forum of scientists involved in the research of the Aral, Dead, Caspian Seas, as well as other critical lakes in USA, Africa, Central Asia, China, and Australia. The review is intended to highlight regional peculiarities of this world-wide problem. Some critical water bodies of Asia are addressed in Section 2, and those of Africa in Section 3. American and Australian critical lakes are discussed in Sections 4 and 5, respectively. Some generalizing conclusions are drawn in Section 6.

2. CRITICAL SEAS AND LAKES OF ASIA

The largest lakes in Central Asia (Fig.1) are the Caspian Sea (in 1970 sea surface was about 370,000 km^2 and maximum depth – 1025 m), the Aral Sea (64,500 km^2 and 69 m in 1960), Lake Balkhash (18,200 km^2, 26 m) and Lake Issyk-Kul (6,300 km^2, 702 m) (Mandych, 1995; Golubev, 1997).

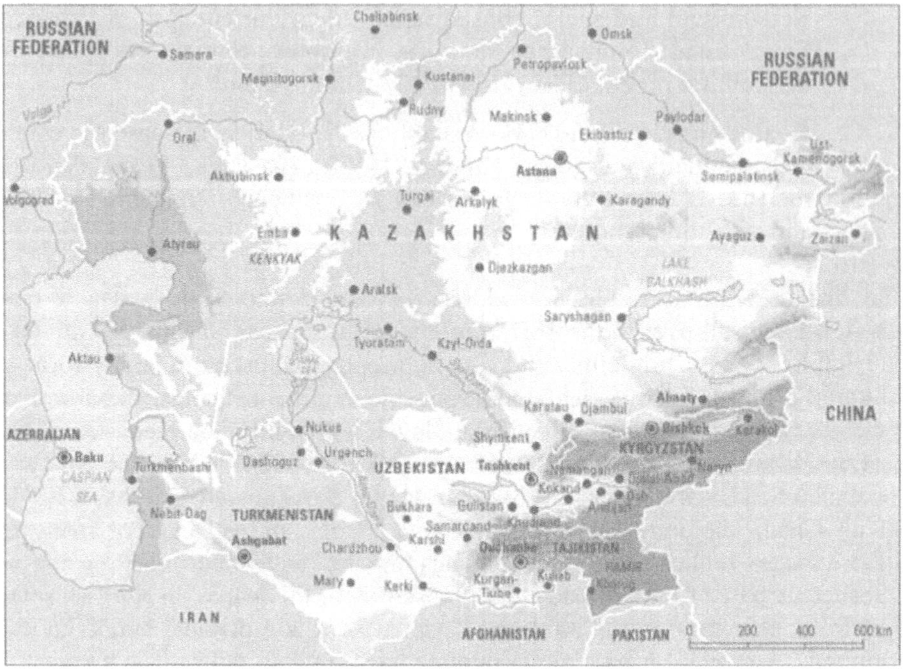

Figure 1. Map of Central Asia.

The Asia's closed and terminal lakes are located in arid or semi-arid regions and therefore sensitive to changes in water balance. Thus, the lake levels and salinity react quickly on any regional climate change or anthropogenic pressure. Unsustainable irrigation in lake basins has already led to very serious environmental and social-economic problems. The lakes will be particularly vulnerable to any future rainfall reduction or temperature increase. We note that according to some global warming studies, e.g., the report published by the Tyndall Centre for Climate Change Research at the University of East Anglia in 2002, the future climate change in Central Asia is likely to be more abrupt than that in other regions.

2.1. Aral Sea

The Aral Sea (Fig.2, 3) is one of the most striking examples of what unsustainable use of water can do to aquatic ecosystems (Micklin, 1988; Glazovsky, 1995a,b; Micklin and Williams, 1996). Once the fourth largest inland water body with the surface area of over 66,000 km^2, the total volume of nearly 1,070 km^3 and the maximum depth 69 m, in 1960, the Aral Sea

Figure 2. Survey of the Sea of Aral by Commander A. Butakoff, Imperial Russian Navy, in 1848-1849.

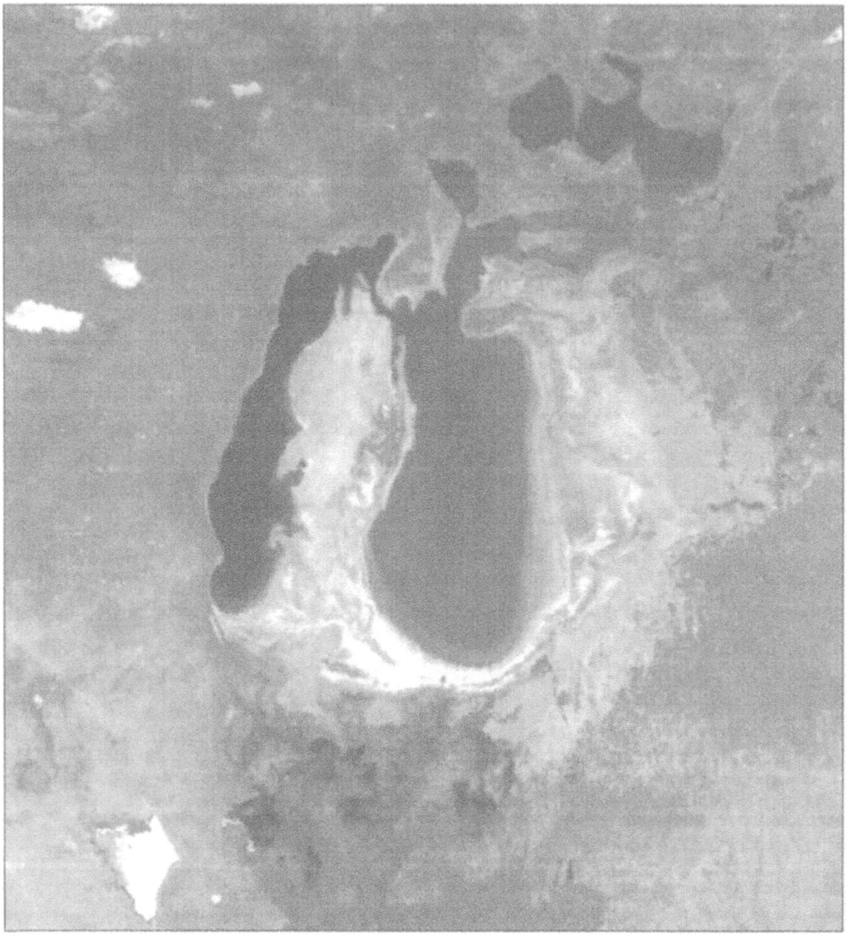

Figure 3. The MODIS image of the Aral Sea on 9 November 2002.

was about the size of the Netherlands and Belgium taken together. Many fish species were living in the brackish (mean salinity about 10 g/l) water, 12 of them were very important for fisheries (yearly catches of 44,000 tons on average). An economy tightly related to the sea (fisheries, sea transport) was formed in the Aral Sea Region. But over the past 40 years, the freshwater discharge into the Aral Sea from the Amu-Darya and Syr-Darya rivers, formerly totaling to 50 km^3 on average, has been decreasing because of diversions for irrigation and ceased almost completely. As a result, the sea surface level (Fig.4) has dropped by almost 23 m (in the Large Aral), the sea has shrunk by a factor of five in its original size, factor of ten in its volume, the salinity exceeded 82 g/l in the western Aral Sea and is even higher in the

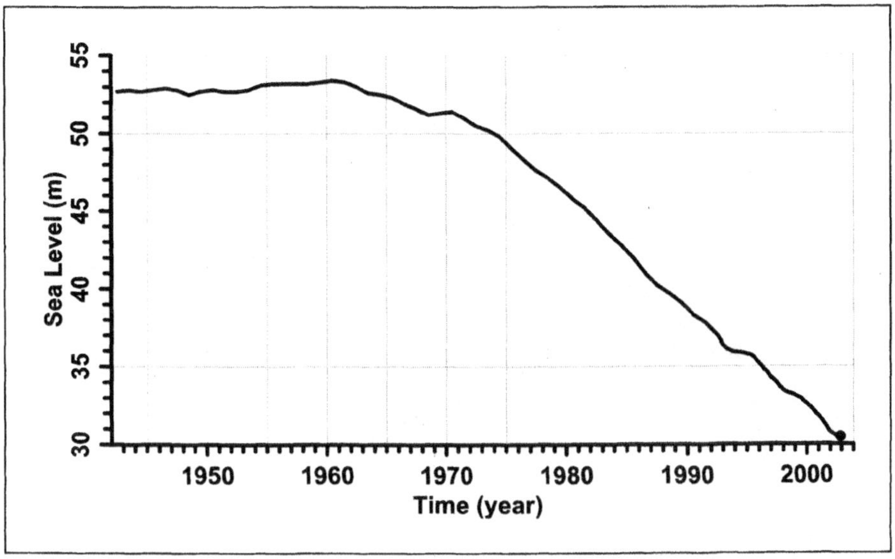

Figure 4. Time variation of the Aral Sea level (1940-2000).

eastern part. Today, the Aral Sea is divided in three almost separate parts (Fig.3, 5). In 1989, the northernmost part and the main body of the sea separated, forming two individual lakes known as the Small Sea and the Large Sea. Presently the Large Sea consists of two distinct basins connected through a narrow (~ 3 km) and shallow (~ 3 m) channel (Fig.5). The western basin (65 km^3) is a trench with a steep bottom slope at the western side where the maximum depth still exceeds 40 m. The eastern basin is a relatively large but shallow with the maximum depth of about 7 m, the estimated total volume of 25 km^3, and salinity of 160 g/l.

The desiccation and salinization of the sea have led to the degradation of the endemic ecosystem and had severe impact on the quality of life and health of the local population. Cancers, lung, kidney and liver diseases and infant mortality (10%) are 30 times higher than they used to be. The local economy is in a deep crisis. Another potential ecological problem for the Aral Sea Region is the former Vozrozhdeniye Island, now a peninsula (Fig.3), which used to be the World's largest bacteriological weapon testing site.

The Aral Sea shallowing is believed to have triggered strong negative impacts on the regional climate, including, but not limited to, the desertification processes. It is of great importance to formulate a sound forecast of the future development of the Aral Sea conditions. Such a forecast could help to elaborate measures aimed to ease the environmental

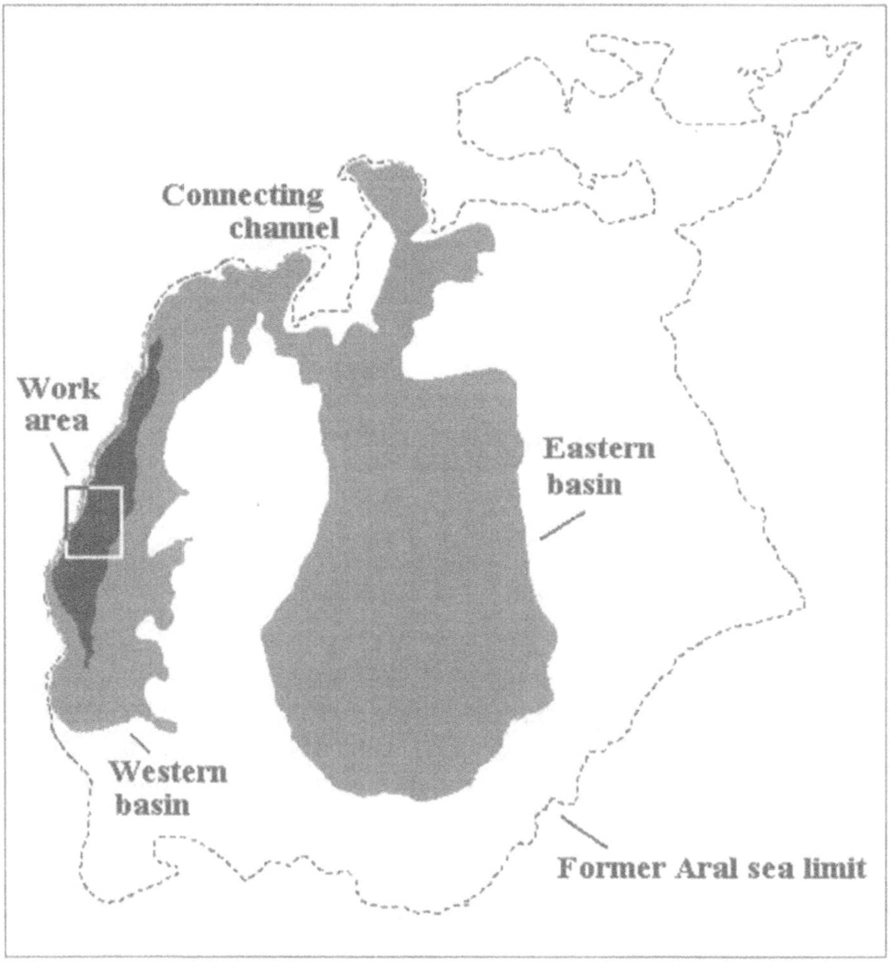

Figure 5. Map of the Aral Sea in November 2002 and the work area. The black shading indicates the horizontal extent of the discovered hydrogen sulphide contamination zone.

and social impacts. A few models focused on this objective have been developed (e.g., Bortnik and Chistyaeva, 1990, Small et al., 2001a,b). Generally, the model predictions released a decade ago or earlier turned out to be inaccurate and the Aral Sea shallowing has actually progressed faster than it had been expected. Obviously, models depend heavily on the "ground truth" data for their correct realization and validation. Such data should include, for example, the Aral Sea salinity and temperature distributions, the density stratification, the chemical composition of salts in the water, etc. The density stratification and depth, in particular, largely control mixing and hence the sea surface temperature (SST) (Ginzburg et al., 2002, 2003). In turn, the summer SST determines the evaporation rate as the key predictor parameter. Despite the growing international interest towards the Aral problem, few *in situ* hydrological, hydrochemical or biological observations have been made in the Aral Sea since early 1990s (Kostianaya et al., 2002). Apart from the known economical and political issues connected with the collapse of the former Soviet Union, this is because the Aral Sea in its present limits is physically difficult to access. For these reasons, the vertical distributions of the thermohaline parameters and the density stratification is poorly known. Recent field surveys gave new interesting results (e.g., Zavialov et al., 2003a,b, see also Friedrich et al., this volume, Mirabdullaev et al., this volume). In the Shirshov Institute expedition of November 2-19, 2002, a hydrophysical and hydrobiological survey in the Uzbekistan deep western part of the Aral Sea was conducted (Fig.5). Using a motor boat, surface-to-bottom CTD profiling was conducted and numerous water samples at different depth levels were collected. A strong vertical stratification in temperature, salinity and density with the upper 20 m well mixed layer (10.6°C, 82 g/l) was observed. Deeper a strong halocline with the salinity increase up to nearly 95 g/l at the bottom was accompanied by a temperature inversion of 4°C. Salinity was almost 30-40 g/l higher than the predicted values. An important consequence of the effective isolation of the lower level from the exchanges with the atmosphere by strong density stratification is the lack of ventilation resulting in the hydrogen sulphide contamination. All water samples taken at depths below 22 m contained hydrogen sulphide. Finally, the absolute lake surface level 30.47 m above the ocean level was directly determined. This value is some 4 m below the predictions published in 1990 (Bortnik and Chistyaeva, 1990).

2.2. Dead Sea

The Dead Sea is a large and deep salty lake located in the Jordan Rift Valley between Israel and Jordan (Fig.6, 7). It is the lowest land point on the Earth surface, and the Dead sea waters are probably the saltiest and densest lake/sea waters in the world. The Dead Sea is a terminal lake, and its level is fully determined by the balance between river runoff, precipitation and evaporation. Historical records indicate that the sea level has oscillated significantly (Fig.8). Since 1960s, the hydrological regime of the Dead Sea has been strongly influenced by anthropogenic pressure, namely by use of the Dead Sea watershed by Israel, Syria and Jordan (e.g., Gertman and Hecht, 2002). As a result, the evaporation exceeded fresh water inflow to the sea. Moreover, Israel and Jordan are using the sea waters for mineral production

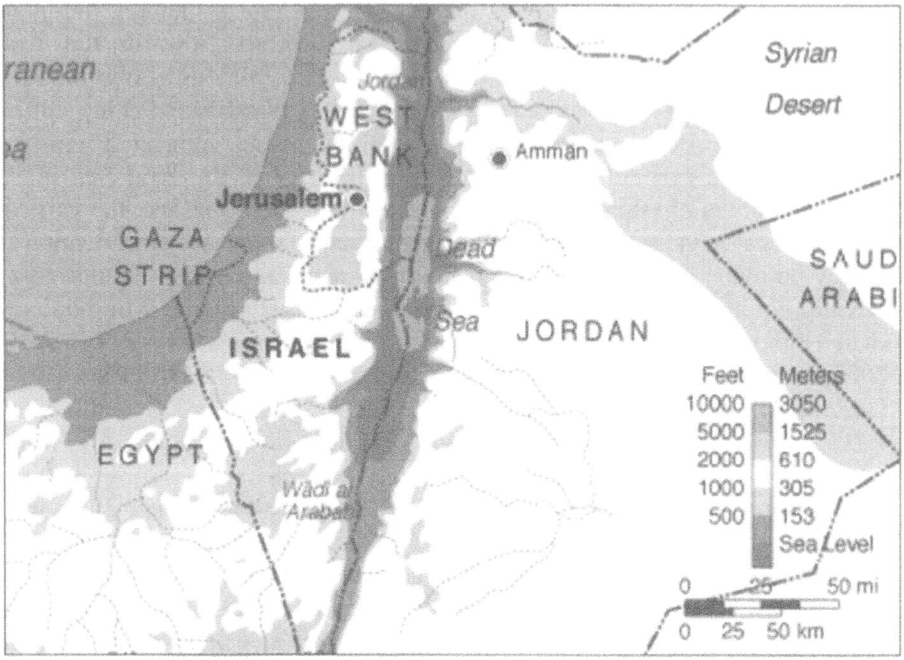

Figure 6. Geographical location of the Dead Sea.

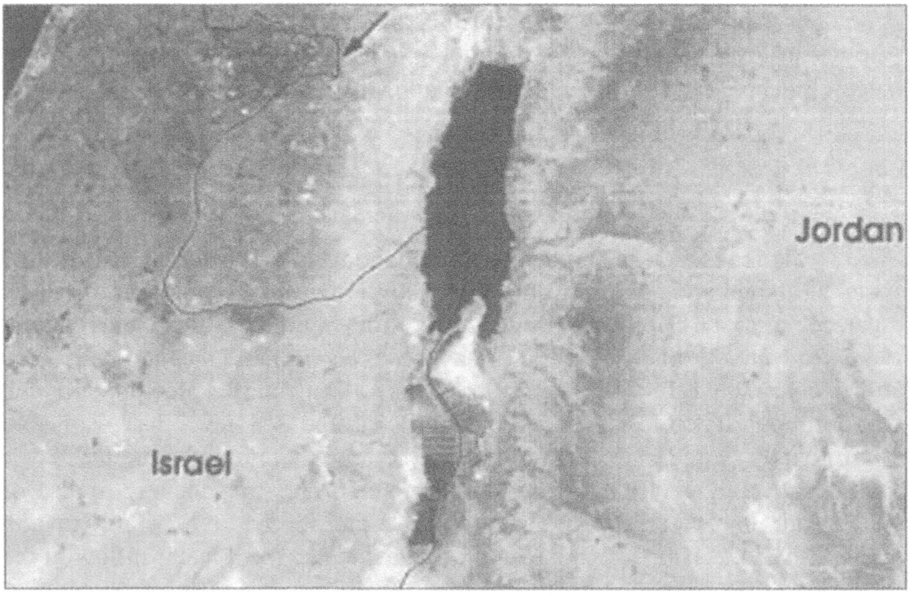

Figure 7. The MODIS image of the Dead Sea on 10 September 2000. Salt evaporation ponds are clearly visible south of the sea.

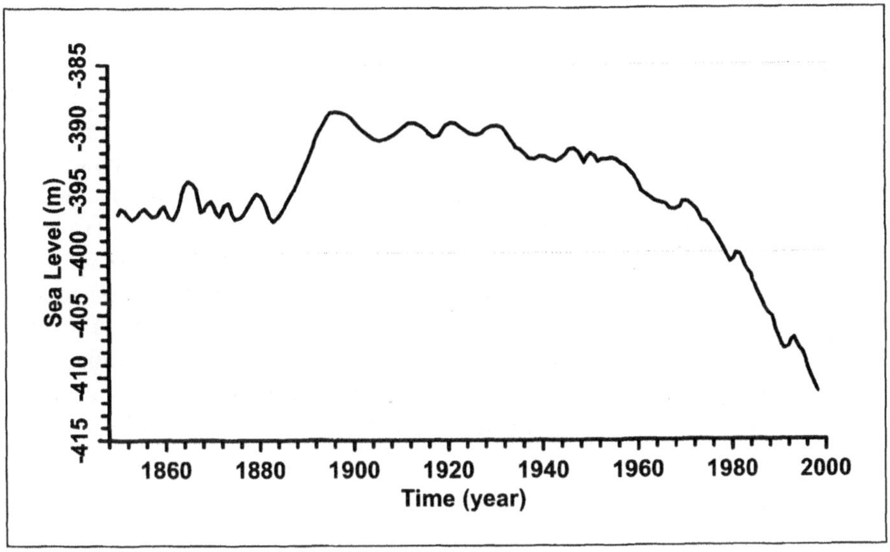

Figure 8. The Dead Sea level variability (*http://exact-me.org/overview/p4144.htm*).

at salt evaporation ponds located southward of the sea (Fig.6), which also contributes to the consumption of the lake waters. The evaporation ponds are suspected to be responsible for 25-30% of the total Dead Sea evaporation. The upper layer salinity has been increasing and the vertical stability of the water column decreasing. During the winter of 1978-1979 the lake waters overturned bringing to the end the long-term stable meromictic hydrological regime. A holomictic period is characterized by stable vertical stratification during the summer season and overturn in winter. A meromictic period is characterized by stable stratification throughout the year. The Dead Sea has been a meromictic lake for several centuries. Since 1977, the length of the Dead Sea decreased from 80 km to 50 km and the level has dropped by 14 m. At the same time, salinity reached 280 g/l (Gertman and Hecht, 2002). In present, the maximum depth is 316 m, the surface area is 815 km^2, and the volume is 146 km^3 . The sea level is about 414 m below the World ocean level (in 1960, the Dead Sea level was -397 m. Since 1977, a continuous level drop was observed at an average rate of 0.6-1.0 m/year. Plans have been developed to construct a pipeline to bring water from the Red Sea into the Dead Sea in order to stabilize or raise the water level. The pipeline, referred to as the "Peace Conduit" Project, would bring about 450 million cubic meters of the Red Sea water into the Dead Sea annually.

2.3. Caspian Sea

Another example of an inland sea suffering from anthropogenic pressure, although of somewhat different nature, is the Caspian Sea (Kosarev, 1975; Kosarev and Yablonskaya, 1994; Rodionov, 1994; Kostianoy and Lobkovskiy, 2003; Kosarev et al., 2004). The Caspian presents the world's largest isolated water body (Fig.9, 10). Currently, its level is at 27 m below the world ocean level (Fig.11). The sea occupies an area of 392,600 km^2, with mean and maximum depths being 208 m and 1025 m, respectively (Kosarev, 1975; Terziev et al., 1992; Kosarev, Yablonskaya, 1994). Caspian's longitudinal spread is three times larger than latitudinal one (1000 km and 200-400 km), resulting in great variability of climatic conditions over the sea. The biologically most productive Northern Caspian is some 1% of the total Caspian water volume (78,600 km^3), while the deep water of the

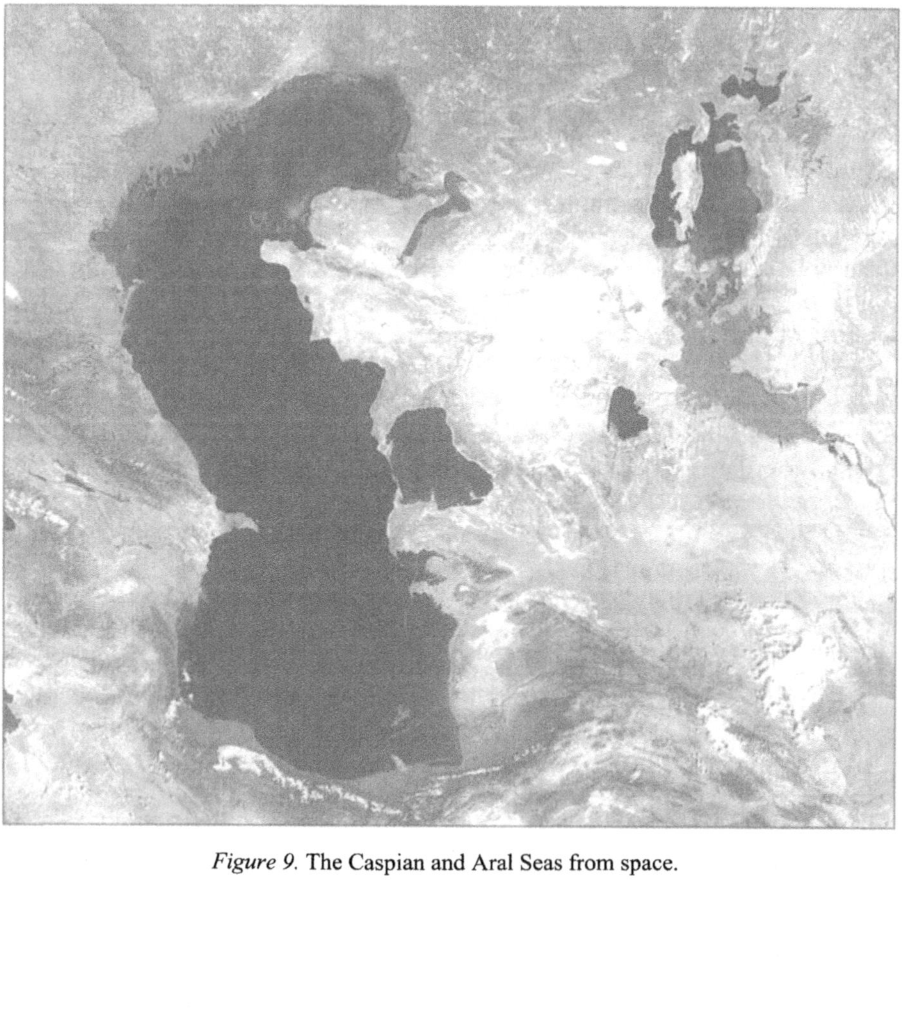

Figure 9. The Caspian and Aral Seas from space.

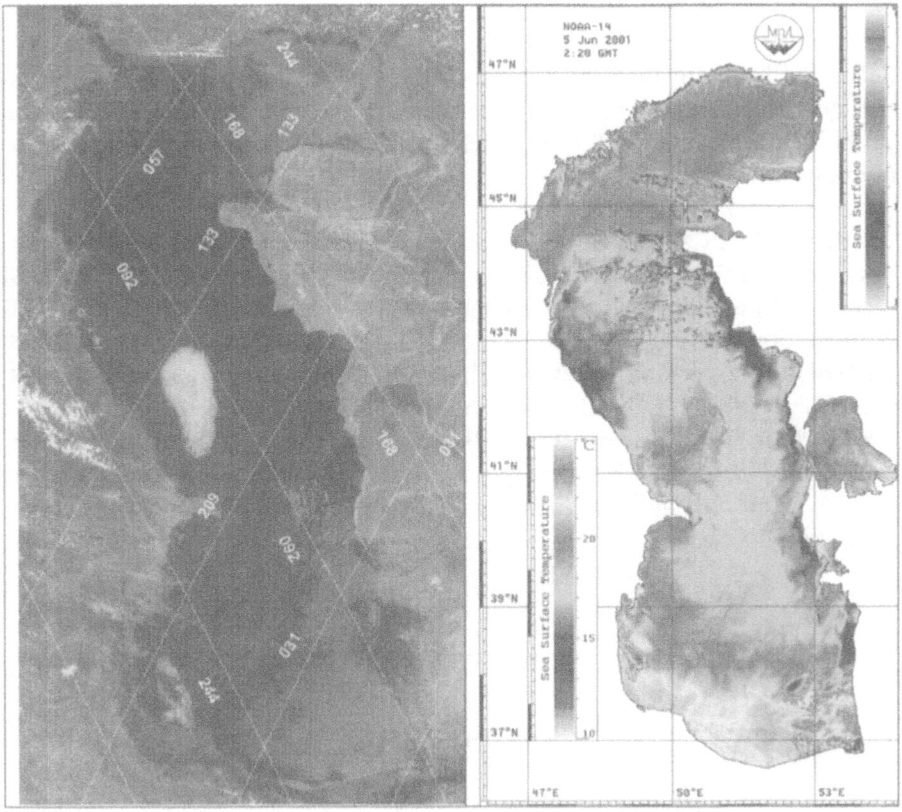

(a) (b)

Figure 10. The Caspian Sea: (a) MODIS image on 28 May 2002 with TOPEX/Poseidon ground tracks and (b) sea surface temperature derived from NOAA-14 on 5 June 2001.

Middle and Southern Caspian accounts for 99%. There has been serious concern over the Caspian Sea level fluctuations (Fig.11). Estimates suggest that these fluctuations as climatically conditioned and show their direct connection with components of the Caspian water budget, especially Volga river runoff contributing 80% of total fresh water flow into the sea (Terziev et al., 1992; Kosarev and Yablonskaya, 1994). Over the past half-century, there was a regression of the Caspian until 1977 when the sea level lowered to -29 m. This drop is considered to be the deepest for the last 400 years. In 1978, the water level started to rise rapidly, and now it is almost stable near the -27 m level (Fig. 11). Today, TOPEX/Poseidon altimetry allows to monitor the Caspian Sea level all over the sea with high accuracy, spatial and temporal resolution (Fig.10a, 12). Since 1997, there was a progressive decrease of the sea level with a rate of 8 cm/year.

Overfishing is another major problem of the Caspian Sea. The most alarming is the situation with beluga sturgeon, whose population is estimated to have decreased by more than 90 percent in the past twenty years. It is believed that the beluga sturgeon does not reproduce in the wild any more. Now, 90% of the fish are caught before they have spawned, denying them a chance to reproduce.

In addition, the Caspian Sea fisheries are at risk because of an invasion of a *Mnemiopsis leidyi* jellyfish (Fig.13) that threatens the food chains. The first record of Mnemiopsis in the Caspian Sea was made in the autumn of 1999 (Vinogradov et al., 2000, 2002; Kideys, 2002). During the summer of 2000, a population explosion followed. This caused the number of kilka, the last exploitable fishery in the Caspian since the decline of sturgeon, to drop by at least 50% in just a few months. Kilka is also the main food of the Caspian seal (*Phoca caspica*). The introduced jellyfish is a serious challenge to their very survival. Oil and gas industry also threatens Caspian ecosystems and endangers the Sea.

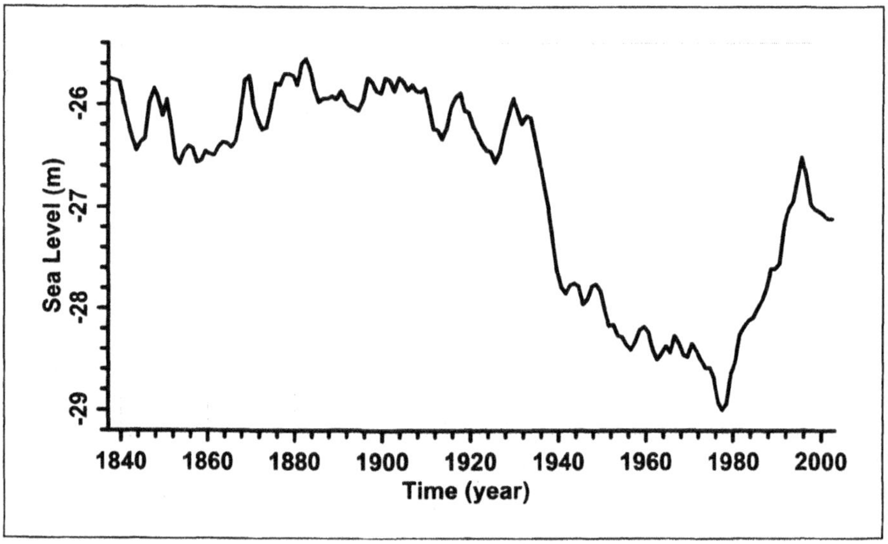

Figure 11. The Caspian Sea level variability by instrumental measurements (1837-2002).

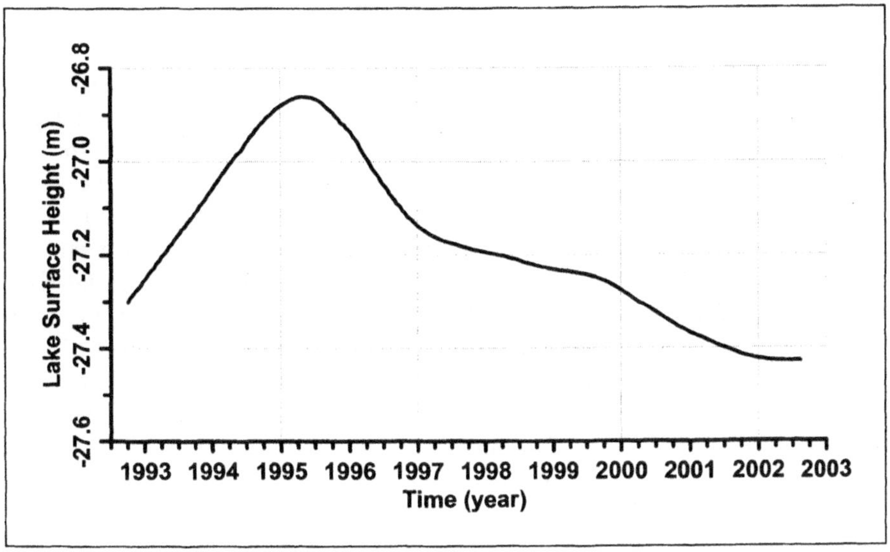

Figure 12. The Caspian Sea level variability based on TOPEX/Poseidon altimetry (1993-2002).

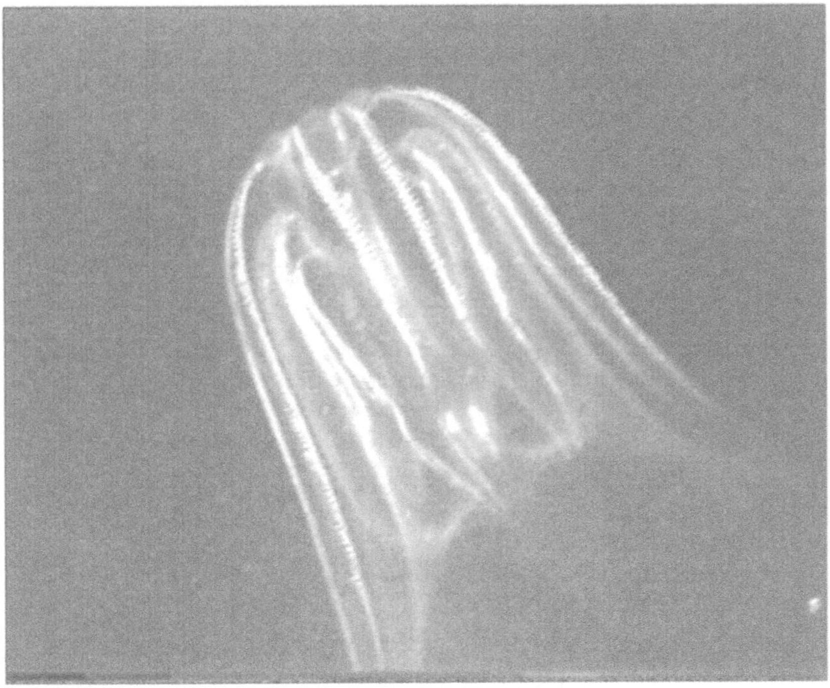

Figure 13. Mnemiopsis leidyi has invaded the Caspian Sea in 1999.

2.4. Kara-Bogaz-Gol Bay

The Kara-Bogaz-Gol (KBG) is a large lagoon eastward of the Caspian Sea (Fig.9, 14), normally about 20,000 km^2 and only a few meters deep. KBG is 2-3 m lower than the Caspian Sea, so water flows from the Caspian through a narrow strait into the KBG, where it evaporates. The KBG is one of the saltiest bodies of water in the world: salt concentrations up to 350 g/l have been documented. The salt in this natural evaporation basin has been used commercially since at least the 1920s. In March 1980, the Caspian-KBG strait was dammed. In response to this anthropogenic intervention, the KBG had already dried up completely by November, 1983 (Fig.14b). In 1992, the dam was exploded. The KBG has been filling up by the Caspian water at a rate of 1.5 m/yr as observed since the beginning of the TOPEX/Poseidon altimetry mission (Fig.15a). Further KBG level evolution with characteristic seasonal oscillations and decreasing tendency with a rate of 6 cm/yr is similar to that of the Caspian Sea (Fig.15b, 12).

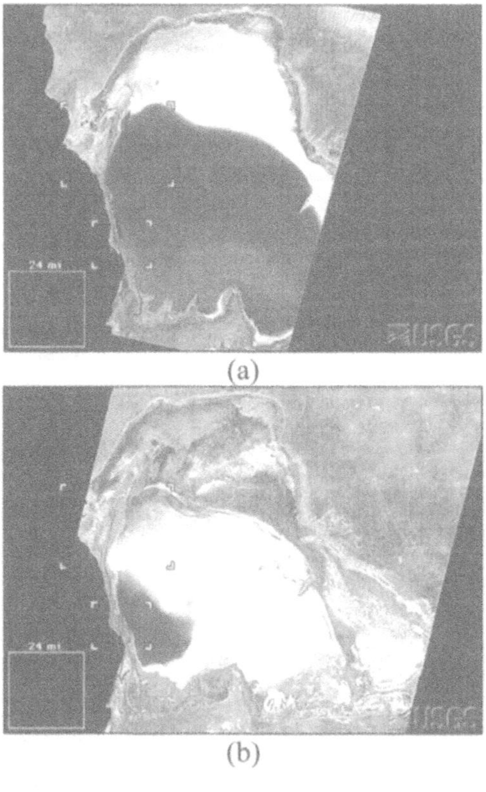

Figure 14. The Landsat-1 images of Kara-Bogaz-Gol Bay on: (a) 4 December 1972 and (b) 25 September 1987. White areas are salt on the dried bottom.

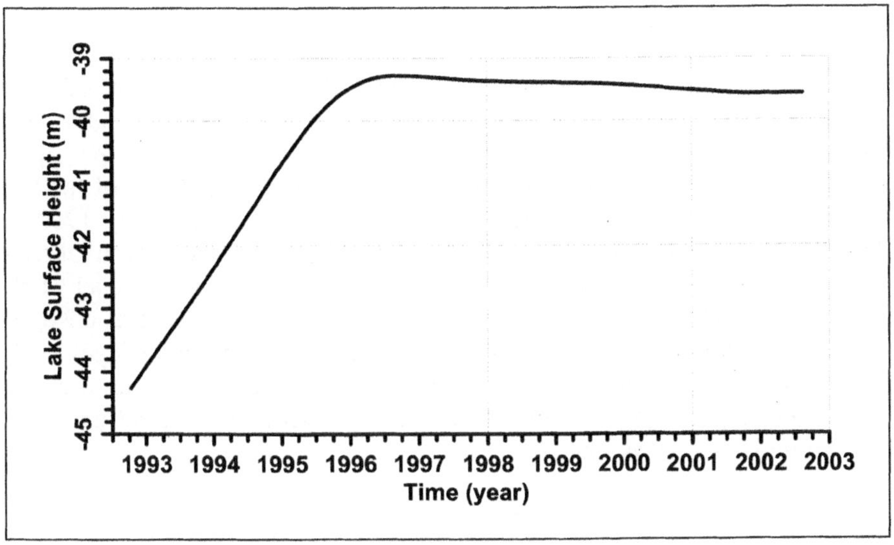

(a)

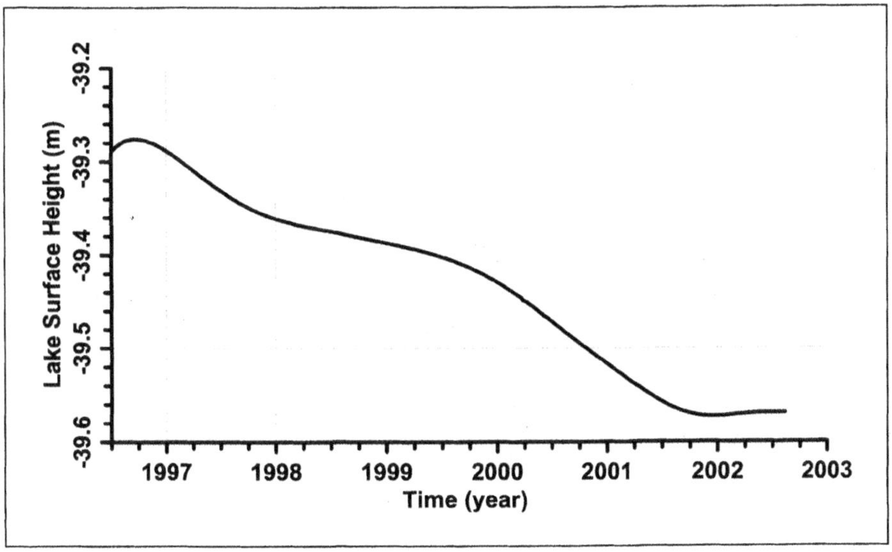

(b)

Figure 15. The Kara-Bogaz-Gol Bay level variability based on TOPEX/Poseidon altimetry: (a) in 1992-2002 and (b) in 1997-2002.

2.5. Lake Sarykamysh

Lake Sarykamysh is a large drainage water body, located eastward of Kara-Bogaz-Gol Bay and southwestward of the Aral Sea (Fig.16), which has been used as a discharge collector of salty irrigation water from the fields. Currently, the lake covers an area exceeding 3,000 km^2. Salinity of the lake waters has been continuously increasing: from 3-4 g/l in the early 1960s to 12-13 g/l in 1987 (Glazovsky, 1995b). Since 1992, the Sarykamysh lake has been progressively increasing in size, reaching its maximum level at the begining of 2000 with an increase of 4 m at a rate of 0.6 m/year as observed since the beginning of the TOPEX/Poseidon altimetry mission (Fig.17). The lake is growing due to water being diverted from the Amu Darya river - water that ought to go into the Aral Sea gets backed up behind dams and instead goes into Sarykamysh.

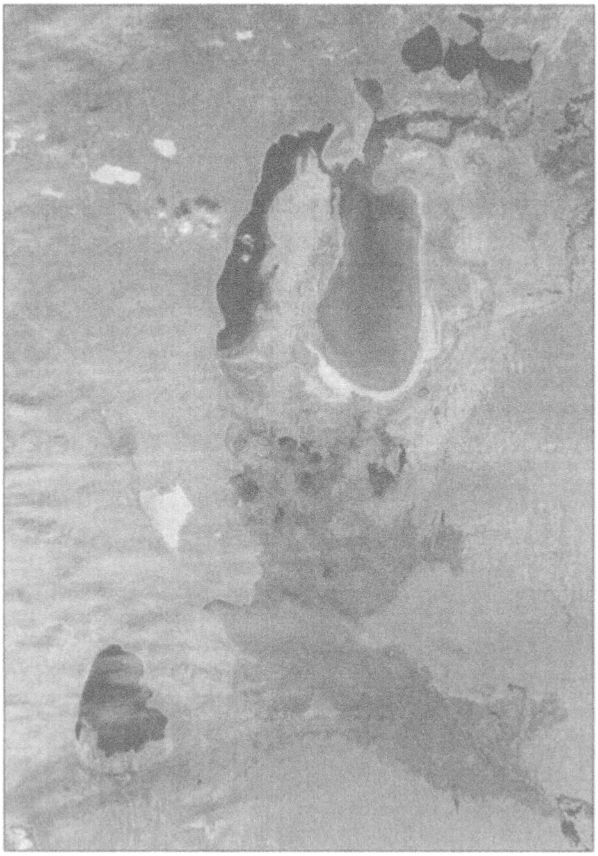

Figure 16. MODIS image of Lake Sarykamysh (seen southwestward of the Aral Sea) taken on November 7, 2002.

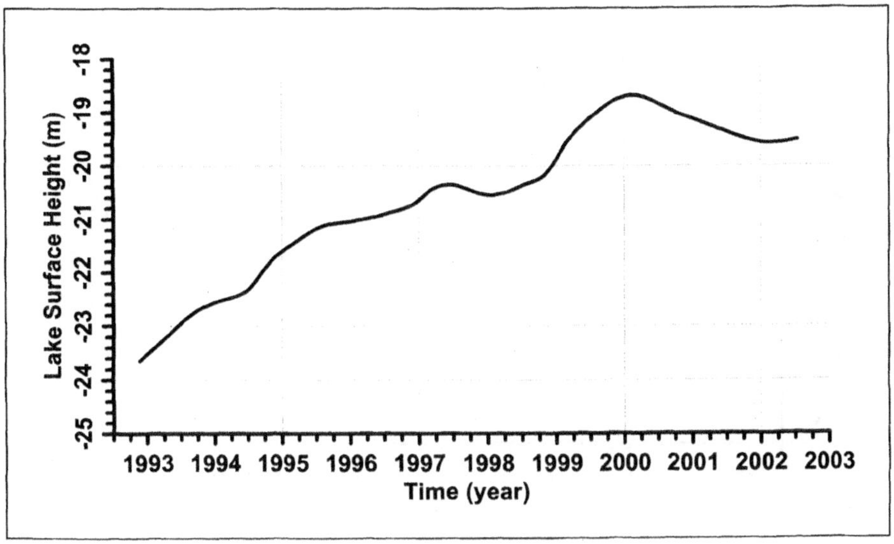

Figure 17. The Lake Sarykamysh level variability based on TOPEX/Poseidon altimetry (1992-2002).

2.6. Lake Balkhash

The Balkhash Lake is located at the altitude of 342 m in the vast Balkhash-Alakul, eastern Kazakhstan (see Fig.1, 18) (Abrosov, 1973; Shaporenko, 1995). Its surface is 18,200 km^2, the volume is 105 km^3, the mean depth is 5.8 m, the maximum depth is 26.5 m. It is 600 km long and 30 km wide. In

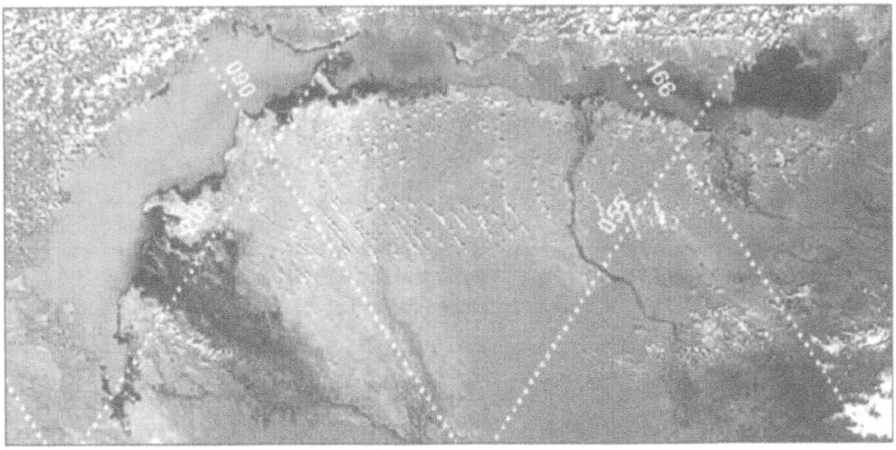

Figure 18. MODIS image of Lake Balkhash on 13 June 2002 and ground tracks of TOPEX/Poseidon.

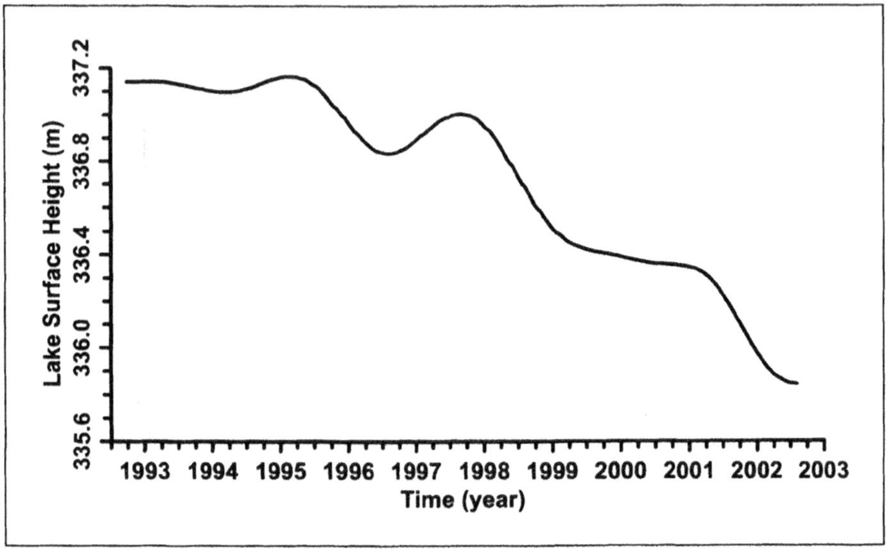

Figure 19. Western Lake Balkhash level variability based on TOPEX/Poseidon altimetry (1992-2002).

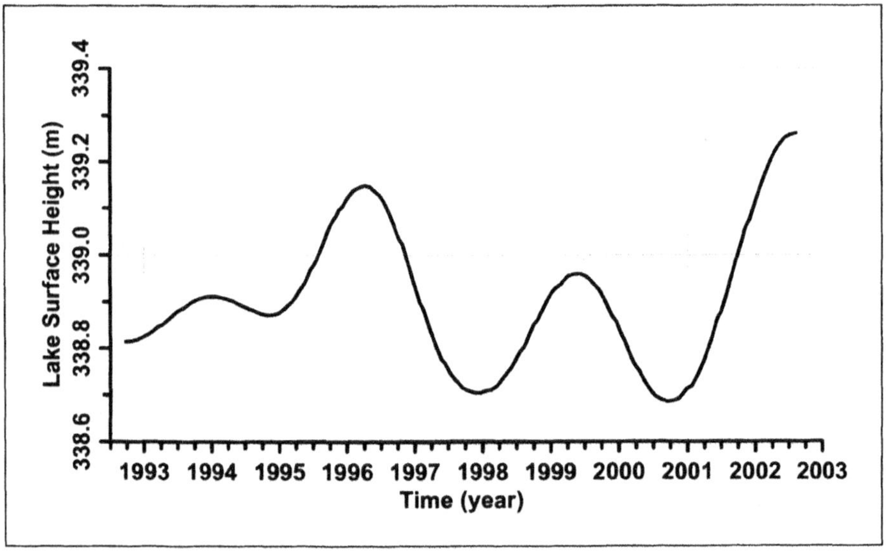

Figure 20. Eastern Lake Balkhash level variability based on TOPEX/Poseidon altimetry (1992-2002).

central part of the lake, large Sary-Isek Peninsula divides Lake Balkhash into western and eastern parts with mean depths 4.6 m and 7.6 m, respectively.

The area and water volume of the lake vary considerably depending on the large amplitude of long-term and secular fluctuations of its water level. For example, during XX century its water surface has changed from 15,730 km^2 (1946) to 23,464 km^2 (1910) and its water volume from 82.7 km^3 to 163.9 km^3. About 80% of the lake inflow (15 km^3/year) is due to the Ili river feeding the western basin (Fig.24). The western part of the lake has fresh or slightly salty water (0.5-1.5 g/l) depending on the secular fluctuation of its water level, while the eastern part is characterized by rather high concentration of dissolved solids (up to 7 g/l). The main reasons for such a difference are the presence of a large river inflow into the western part and the reduced exchanges between the two basins.

The water resources of the lake and its tributary rivers are used for irrigation, municipal and industrial water. Since 1970, the Ili run-off has been decreased by a filling of Kapchagay Reservoir (with a volume of 29 km^3), which was built in the middle part of the river for irrigation and hydropower generation (Shaporenko, 1995). This resulted in negative consequences on the water level, salt balance and ecology of the lake. Satellite altimetry for the last 10 years shows that a progressive drop of 1.4 m. was observed has taken place in the western basin (Fig.19, 20). In terms of fishery, the Balkhash Lake has been effectively managed through introduction of a number of fish species and their food organisms (Petr, 1992). Damming of the Ili River has contributed to fishery problems in Lake Balkhash. This problem is largely associated with the gradual desiccation of the deltaic lakes in which many fish species previously spawned and found excellent nursery areas (Petr and Mitrofanov, 1995).

2.7. Lake Issyk-Kul

The Issyk-Kul Lake (Fig.21, 22), one of the world's largest lakes, is located at the altitude of 1,607 m in a depression between the Kungey- and Terskey-Alatau ridges of the Northern Tien Shan in Kyrgyzstan (Tsigelnaya, 1995). Its surface is 6,292 km^2, volume is 1,738 km^3, the mean depth is 280 m, maximum depth is 678 m. It is 179 km long and 60 km wide. Issyk-Kul never freezes, even in severe winters. Cyclonic water circulation in the lake causes an extremely high transparency at the center where the Secchi disc reading reached 53 m in the stormy winter of 1985 (Tsigelnaya, 1995). The total river discharge into the lake from as many as 118 tributary rivers amounts to 3.7 km^3/year.

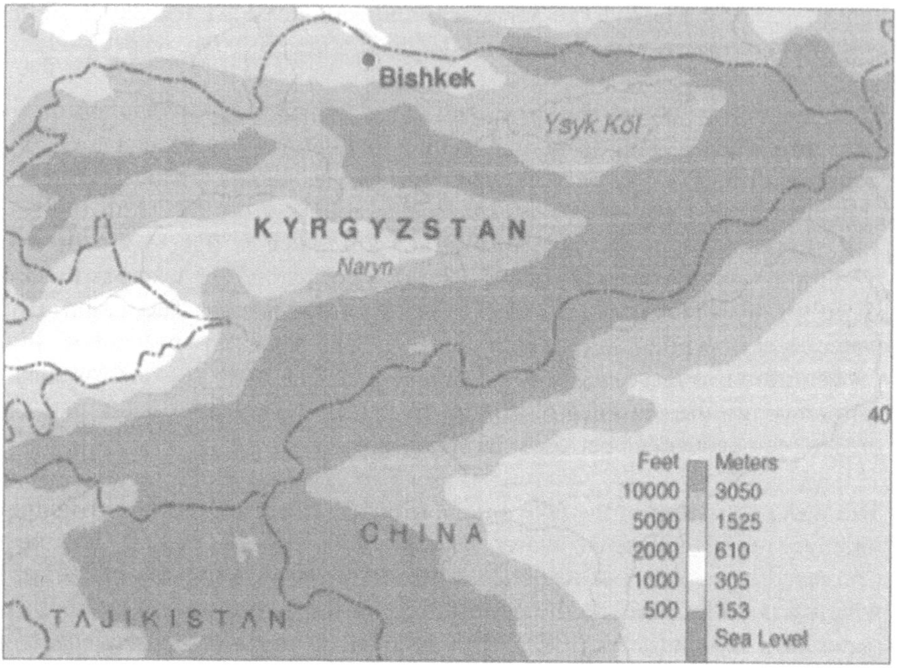

Figure 21. Geographical map of Kyrgyzstan.

Figure 22. The MODIS image of Lake Issyk-Kul on 13 June 2002 and ground tracks of TOPEX/Poseidon.

The Issyk-Kul Lake level drop was about 10 m since 1876. Since 1927 till the early 1990s the lake level has decreased by 2.64 m with short-term periods of increase observed in 1929, 1935-1936, 1941-1942, 1956-1960, 1970-1971, and 1989 (Tsigelnaya, 1985). In 1980s, the lake level was progressively declining by 5 cm/year (Savvaitova and Petr, 1992), reflecting diversions for irrigation and other causes. The area already dried up during the last 25 years amounted to more than 55 km^2. At present, about 90% of irrigated land involves furrow irrigation in which part of the water evaporates and the rest drains into the lake but contains fertilizers and harmful agrochemical compounds. Since 1992, satellite altimetry monitoring of the lake level shows that there were no significant trends, but its oscillations reach 5 m in amplitude (Fig.23). By the 1970s, an intensive fishery has drastically reduced the stocks of native fish. These were further affected by introducing two predators, pikeperch and the Sevan trout (*Salmo ischchan issykogegarkuni*) (Savvaitova and Petr, 1992). Expected increase of the Northern Hemisphere temperature by the mid of XXI century by 2°C and decrease of precipitation by 10% would have great effect on river runoff from the Issyk-Kul basin that is fed by melting snow and glaciers. Calculations show that inflow to the lake would not exceed 0.5 km^3 and Issyk-Kul will progressively disappear (Tsigelnaya, 1995).

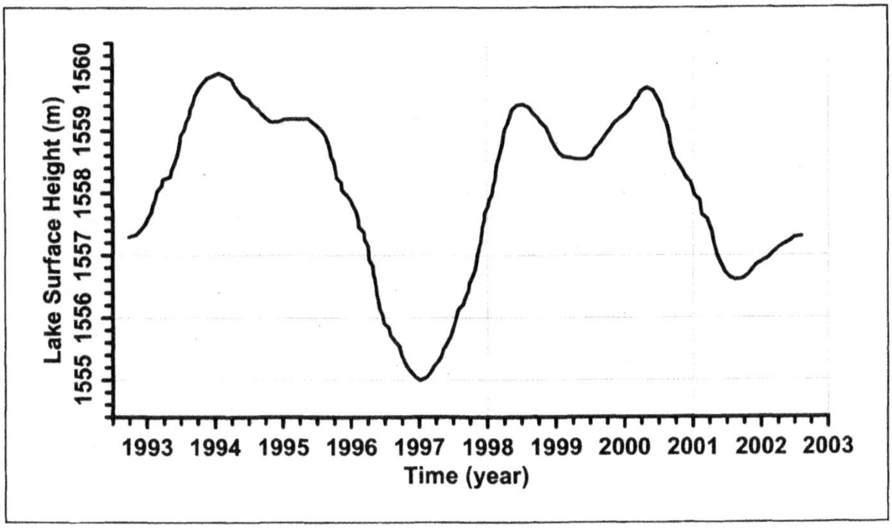

Figure 23. The Issyk-Kul lake level variability based on TOPEX/Poseidon altimetry (1992-2002).

2.8. Lake Lobnor and other critical lakes in China

Northern and northwestern China has been experiencing a desiccation process since 1950s. As a result, for example, the depth of the lake Ohlin at the head of the Yellow River has been dropping by over 2 cm annually (Wang Hongchang, 1993). Since 1950s, the average annual precipitation has decreased by at least one third. The lake Lobnor vanished completely in 1972 (Fig.24). Lobnor became nuclear testing site in China. Consequences of test explosions potentially threaten not only to Northwestern China lands and population but also to all Asia. Nowadays, due to nuclear testing and unsustainable of the region's natural resources, an ecological catastrophe may be approaching.

The Qinghai Hu Lake (Fig.25) is shrinking due to a decrease in rainfall, underground water supply and unsustainable use of the water for irrigation. According to the lake's water resources protection bureau, the water level has been decreasing at an average of 10 cm per year between 1959 and 1982. Emergency measures taken then led to the lake rise at an average of 10 cm per year from 1983 to 1989. But for the last decade, the lake's water level has again dropped off. The total drop since 1908 reached 11.7 m. The lake salinity has increased from 5.6 to 12 g/l since 1950.

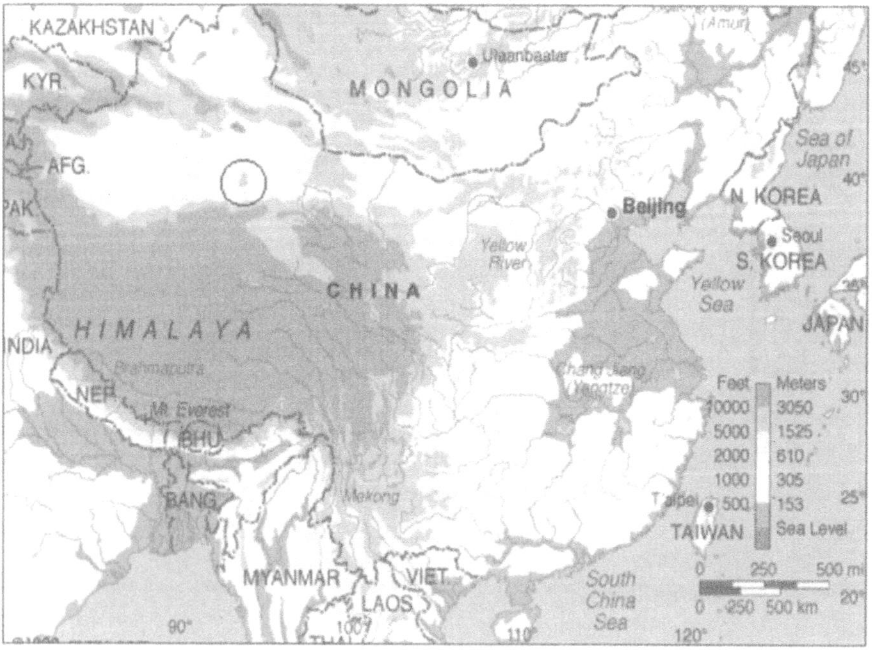

Figure 24. Circle denotes the Lobnor Lake location in China.

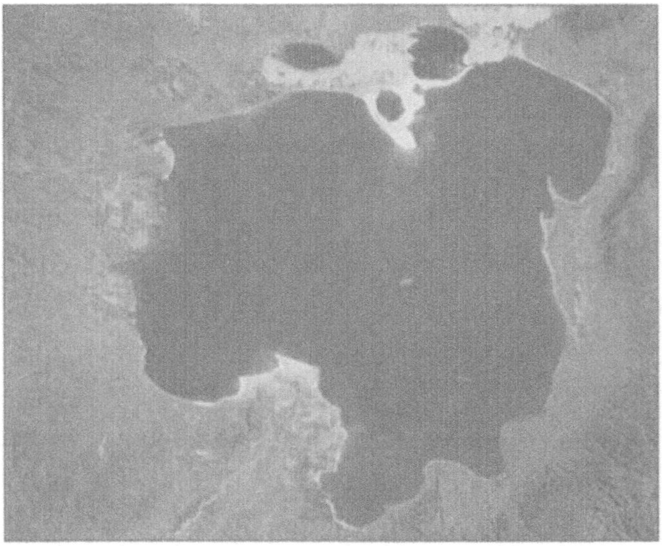

Figure 25. MODIS image of Qinghai Hu Lake taken on November 7, 2002.

Because of rapid growth of population, the surface of Ebinur Lake, the largest salt lake in northwest China's Xinjiang Uygur Autonomous Region, has shrunk to 530 km^2 in the past five decades (its surface was 1,200 km^2 in the 1950s). As a result, many of plant and animal species living in and around the lake have been extinct. Estimates suggest that the wind blows about 4.8 million tons of dust and sand away from the region annually. To prevent the lake from further shrinking, the regional authorities have taken a series of measures such as planting trees around the lake to preserve it.

3. CRITICAL LAKES OF AFRICA

The global warming, induced by the increase of carbon dioxide in the atmosphere, and withdrawal and/or diversion of water from inflowing rivers are the reason for the water level drop in several African lakes (Fig.26). Decreasing snow cover and land-ice extent continue to be highly positively correlated with increasing land-surface air temperatures (IPCC WG I, 2001). During XX century, the mean land surface temperature in Africa has increased by 0.9°C (Fig.27). This resulted in desiccation of rivers and lakes, as well as in snow and ice melting in mountains. For example, the total area of ice on Kilimanjaro decreased significantly. At the beginning of last century it amounted to 12.5 km^2, but by 2000, the ice cover is smaller than 2 km^2. Estimations show that snow and ice on Kilimanjaro could vanish completely by 2015 (Fig.28).

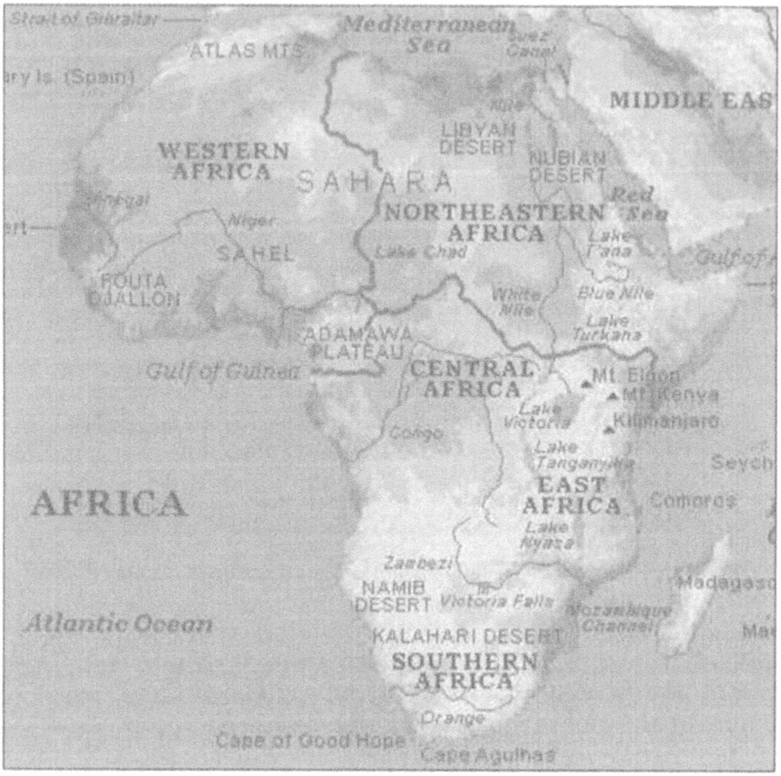

Figure 26. African lakes.

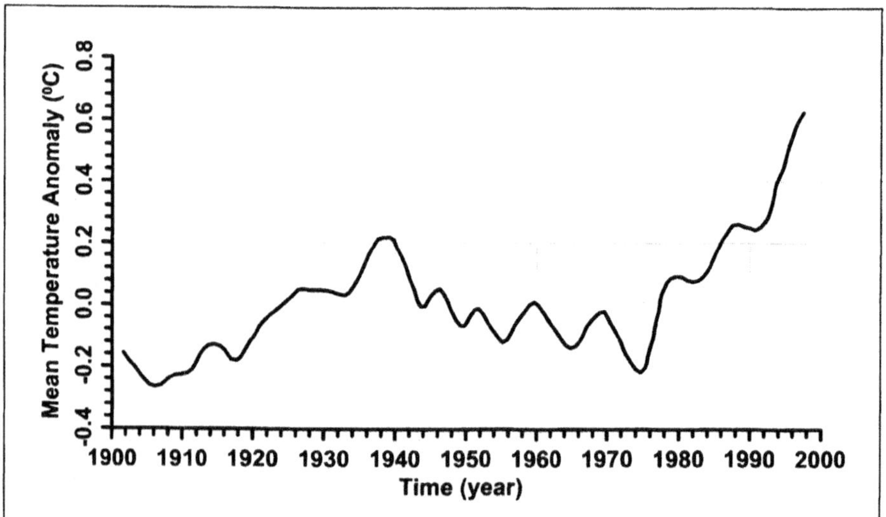

Figure 27. Mean temperature anomaly (°C) in Africa during the last century.

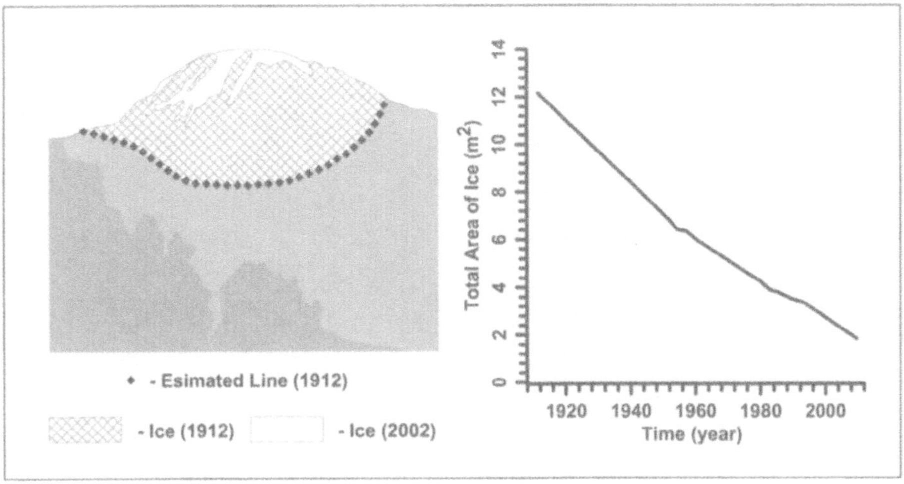

Figure 28. The melting snows of Kilimanjaro, 1912-2002 (UNEP).

3.1. Lake Chad

Lake Chad (Fig.29), once one of the largest on the Earth, has been dramatically decreasing since the 1960s (Fig.30, 31). Lake Chad has been a source of freshwater for irrigation projects in Chad, Niger, Nigeria and Cameroon in West Africa. Maps drawn from a series of satellite images show a dramatic decrease in the size of the lake over the past 30 years (Fig.30). Since 1963, the lake has shrunk to nearly a twentieth of its original size, due both to climatic changes and to high demands for agricultural water. Since 1963, the surface area of Lake Chad has decreased from approximately 25,000 km^2 to 1,350 km^2 (Scientific American, 2001). Before 1980s, irrigation did not have a strong impact on the Lake Chad ecosystem. Between 1983 and 1994, irrigation water use increased fourfold. About 50% of the decrease in the lake's size since the 1960s is attributed to human massive water use, with the remainder attributed to the natural climate variability. Since the early 1960s, the region has been experiencing a significant decline in rainfall.

Research carried out over the past 40 years indicates that the main factors in the shrinking of the lake have been:

(1) Unsustainable use of water resources by Niger; Nigeria, Cameroon and Chad, which have diverted water from both the lake and the tributary rivers ;
(2) Major overgrazing in the region (Coe and Foley, 2001), resulting in the loss of vegetation and hence a drier climate.

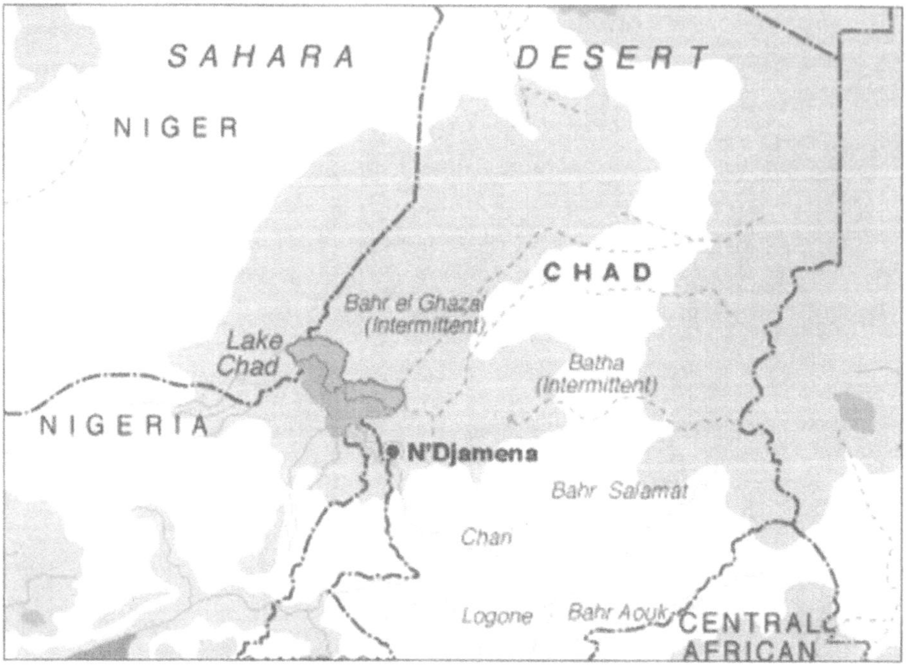

Figure 29. Geographical location of Lake Chad.

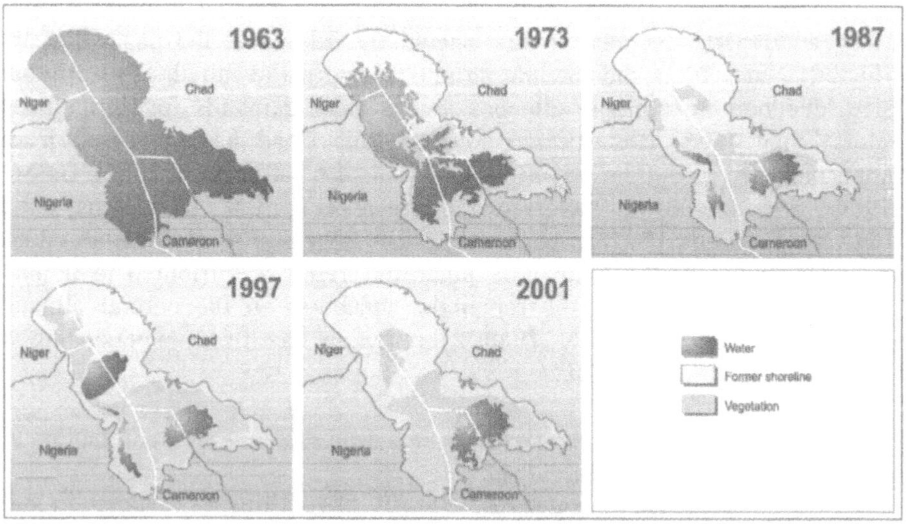

Figure 30. Desiccation of Lake Chad (1963-2001)
(*http://www.gsfc.nasa.gov/gsfc/earth/environ/lakechad/chad.htm*).

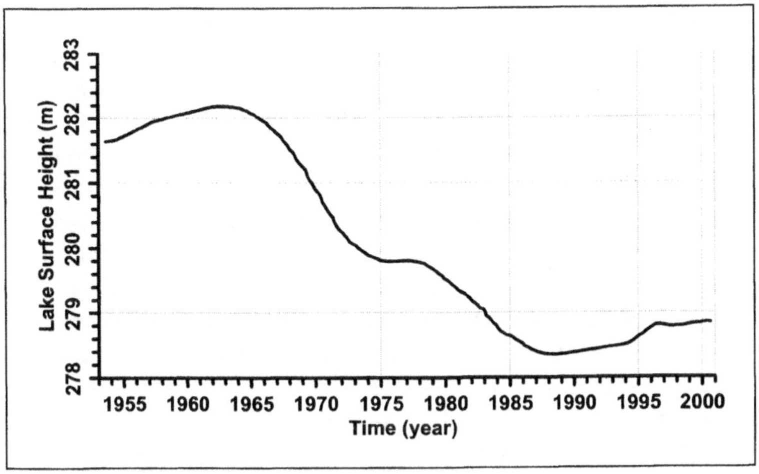

Figure 31. Variability of the Lake Chad level (gauge measurements in 1954-2001).

The changes in the lake have contributed to potable water shortages, collapsed fisheries, soil salinization, and increasing poverty throughout the region. Another ecological problem is related to the invasive plant species which currently cover about 50% of the remaining surface of the Lake Chad.

The monitoring of the lake level by TOPEX/Poseidon shows that the dramatic decrease of the level has apparently stopped and a rise of the water level of about 1 meter since 1993 is observed (Fig.32).

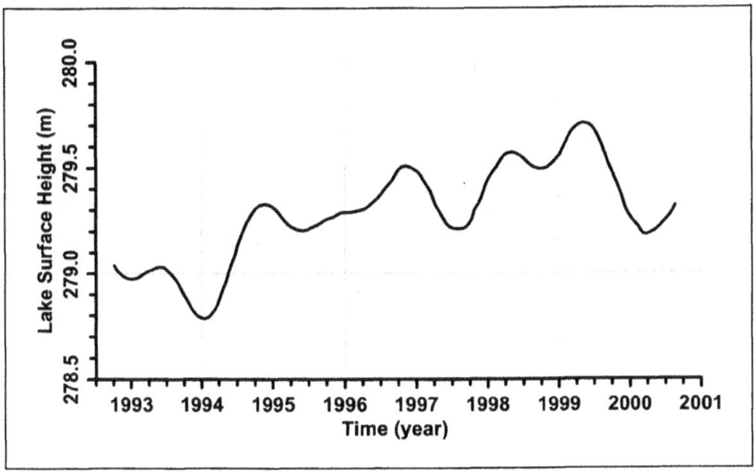

Figure 32. Variability of the Lake Chad level by TOPEX/POSEIDON satellite altimetry in 1992-2001 (AVISO).

3.2. Lake Victoria

Lake Victoria, with a surface area of 68,800 km^2, is the world's second largest, and the largest in Africa body of fresh water in terms of surface area, (Serruya and Pollinger, 1983; Payne, 1986). The lake is shared by Kenya (6%), Uganda (45%), and Tanzania (49%) (Fig.33). The lake occupies a wide depression near the equator, between the East and West Great Rift Valleys, but its drainage basin is relatively small (184,000 km^2). Lake Victoria is relatively shallow, with the maximum depth of about 84 m, and the average depth of about 40 m. Because the lake is shallow, its volume (2,750 km^3) is smaller than that of many other lakes with much smaller surface area. For example, the volume of the lake is only 15% that of Lake Tanganyika, even though the latter has less than half the surface area of Lake Victoria. The lake water is drained at a rate of about 600 m^3/s, at Jinja on the northern shore, into the Victoria Nile, the only outlet of the lake. Lake Victoria is of great socio-economic importance for 20 million people living in the basin (Serruya and Pollinger, 1983; Payne, 1986).

Over the past few decades, Lake Victoria has been a subject to drastic changes in its water quality and ecology due to anthropogenic activities (Serruya and Pollinger, 1983; Payne, 1986). Pollution to the lake has increased alongwith rapid growth of population. Sources of pollution to the lake include sewage discharges, agricultural runoff; sediments resulting from soil erosion in the catchment area due to deforestation and overgrazing; and industrial pollution from many local industries. All these factors have resulted in the eutrophication of the lake because of the increase in nutrient supply to the lake. Algae blooms and massive fish deaths in the lake have been reported. Yet another problem the lake is facing is the extinction of native species of fish in the lake due to introduction of two foreign fish species, the nile perch and nile tilapia, in the 1950s. According to the "World Resources 2000-2001" report, prior to 1970, Lake Victoria had more than 350 species of fish from the cichlid family. Ninety percent of these were unique to the lake. The introduction of Nile perch and tilapia has caused a collapse in the lake's biodiversity, as most of the native species have been extinct. Several projects are underway aimed at addressing the Victoria lake environment issues, e.g., the World Bank funded Lake Victoria Environment Management Project (LVEMP).

The TOPEX/Poseidon satellite altimetry shows that there was no significant trend in the Victoria Lake level in the last 10 years, but interannual oscillations may reach 1.5 m in amplitude (Fig.34). The notable drop of the lake level by 1997 accompanied by a pick in 1998 and the following

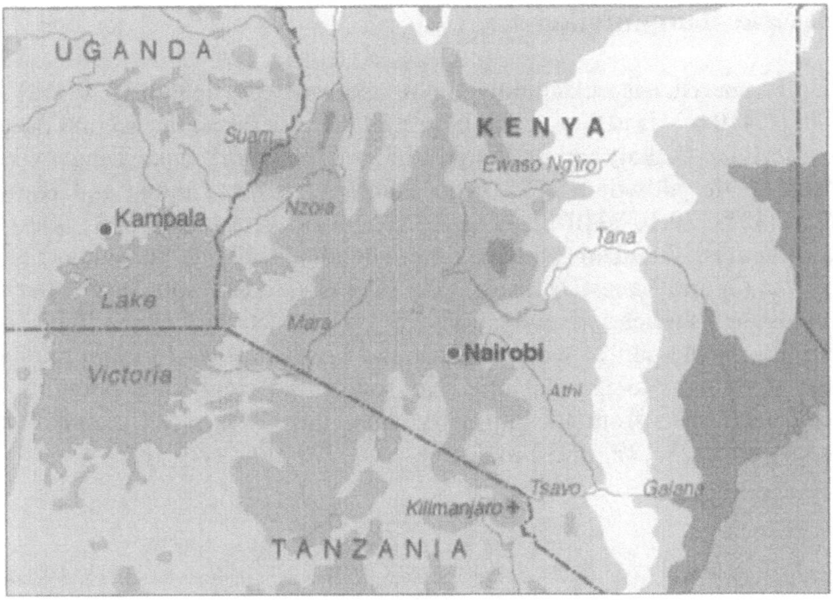

Figure 33. Map of the Lake Victoria region.

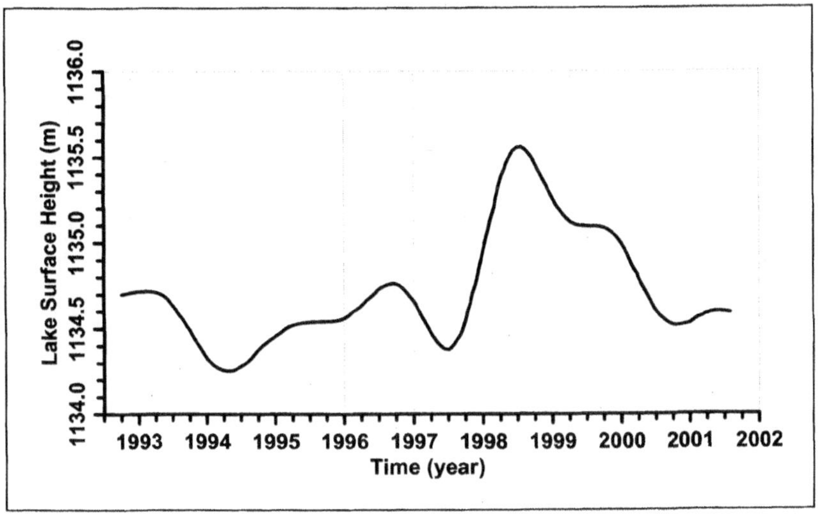

Figure 34. The Lake Victoria level variability based on TOPEX/Poseidon altimetry (AVISO).

decrease of the level by about of 1 m is a characteristic feature for other African lakes also, as shown in figures below. A sharp level rise in 1998 that observed all lakes in this part of the continent was brought on by increased rainfall in that year (AVISO).

3.3. Lake Tanganyika

Lake Tanganyika is outstanding for its extraordinary length (670 km) and depth (1,470 m). It is the second largest of African lakes, the second deepest (next to Lake Baikal) and the longest lake of the world. Lake Tanganyika is located at the altitude of 773 m in Tanzania, Zaire, Zambia and Burundi (3.25-8.45°S, 29.10-31.10°E) (Fig.35). The lake surface area is 32,000 km^2, the volume is 17,800 km^3, the maximum depth is 1,471 m, the average depth is 572 m (World Lakes Database). The lake is used as a source of water and grounds for fisheries and navigation. The only outlet, the Lukuga river, starts from the middle part of western coast and flows westward to join the Zaire River draining into the Atlantic. No sources of significant pollution are present. As seen from the altimetry data, there was a 1.5 m lake level decrease in 1993-1997 and 1 m decrease in 1999-2001 (Fig.36).

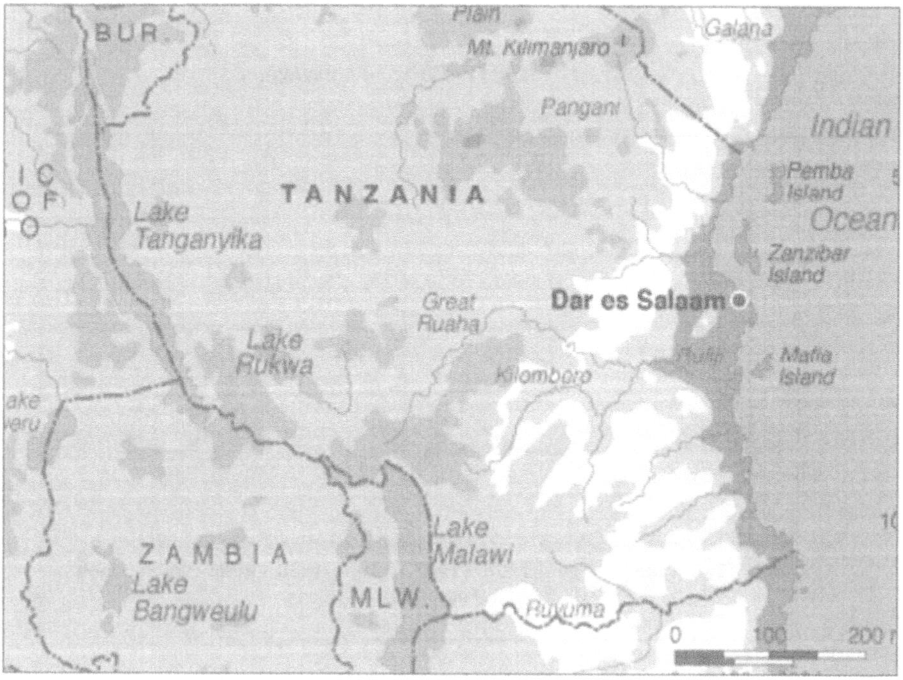

Figure 35. Map of the Lake Tanganyika region.

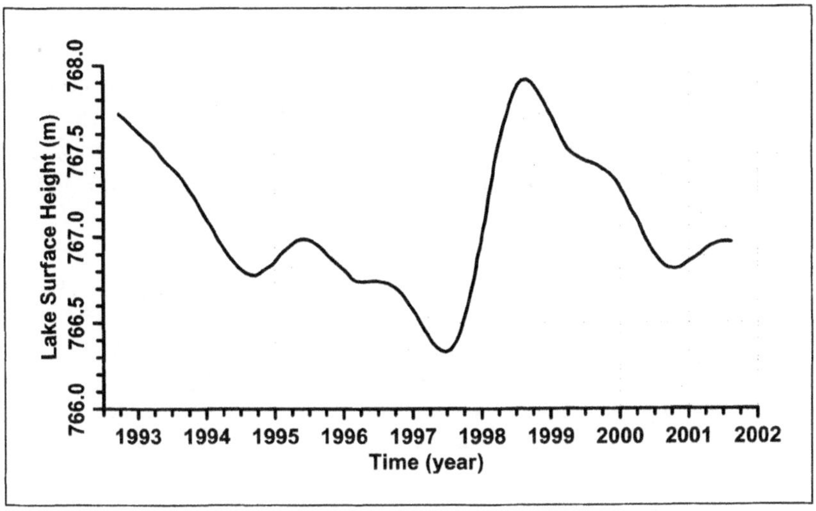

Figure 36. The Lake Tanganyika level variability based on TOPEX/Poseidon altimetry (AVISO).

3.4. Lake Rukwa

Lake Rukwa (Fig.35) is a shallow lake located at the altitude of 793 m in Tanzania (8°S, 32.25°E). The lake surface area has is 3,000 km^2 (World Lakes Database). There was a 2.5 m lake level decrease in 1993-1997 and 1.5 m level drop in 1999-2000 (Fig.37).

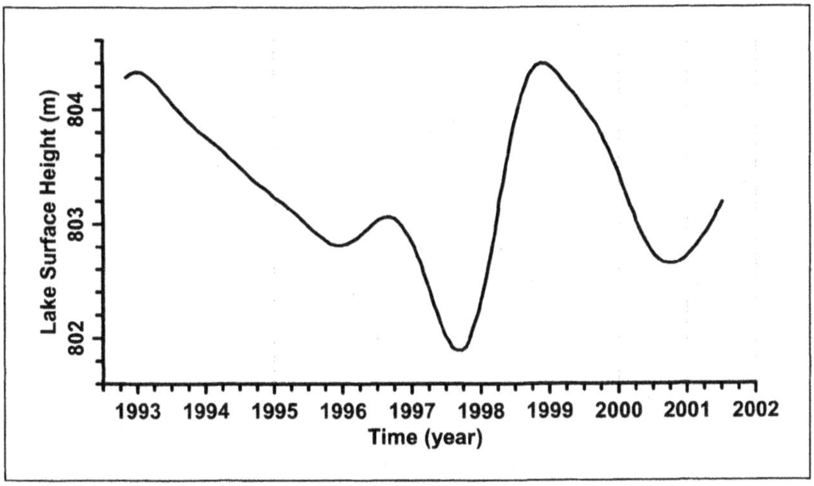

Figure 37. The Lake Rukwa level variability based on TOPEX/Poseidon altimetry (AVISO).

3.5. Lake Tana

Lake Tana (Fig.38) is located at the altitude of 1,788 m in Ethiopia (11.4°N, 37.2°E). The lake surface area is 3,600 km^2, the volume is 28 km^3, the maximum depth is 14 m, the average depth is 9 m (World Lakes Database). There was a continuous lake level decrease of 2 m in 1992 through 2002 (Fig.39).

Figure 38. Map of the Lake Tana region.

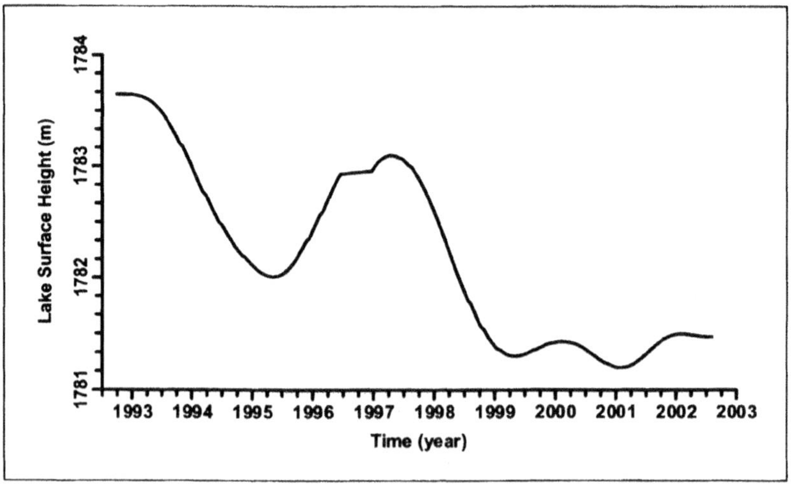

Figure 39. The Lake Tana level variability based on TOPEX/Poseidon altimetry.

4. CRITICAL LAKES OF NORTH AMERICA

4.1. Great Salt Lake

The Great Salt Lake (Fig.40) located at the altitude of 1,283 m in Utah, USA (41.1°N, 112.4°W) is the fourth largest terminal lake in the world. It is one of the largest migratory bird sites in North America. The lake has surface area of 5,000 km^2 and maximum depth of 12 m (World Lakes Database). The local topography is such that small changes in the lake surface level lead to enormous changes in the land surface dried up when the lake level drops (Fig. 41). In 1963 when the level was record low (1,277.5 m), the lake covered 2,460 km^2, while in 1986 with 1,283.8 m, it covered about 5,960 km^2 (Arnow and Stephens, 1990). The history of the water-level changes (Fig. 42) shows energetic variability at temporal scales from years to decades. Now, the lake level is the lowest since 1980. The salinity of the lake waters is remarkably high, namely up to 280 g/l. The Great Salt Lake salinity varies considerably depending on the site and lake level (Fig.43). The lake level increase from 1982 to 1986 (Fig.42) was, at least partly, related to the enhanced precipitation during the 1982-1983 El Nino (Arnow and Stephens, 1990).

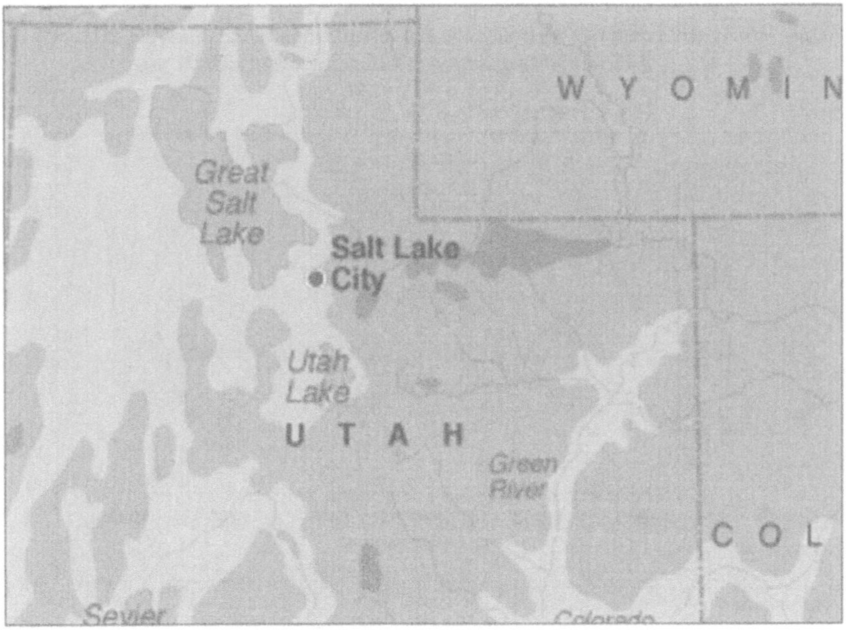

Figure 40. Map of the Great Salt Lake region.

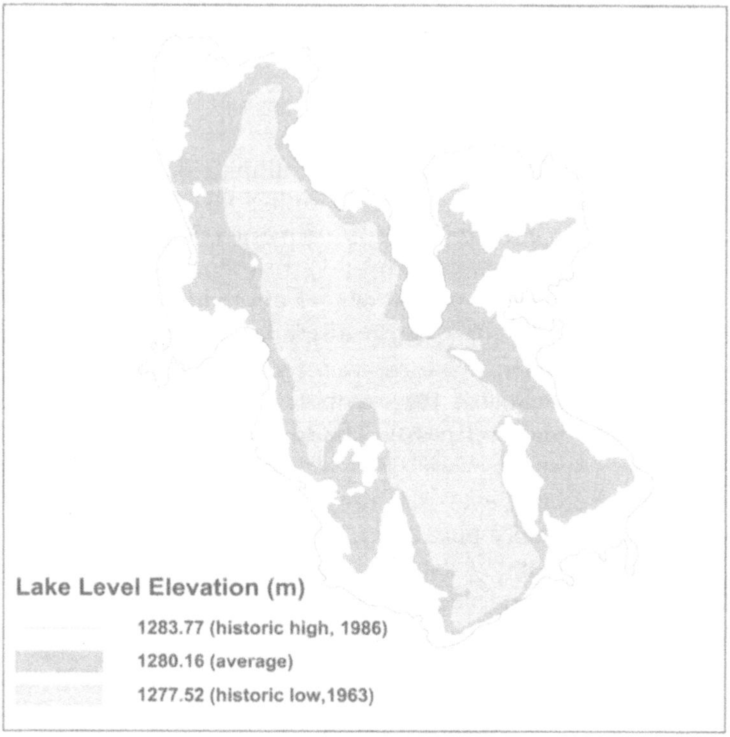

Figure 41. Area covered by Great Salt Lake at historic high, low, and average levels, 1847-1986 (modified from Arnow and Stephens, 1990).

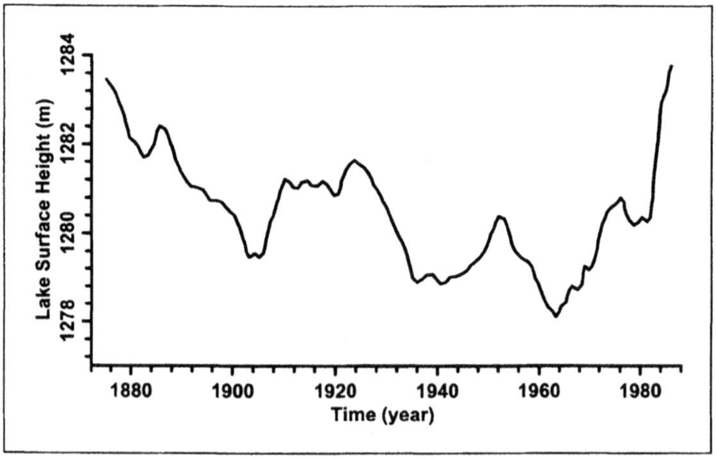

Figure 42. Annual peak water levels of Great Salt Lake, 1875-1986 (from Arnow and Stephens, 1990).

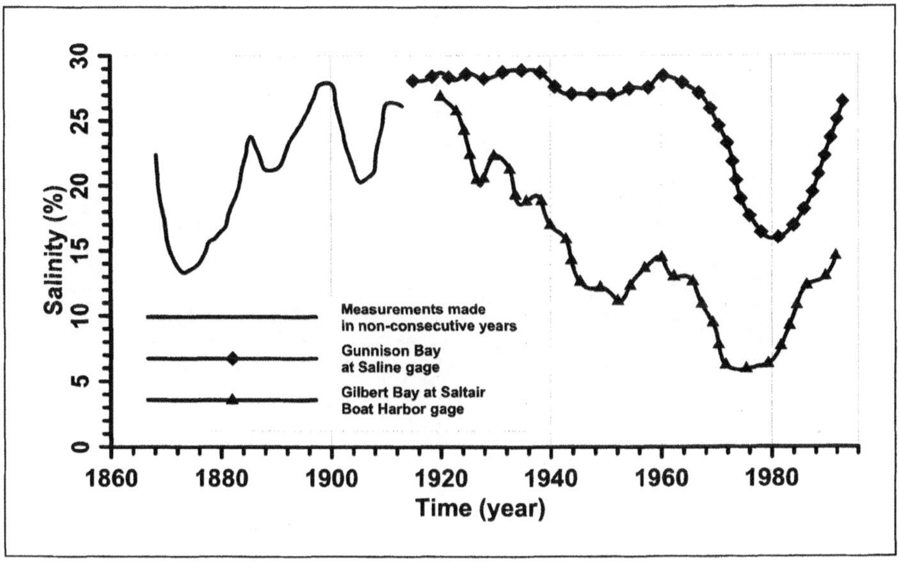

Figure 43. Salinity (%) in Great Salt Lake at historic period, 1850-1997, according to U.S. Department of the Interior, U.S. Geological Survey (*http://ut.water.usgs.gov/salinity/index.html*)

4.2. Pyramid Lake

The Pyramid Lake (Fig.44) is the deepest terminal salty lake in the Western Hemisphere. It is located at the altitude of 1,160 m in Nevada, USA (40.0°N, 119.5°W). The lake surface area is 453 km^2, the volume is 27 km^3, the maximum depth is 105 m, the average depth of 60 m (World Lakes Database). There was a 21 m lake level decrease since 1910 and a salinity build-up from 3.8 to 5.5 g/l between 1933 and 1980 (Wheeler, 1974). Pyramid Lake receives about 85% of its annual water input from Truckee River. The lake is a source of water and grounds for sightseeing and tourism, recreation and fisheries. As the lake level declined, a delta formed at the mouth of the Truckee River, preventing spawning migrations of fishes. This delta and the Derby Dam impeding upstream migration resulting in decline of the Lahontan cutthroat trout in 1938 (reintroduced in the 1950s). Other environmental concerns include water quality issues as the population upstream rapidly grows (World Lakes Database).

Figure 44. Map of the Pyramid Lake region.

4.3. Mono Lake

The Mono Lake (Fig.45, 46) is located at the altitude of 1,945 m in California, USA (38°N, 119°W). The lake surface area is 180 km^2, the volume is 2,97 km^3, the maximum depth is 43 m, and the average depth is 17 m (World Lakes Database). There was a remarkable 17 m lake level decrease between 1920 and 1980 (Fig.47) because of diversions of Mono's tributaries. Currently, the salinity of the Mono Lake is 79 g/l (as of 2002), which is almost twice the value of 1920.

Figure 45. Map of the Mono Lake region.

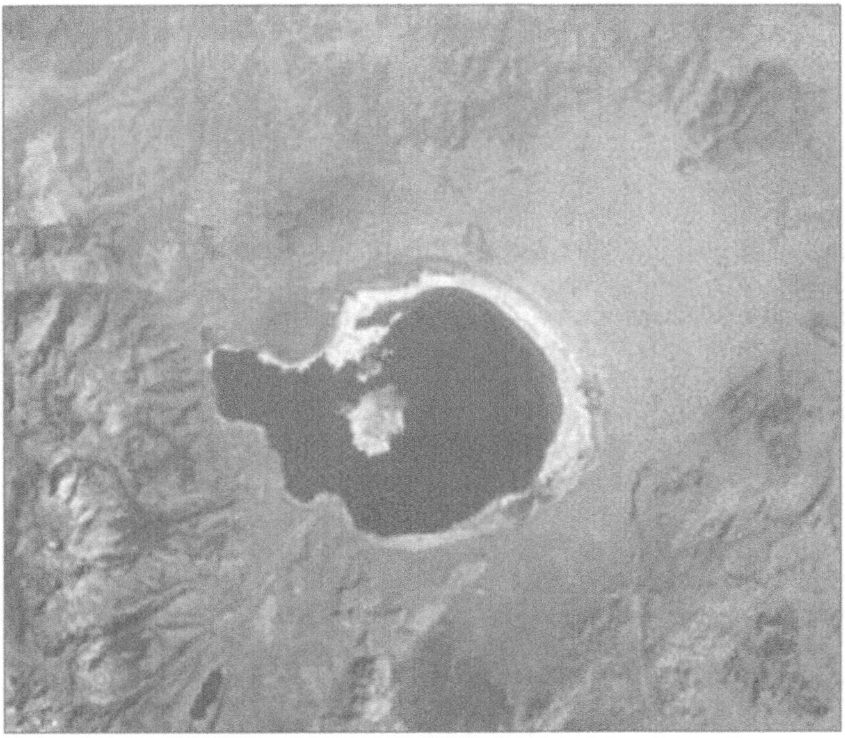

Figure 46. The MODIS image of Mono Lake on 15 September 2002.

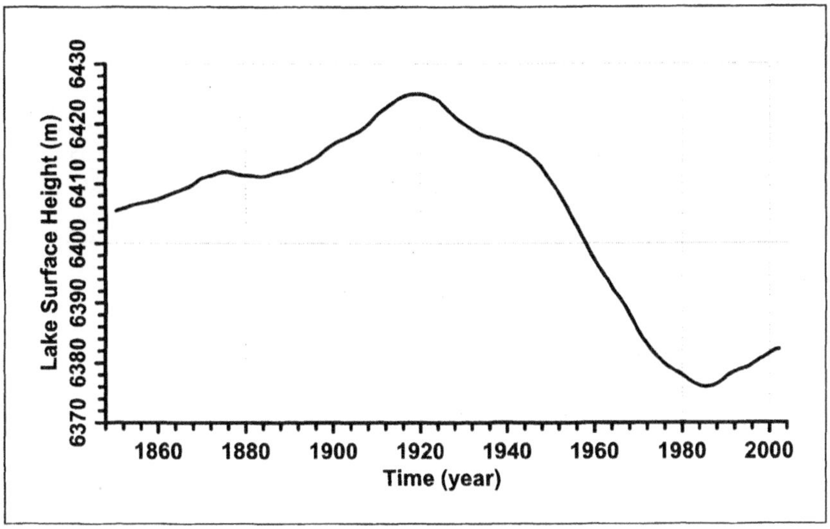

Figure 47. The Mono Lake level (feet) variability since 1850 to 2002.

4.4. Salton Sea

The Salton Sea is California's largest lake (33.12°N; 115.51°W) (Fig.48, 49). At the surface elevation of 70 m below the sea level, it has a surface area of 950 km^2. The maximum depth of the Sea is about 17 m and the average depth is 10 m. The annual inflow to the Sea averages about 1.5 km^3 (www.sci.sdsu.edu/salton//Salton%20Sea%20Description.html). The Salton Sea was formed in 1905 when a strong flooding caused the Colorado River flow into the Salton Basin for 18 months. Since then, the lake level has been maintained mainly by agricultural return flows from the surrounding valleys. The Salton Sea has several serious problems. Salinity of this terminal lake progressively increases, now reaching 44 g/l. This salinity increase has led to decline in the lake's ecosystems, including fishes (www.sci.sdsu.edu/salton//Salton%20Sea%20Description.html).

High concentrations of nutrients in Salton's inflows lead to eutrophication of the Sea. In addition, about 15% of the discharge polluted by industrial chemicals and sewage. In particular, selenium that is believed to have caused mortality among wildlife at some irrigated areas in the western U.S. has been investigated in the Salton basin by the U.S. Department of the Interior (www.sci.sdsu.edu/salton)

Figure 48. Map of the Salton Sea region.

Figure 49. The MODIS image of the Salton Sea on 9 October 2002.

4.5. Great Lakes

The Great Lakes (Fig.50) contain about 20% of Earth's surface fresh water. General physical characteristics of the lakes are shown in Table 1. Pollution to all of the Great Lakes, may be except Lake Superior, has increased significantly over the past few decades. This has resulted in increased concentrations of nutrients, chlorides, sulphates, and other compounds in Lakes Michigan, Huron, Erie and Ontario (World Lakes Database).

Lake Superior is the second largest lake in the world after the Caspian Sea. The water of the lake is still of excellent quality and oligotrophic. Lake Huron is the second largest of the Great Lakes and the fifth largest in the

Figure 50. Map of the Great Lakes region.

Table 1. Physical dimensions of the Great Lakes (World Lakes Database).

	Superior	Huron	Michigan	Erie	Ontario
Surface area, km^2	82,367	59,570	58,016	25,821	19,009
Volume, km^3	12,221	3,535	4,871	458	1,638
Maximum depth, m	406	228	281	64	224
Mean depth, m	148	53	84	17.7	86
Water level	Regulated	Unregulated	Unregulated	Unregulated	Unregulated
Length of shoreline, km	4,768	5,088	2,656	1,369	1, 161
Residence time, yr	191	22.6	99.1	2.6	7.9
Catchm. area, km^2	124,838	128,464*	117,845	78,769*	75, 272*

* Not including the catchments of the upstream Great Lakes.

world, with the most irregular shape of any of the Great Lakes. The water quality of the lake is still excellent and oligotrophic. Lake Michigan is the third largest of North America's Great Lakes. Since 1970s, the lake water along the southern coast has been increasingly eutrophicated because of urban and industrial pollution. To prevent the lake from further pollution, the wastewater was diverted to the Mississippi River instead of returning to the lake. As the result, the quality of the lake water appears to be recovering (World Lakes Database). Lake Erie is the fourth largest and shallowest of the five Great Lakes. The western side of the lake is a heavily industrialized and populated area. The lake is a subject to considerable eutrophication similar to that of Lake Ontario before the 1970s. Lake Ontario is the smallest of the Great Lakes. The most populated and industrialized area of Ontario Province is located on the western bank of the lake. The eutrophication of the lake started at the beginning of the past century and continued until mid-1970s.

The aggregate level of the Great Lakes is currently decreasing. Since 1997, Lakes Huron, Michigan and Erie have dropped over 1 m, and further drop of 0.5-1 m has been predicted for the next few years. Also, a dramatic shift in the seasonal changes in water levels on the Great Lakes has been reported (ENN, 2000). Analysis based on 139-year-long (1860-1998) record from four stations around the Great Lakes demonstrated that the annual highs and lows of lakes Ontario and Erie are now occurring earlier by about a month. Also, the range of Lake Ontario's level annual cycle has increased by a factor of 1.5.

It is believed that the principal mechanisms responsible for all these changes are: (1) Decrease of rainfall and snowfall; (2) Elevated temperatures and thus enhanced evaporation and transpiration rates and reduced ice cover; (3) Irreversible loss of water for urban and industrial uses – for example, Chicago sends its used water taken from the lakes to the Mississippi after treatment instead of the lakes (Mitchell, 2002).

5. SOME CRITICAL LAKES OF AUSTRALIA

Lake Eyre (Fig.51) is Australia's largest salty lake, located in the driest region in the country. The Eyre's drainage basin is as large as 1,140,000 km^2. The lake consists of two parts - Lake Eyre North, 144 km long and 77 km wide, is connected by the Goyder Channel with Lake Eyre South, which is 64 km long and 24 km wide (Fig.51). The lake bottom is usually dry and encrusted with salt. However, about twenty flood events have occurred over the last forty years. During such events, the lake becomes temporarily the

Australia's largest lake with the surface area of up to 9,500 km^2. The strongest floodings were registered in 1950, 1974 and 1984 (Fig.51). In 1974 water flowed from Lake Eyre North to Lake Eyre South until an equilibrium level was obtained. Over 30 million tons of salt, or about 7.5% of the total Lake Eyre North salt content, were transferred into Lake Eyre South during this event, creating a salt crust of up to 29 cm. The broad basins of the lakes are very sensible to even slight variations of precipitation. Considering the trends of climate change is therefore essential for understanding the regime and variability of these lakes (World Lakes Database).

An important example of a salty lake endangered by water diversion is Lake Corangamite. This largest permanent water body in continental Australia is a subject to rapid shallowing and salinization because its main tributary, the Woady Yaloak Creek, has been diverted into the Barwon River. From 1959 to 1990, the lake level has decreased by over 2 m and the salinity has increased almost twofold from 35 to 60 g/l.

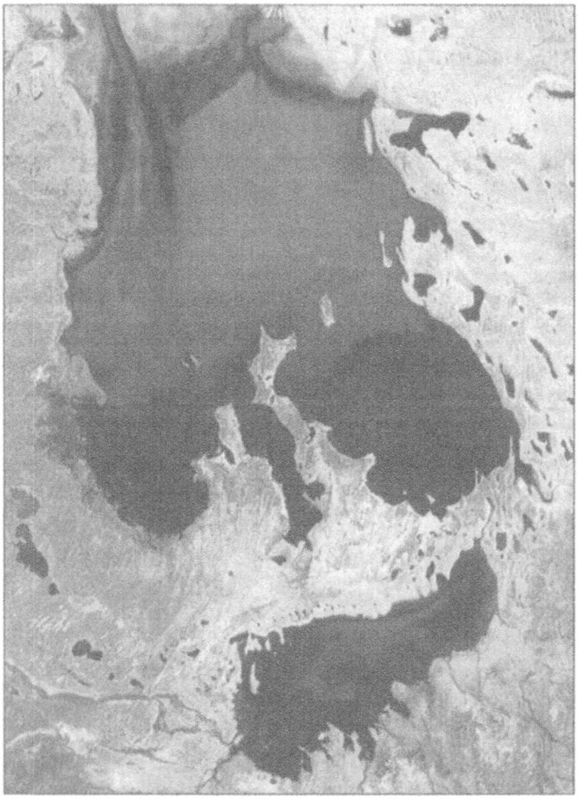

Figure 51. Satellite image of Lake Eyre partially filled, 22 February 1984.

6. CONCLUSIONS

In 1986, the International Lake Environment Committee (ILEC) and UNEP have started a project called "*Survey of the State of World Lakes*", aimed at collecting and analyzing of environmental data on 217 lakes including 64 from Asia, 61 from North America, 56 from Europe, 20 from Africa, 12 from South America and 4 from Oceania (Kira, 1997). The results have indicated that various environmental disruptions are common for the lakes in all continents. Common environmental problems may be classified in the following categories (Kira, 1997):

1) Lake shallowing due to over-use of water from lakes and/or tributary rivers, resulting in a degradation of water quality and lake ecosystems. Rapid salinization is a common feature for originally fresh or brackish inland water bodies.
2) Accelerated siltation in lakes and reservoirs resulting from anthropogenic or natural soil erosion.
3) Lake water acidification resulting from acid precipitation which may resulting in the extinction or of ecosystems. Contamination of water with toxic agricultural and/or industrial chemicals.
4) Eutrophication due to inflow of nitrogen and phosphorus compounds or other nutrients in the discharged water waste water, strongly affecting the biodiversity.
5) In extreme cases, a complete collapse of aquatic ecosystems and desiccation of lakes.

One of the principal tasks for future research is the delimitation of the anthropogenic and natural climate change impacts. Apparently, the climate change is the reason for the water level drop in only a few of the examples discussed in this article. However, if the present climatic trends persist, in the near future it could have a much stronger impact on desiccation of lakes. The ongoing global warming could trigger climate changes difficult to adapt for people and aquatic ecosystems.

Degradation of many inland water bodies as a global environmental problem and its social and economical implications have attracted growing attention of many individuals and organizations. This resulted in a number of relevant national and international research projects or practical programs initiated in the last decades. The 35th International Liege Colloquium on Ocean Dynamics and NATO Advanced Research Workshop "*Dying and Dead Seas*" held in Liege, Belgium, on 5-10 May 2003 concluded, that despite our best efforts, environmental destruction of the lakes in the world continues and the levels of human impact continue to grow. We may be running out of time if we wish to prevent a large scale environmental catastrophe already underway. New ways of thinking that will help to improve policies and practices supporting the vitality of the critical lakes and seas are needed.

Acknowledgments. This work was supported by INCO-Copernicus Project ICA2-CT-2000-10023 *"Desertification in the Aral Region: A Study of the Natural and Anthropogenic Impacts*"; NATO Cooperative Linkage Grant *"Climate Change in the Aral Region: Past Variability, Present Challenges and Future Scenarios"*; Russian Ministry of Industry, Science, and Technologies through Federal Project *"Caspian Sea"* and Grant MD-315.2003.05; and Russian Foundation for Basic Research (Grants 03-05-96630 and 03-05-64926). The satellite altimetry data were obtained from the NASA Physical Oceanography Distributed Active Archive Center (PODAAC JPL), Ocean Altimeter Pathfinder Project (GSFC) and AVISO. Data have been processed using the Integrated Satellite Altimetry Data Base software of the Geophysical Center, Russian Academy of Sciences. The ILEC World Lakes Database was very useful for preparing this review. A.K. and P.Z. are grateful to NATO for supporting their participation in the *"Dying and Dead Seas"* ARW as invited speakers.

7. REFERENCES

Abrosov V.N. *Balkhash Lake*. Leningrad, Nauka. 1973, 180 pp. (in Russian).

Arnow T., Stephens D. *Hydrologic characteristics of the Great Salt Lake, Utah: 1847-1986.* U.S. Geological Survey Water-supply Paper 2332, 1990, 32 pp.

AVISO. www-aviso.cls.fr/html/applications/niveau/global_regional_uk.html.

Bortnik V.N., Chistyaeva S.P. (Eds.). *Hydrometeorology and Hydrochemistry of the USSR Seas, Vol. VII: The Aral Sea.* Leningrad, Gidrometeoizdat, 1990, 196 pp. (in Russian).

Coe T., Foley J.A. Lake Chad has shrunk into 'a ghost'. Afrol News, April 2001. Viewed at www.afrol.com/categories/environment/env062_lake_chad.htm.

ENN. Great Lakes signal great shift in seasonal changes. June 13, 2000.

Gertman I., Hecht A. The Dead Sea hydrography from 1992 to 2000. J. Marine Systems, 2002, 35: 169-181.

Ginzburg A.I., Kostianoy A.G., Sheremet N.A. Thermal regime of the Aral Sea at present time (1982-2000) based on satellite data. Issledovanie Zemli iz kosmosa, 2002, 4: 62-69 (in Russian).

Ginzburg A.I., Kostianoy A.G., Sheremet N.A. Thermal regime of the Aral Sea in the modern period (1982-2000) as revealed by satellite data. J. Marine Systems, 2003 (in press).

Glazovsky N.F. "Aral Sea". In *Enclosed Seas and Large Lakes of Eastern Europe and Middle Asia,* A.F. Mandych (Ed.), SPB Academic Publishing, Amsterdam, The Netherlands, 1995a, 119-154.

Glazovsky N.F. "The Aral Sea Basin". In *Regions at Risk: Comparisons of Threatened Environments,* Jeanne X. Kasperson, Roger E. Kasperson, and B. L. Turner II (Eds.), United Nations University Press, TOKYO - NEW YORK – PARIS, 1995b.

Golubev G.N. "Environmental Problems of Large Central Asian Lakes". In *Guidelines of Lake Management. Vol.8: The World's Lakes in Crisis.* S.E. Jorgensen and S. Matsui, Eds. ILEC Foundation and UNEP, 1997, 67-78.

IPCC WG I. *Climate Change. B.3 Observed Changes in Snow Cover and Land- and Sea-Ice Extent*, Technical Summary, 2001, 83 pp.

Kideys A.E. Fall and rise of the Black Sea Ecosystem. Science, 2002, 297 (30 August), 1482-1484.

Kira T. "Survey of the State of World Lakes". In *Guidelines of Lake Management. Vol.8: The World's Lakes in Crisis.* S.E. Jorgensen and S. Matsui, Eds. ILEC Foundation and UNEP, 1997, 147-155.

Kosarev A.N. *Hydrology of the Caspian and Aral Seas.* Moscow, Moscow State University, 1975, 272 pp. (in Russian).

Kosarev A.N., Tuzhilkin V.S., Kostianoy A.G. Main features of the Caspian Sea hydrology. J. Marine Systems, 2004 (submitted).

Kosarev A.N., Yablonskaya E.A. *The Caspian Sea.* Academic Publishing, The Hague, 1994, 259 pp.

Kostianaya E.V., Ginzburg A.I., Mikhailov V.N., Gregoire M. *The Aral Sea: Selected Bibliography,* (Eds.) J.C.J. Nihoul, A.N. Kosarev, A.G. Kostianoy, I.S. Zonn. Moscow: Noosphere, 2002, 232 pp.

Kostianoy A.G., Lobkovskiy L.I. Satellite monitoring of the Caspian Sea. Vestnik Kaspiya, 2003 (in press, in Russian).

Mandych A.F. *Enclosed Seas and Large Lakes of Eastern Europe and Middle Asia.* SPB Academic Publishing, Amsterdam, The Netherlands, 1995, 270 pp.

Micklin P.P. Desiccation of the Aral Sea: A water management disaster in the Soviet Union. Science, 1988, 241: 1170-1176.

Micklin P.P., Williams W.D. *The Aral Sea Basin.* Proc. NATO ARW "Critical Scientific Issues of the Aral Sea Basin: State of Knowledge and Future Research Needs", Tashkent, Uzbekistan, May 2-5, 1994. Springer-Verlag, Berlin, Heidelberg, New York, 1996, 187 pp.

Mitchell J.G. Down the drain. National Geographic, September 2002, 34-51.

Montaigne F. Water pressure. National Geographic, September 2002, 2-33.

Payne A. I. *The Ecology of Tropical Lakes and Rivers.* John Wiley and Sons Ltd., New York, 1986.

Petr T. Lake Balkhash, Kazakhstan. Int.J.Salt Lake Res., 1992,1(1):21-46.

Petr T., Mitrofanov V.P. "Fisheries in Arid Countries of Central Asia and in Kazakhstan Under the Impact of Irrigated Agriculture". In: *FAO Fisheries Report (Suppl.)* No.512, Rome, FA0, 1995, 40-79.

Rodionov S.N. *Global and Regional Climate Interaction: The Caspian Sea Experience.* Kluwer Academic Publishers, Dordrecht, the Netherlands, 1994, 243 pp.

Savvaitova K., Petr T. Lake Issyk-kul, Kirgizia. Int.J.Salt Lake Res. 1992, 1(2): 21-46.

Scientific American. Lake Chad is Disappearing, March 1st 2001. Viewed at www.sciam.com/news/030101/4.html.

Serruya C., Pollinger U. *Lakes of the Warm Belt.* Cambridge University Press, Cambridge, 1983, 569 pp.

Shaporenko S.I. "Balkhash Lake". In *Enclosed Seas and Large Lakes of Eastern Europe and Middle Asia,* A.F. Mandych (Ed.), SPB Academic Publishing, Amsterdam, The Netherlands, 1995, 155-198.

Small E.E., Giorgi F., Sloan L.C., Hostetler S. The effects of desiccation and climatic change on the hydrology of the Aral Sea. J. Climate, 2001a, 14: 300-322.

Small E.E., Sloan L.C., Nychka D. Changes in surface air temperature caused by desiccation of the Aral Sea. J. Climate, 2001b, 14: 284-299.

Terziev F.S., Kosarev A.N., Kerimov A.A. (Eds.). *Hydrometeorology and Hydrochemistry of the Seas. Volume VI: The Caspian Sea: Issue 1: Hydrometeorological Conditions*, Hydrometeoizdat, St. Petersburg, 1992, 360 pp. (in Russian).

Tsigelnaya I.D. "Issyk-Kul Lake". In *Enclosed Seas and Large Lakes of Eastern Europe and Middle Asia,* A.F. Mandych (Ed.), SPB Academic Publishing, Amsterdam, The Netherlands, 1995, 199-230.

Vinogradov M.E., Shushkina E.A., Anokhina L.L., Vostokov S.V., Kucheruk N.V., Lukashova T.A. Mass development of the ctenophore Beroe ovata Eschscholtz near the northeastern coast of the Black Sea. Okeanologiya, 2000, 40(1): 46-49 (in Russian).

Vinogradov M.E., Shushkina E.A., Vostokov S.V., Vereshchaka A.L., Lukasheva T.A. Population dynamics of the ctenophores Mnemiopsis leidyi and Beroe ovata near the Caucasus shore of the Black Sea. Okeanologiya, 2002, 42(5): 693-701 (in Russian).

Wang Hongchang. Deforestation and desiccation in China. A Preliminary Study. 1993. Viewed at www.library.utoronto.ca/pcs/state/chinaeco/forest.htm.

Wheeler S.S. *The Desert Lake: the Story of Nevada's Pyramid Lake.* Caxton Printers, Ltd., Caldwell, Idaho, 1974.

World Lakes Database. www.ilec.or.jp/database/database.html.

Zavialov P.O., Kostianoy A.G., Emelianov S.V., Ni A.A., Ishniyazov D.P., Khan V.M., Kudyshkin T.V. Hydrographic survey in the dying Aral Sea. Geophys. Res. Lett., 2003a, 30: doi: 10.1029/2003 GL017427.

Zavialov P.O., Kostianoy A.G., Sapozhnikov Ph.V., Scheglov M.A., Khan V.M., Ni A.A., Kudyshkin T.V., Pinkhasov B.I., Ishniyazov D.P., Petrov M.A., Kurbaniyazov A.K., Abdullaev U.R. Modern hydrophysical and hydrobiological state of the western Aral Sea. Okeanologiya, 2003b, 43(2): 316-319 (in Russian).

Chapter 2

WATER MANAGEMENT ASPECTS OF AMU DARYA

Options for future strategies

Jochen Froebrich,
Fachgebiet Gewässergütemodellierung - Prof. Herwig. Lehmann,
Water Quality Protection and Management Section,
University of Hannover, Germany

Odil Kayumov
Scientific Technical Center Tosza Darya, Tashkent, Uzbekistan

1. INTRODUCTION

"Where the water ends, the land ends". This statement from Central Asia makes clear, better than any numbers, the vital dependence of local people on the fresh water of the Aral Sea tributaries. The use of this water for irrigation has a long history. Numerous oases have existed since ancient times and these were supported by the favourable flow regime of the Amu Darya, including the low saline summer flood.

J.C.J. Nihoul et al. (eds.), Dying and Dead Seas, 49–76.

Today, the problem lies in the scale of irrigation , the choice of cultivated crops and the decline in water quality due to the use of agro-chemicals. The greatest deterioration occurred from the 1960s to the 1980s. Since then, the extensive use of large amounts of pesticide has been reduced owing to economic constraints, but accumulated pesticides in soils and the salinity of water remain serious issues. Past environmental destruction has led not only to a reduction in the level of the Aral Sea, but is also endangering human existence in the Aral Sea deltas. Child mortality is very high compared to other NIS and diseases such as typhoid fever, viral hepatitis, anaemia, respiratory diseases, and different types of cancer are widespread (ELPINER, 1999). More than 3 million people have no access to a satisfactory drinking water supply. In addition to huge irrigation water withdrawals and water losses, the quality of return water also causes major ecological problems.

The paper presented here will describe selected aspects of water quality and quantity problems in the Amu Darya basin. Since the Aral Sea crisis cannot be understood without reference to the Amu Darya and Syr Darya tributaries, the paper should generate a greater appreciation of the complexity and interrelationships influencing the hydrology of the region.

Urgent action is needed to achieve at least the bare minimum conditions for the survival of both ecosystems and communities in the region.. Using revised operating rules for existing reservoirs the local population might be supplied with safe potable water in future. Therefore a particular focus is given to the operation of the Amu Darya reservoirs, primarily the dams of the Tuyamuyn Hydroengineering Complex.

2. HYDROGRAPHY OF THE AMU DARYA BASIN

The Amu Darya is part of the Aral Sea basin and its watershed stretches across the territories of Turkmenistan, Afghanistan and the Republics of Tadjikistan, Uzbekistan and Kyrgyzstan. The total Amu Darya basin area covers 612,300 km^2. The river length amounts to 1,415 km (GRDC/UNH) from the confluence of the Pyandj and Vaksh to the Aral Sea. After passing the mountainous upstream part, the Amu Darya meanders through a desert region in the 880 km-section from 100 km east from the city of Termez until it reaches the Tuyamuyn Hydroengineering Complex (THC). The lower Amu Darya region ranges from the THC downstream to the dying Aral Sea (Figure 1). This region, which was a natural river delta in former times, is today characterised by intensive irrigation activities and an extended canal and collector system.

The Amu Darya has the greatest discharge and the largest catchment area of all the rivers in Central Asia. About 80 – 85 % of the Amu Darya flow volume is formed by the Vahsh and Pyandj rivers and the remaining 20 – 15% come to from the Surchandarya, Kafirnigan and Kunduz rivers. The total runoff formation area of the Amu Darya basin is about 226,800 km², of which 7,400 km² are located at glaciated areas.

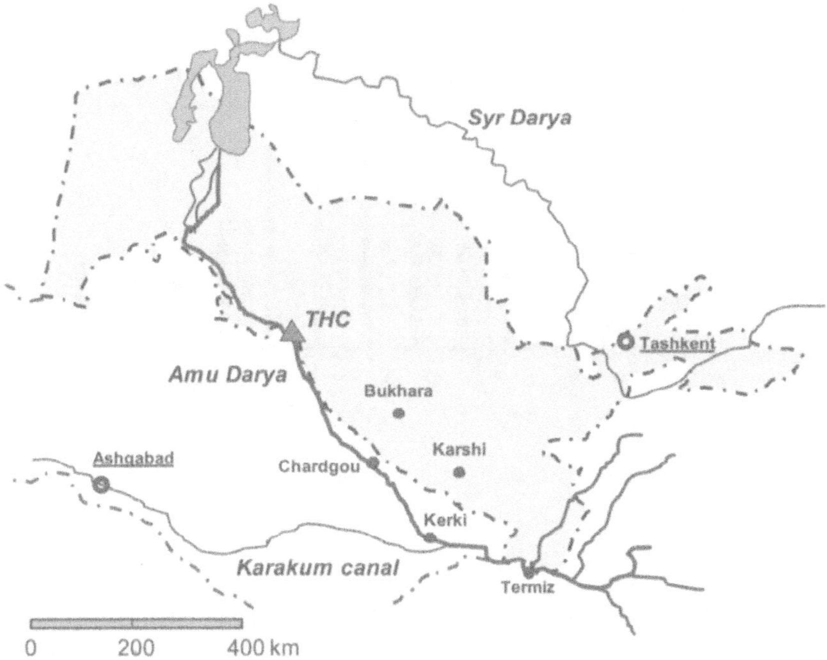

Figure 1. Amu Darya and Syr Darya – the Aral Sea tributaries and location of THC

There is a characteristic flow regime. About 80 % of the annual run-off occurs during April – September and is generated by snow and glacier melting in the Pamir mountains.

Figure 2 presents the Amu Darya flow regime at Kerki station, showing monthly averages for the period 1981 – 1997 (KAYUMOV, 2003). Here the rise in water level and run-off starts in May and the flow maximum is reached in June. Annual water flow of the Amu Darya ranges from 58.6 km³ up to 109.9 km³, these being 5 and 95 percentiles of flow. The average flow volume is 78.4 km³/a.

Tributaries of the Amu Darya are only located at the first 180 km - including the confluence of Pyandj and Vahsh rivers at the upper limit of the Amu Darya. The lack of tributaries in the downstream part, high evaporation (Urgench, average: 1364 mm/y), low precipitation (Urgench, average: 110 mm/y), and infiltration in the groundwater cause a natural decrease in the river discharge. Additional abstractions of water for irrigation, industrial and municipal supply, intensify this decline.

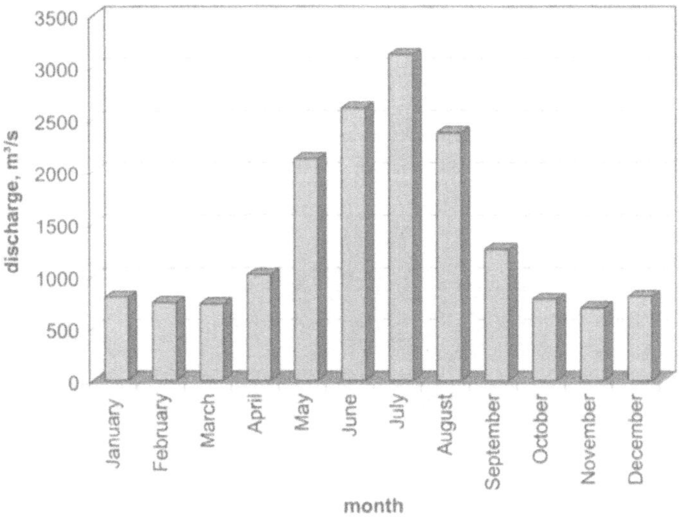

Figure 2. Characterisation of the general flow variability of the Amu Darya, Station Kerki (37°50′N/65°15′E), monthly average of the period 1981 – 1997)

The most important land use in the Amu Darya basin is agricultural cultivation (Figure 3), which depends entirely on irrigation. This irrigation areas has increased during the last 50 years. Major water losses account for 50-80 % of the total water withdrawals, these being due to old and inefficient irrigation channel networks, inefficient irrigation methods, and no economic incentives for water saving.

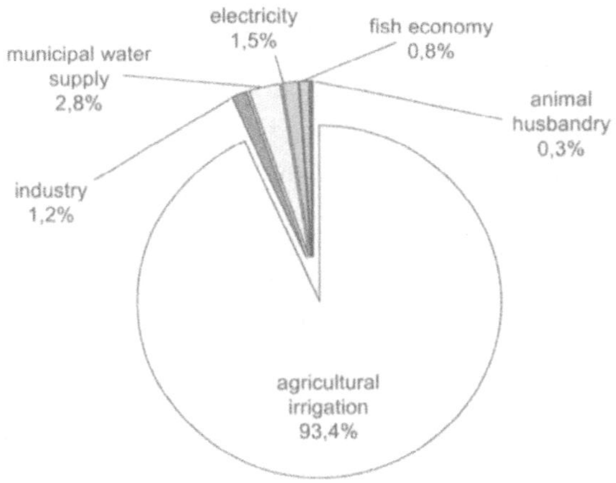

Figure 3. Main water consumers in the Amu Darya basin (after MICKLIN in RESSEL, 1999)

Since 1918 the construction of agricultural irrigation projects has been carried out successively in Turkmenistan and Uzbekistan. A significant part of Amu Darya water is abstracted by the Karakum canal. This 1300 km long channel crosses the Karakum desert and secures the water supply to many regions of Turkmenistan. It is the biggest irrigation-channel in the Amu Darya basin and in Central Asia. Up to now there has been no exact information about the average annual withdrawal from Amu Darya; different estimations range from 8 to 12 km^3/a . Other scources refer to an estimated abstraction of 15 km^3/a in 1990 (= approx. 500 m^3/s) (LIEBMANN, 1990). Although there is an agreement between Uzbekistan and Turkmenistan that limits Turkmenistan's withdrawal to about 13 km^3 per year, the compliance is not assured or enforced (HANNAN AND O'HARA, 1998). Additionally the water losses of the Karakum canal are not identified. The channel is built mostly on desert sand without sufficient sealing to prevent high seepage losses (ORLOVSKY, 1999).

Apart from the Karakum-canal, there are a number of other major intake points downstream of Termiz:

- large parts of the Surchandarya irrigation region are supplied by Amu Darya water.

- Irrigation water is removed near Kerki. This water flows to the city of Karshi through the Karshi canal.
- The Amu-Bukhara canal, for water supply to the irrigation areas in the Bukhara region.
- Irrigated agricultural lands adjacent to the Amu Darya near Kerki and Chardgou, Turkmenistan.

The Amu Darya flow regime is also influenced by anthropogenic impacts, such as the Nurek reservoir in Tajikistan and the dams of the Tuyamuyn Hydroengineering Complex, which are described more in detail below.

3. WATER QUALITY OF THE AMU DARYA

3.1 Sources of pollution

Return water from irrigated fields, containing salts, pesticide residues, defoliants, and nutrients are the major pollution source in the Amu Darya basin. In addition to pollution from return flow, there is also contamination by industrial and municipal wastewaters. Significant longitudinal and temporal water quality variations are a characteristic of the Amu Darya region. These are due to the seasonal flow variation of the less polluted headwater tributaries and the seasonality of return water quantity and quality at different inflow locations.

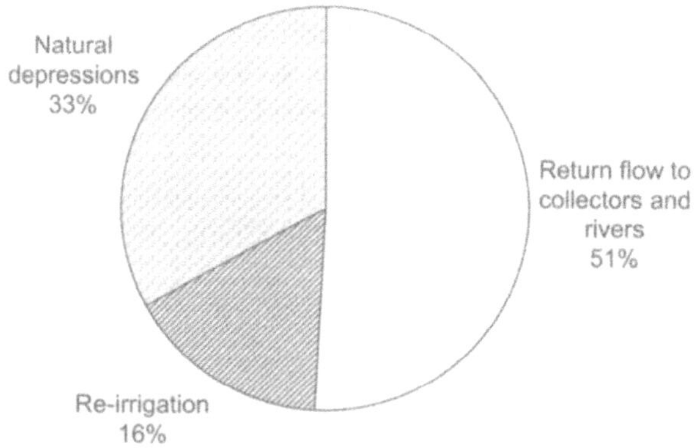

Figure 4. Apportionment of return water in the Amu Darya basin (FROEBRICH AND KAYUMOV, 2003)

The development of irrigation and the construction of required drainage systems have produced an increase in return water volume, especially during the years 1960-1990. Return water is partly collected in artificial parallel channels and enters the Amu Darya either directly or by natural tributaries such as the Surchandarya. The annual total return water volume in the Amu Darya ranged between 16 and 19 km^3 per year during1990-1999. Nevertheless, this quantity corresponds to just 51 % of the total return water volume which is accrued at the irrigated lands. Of the reminder, 33 % is lost to natural depressions and 16 % is used for re-irrigation (Figure 4) (FROEBRICH AND KAYUMOV, 2003). According to estimates by RUBINOVA AND GORELKIN (1998) the annual export of salts with return flow amounts to 87 million tons in the Amu Darya basin.

With the end of the Soviet Union, the extension of irrigation infrastructure has stopped. Difficulties in maintaining the drainage system during the transition period and the commencement of water-conserving measures have resulted in a decrease in return water volume.

Major irrigation areas, which are pollution sources are located in the Bukhara, Karshi and Surchandarya regions, the irrigated fields on the banks of the Amu Darya between Termez and THC and irrigation areas in the Vaksh and Pyandj catchments. These areas affect water quality as far as THC. Downstream of THC the water quality is affected by return water inflow from Khorezem, Darshauz and Karakalpakstan. The proportion of salt loads from different catchments to the river section between Termez and THC is illustrated in Figures 5 and 6. These show that the Bukhara region generates nearly 25% of the salt load of the middle Amu Darya irrigation areas (REDER AND FROEBRICH, 2003).

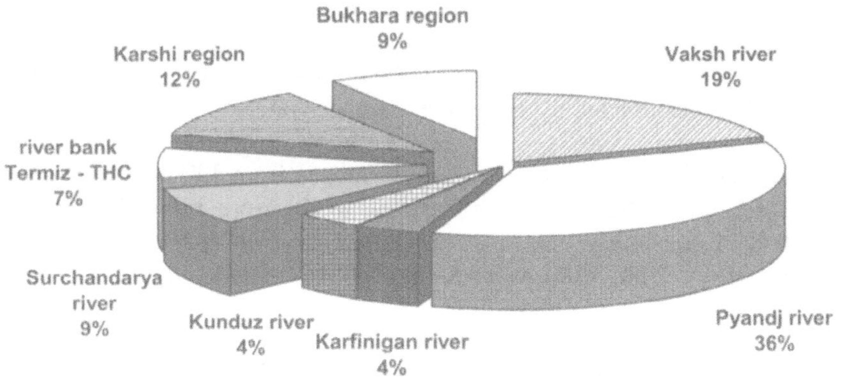

Figure 5. Approximated and estimated percentage distribution of salt load sources in the Amu Darya basin: consideration of the annual salt load at Lebap station (100%)

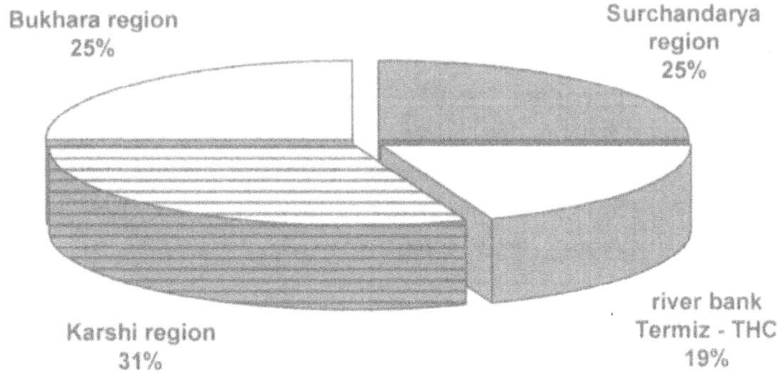

Figure 6. Percentage distribution of salt loads with origin at the irrigation regions of the middle Amu Darya reach (Bukhara, Karshi and Surchandarya regions, and the lands directly of the Amu Darya the banks between Termez and THC)

3.2 Water quality status

Before irrigation activities were greatly expanded, the annual Amu Darya salinity ranged from 200 mg/l to 500 mg/l (Alekin, 1955). Nowadays salinity is significant higher and is still increasing.

The rise of salinity along the river longitudinal profile reveals in particular the effects of the washing out of mineral loads when fields leach (Figure 7). Present data and results also indicate that even the fairly low saline summer flood is affected by inflowing drainage waters.

The monthly average value of salinity ranges from 350 mg/l to 800 mg/l at Termez station, from 350 to 900 mg/l at Kerki, from 350 to 1100 mg/l at Chardgou, from 500 to 1500 mg/l at Darganata, from 600 to 1500 mg/l at Lebap station, from 700 to 2600 mg/l at Samanbay and from 700 to 2800 mg/l at Kiziljar (FROEBRICH AND KAYUMOV, 2003).

The seasonal variation in water salinity is affected by the temporal variation in river discharge. Before the beginning of the growing season in March, the fields are leached and a huge quantity of salts are dissolved and transported with the return water to the collectors and partly to Amu Darya. Due to low flow conditions the inflow of drainage waters leads to Amu

Darya salinity peaks in March, April and partly in May. During the summer flood, the dilution prevents further salinity increase and the lowest salinity concentrations are found between May and September.

Uzbek water quality standards give a threshold of 1000 mg/l salinity both for domestic and industrial water use and for fishery purposes (GOST, 1983). In the upstream section (from the confluence of Vaksh and Pyandj downstream to Chardgou) salinity remains below the limit at all times. Downstream of Chardgou, the limit is exceeded from October to April. Here, it is only in the summer months of flooding (May until September) that there is an acceptable water quality standard.

In the lower river section (downstream of Samanbay) there is very high salinity throughout the year. The high sulphate concentration during the winter is particularly alarming in view of the impact of potable water on human health. The ions exceeding their maximum thresholds are sulphate, magnesium, sodium, and potassium.

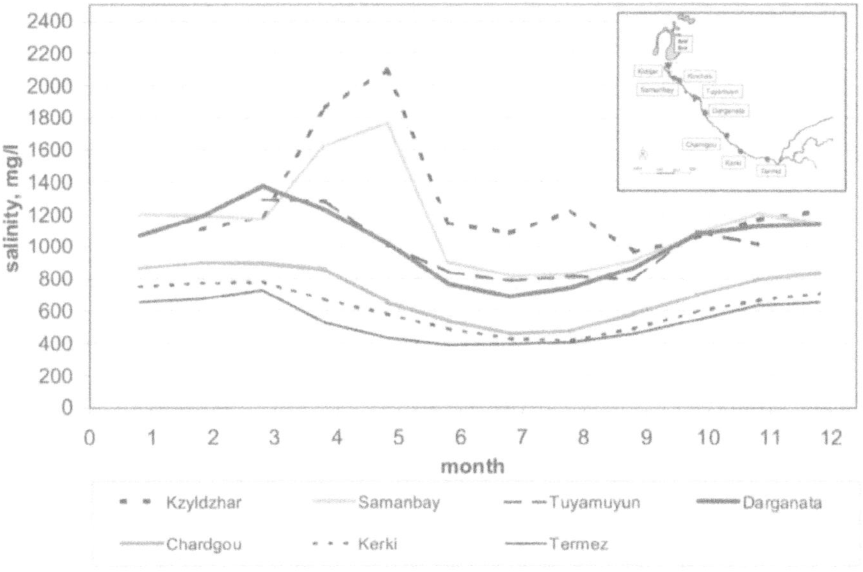

Figure 7. Monthly averaged salinity at seven sampling points on the Amu Darya (average of the period 1991-2001) (FROEBRICH AND KAYUMOV, 2003)

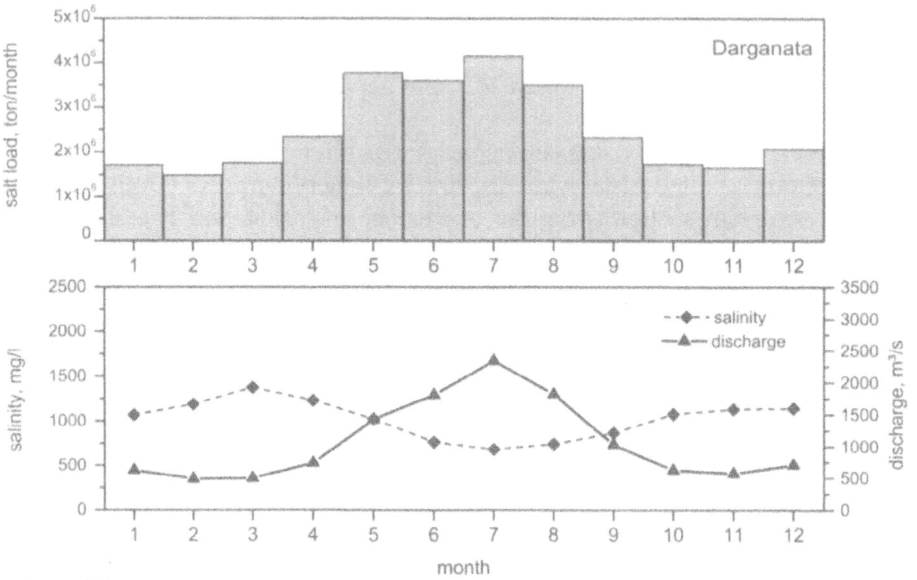

Figure 8. Darganata; upper plot: average monthly salt loads; bottom plot: seasonal variation of averaged monthly flow and averaged monthly salinity (discharges data 1981-2000, salinity data 1991-2001) (FROEBRICH AND KAYUMOV, 2003)

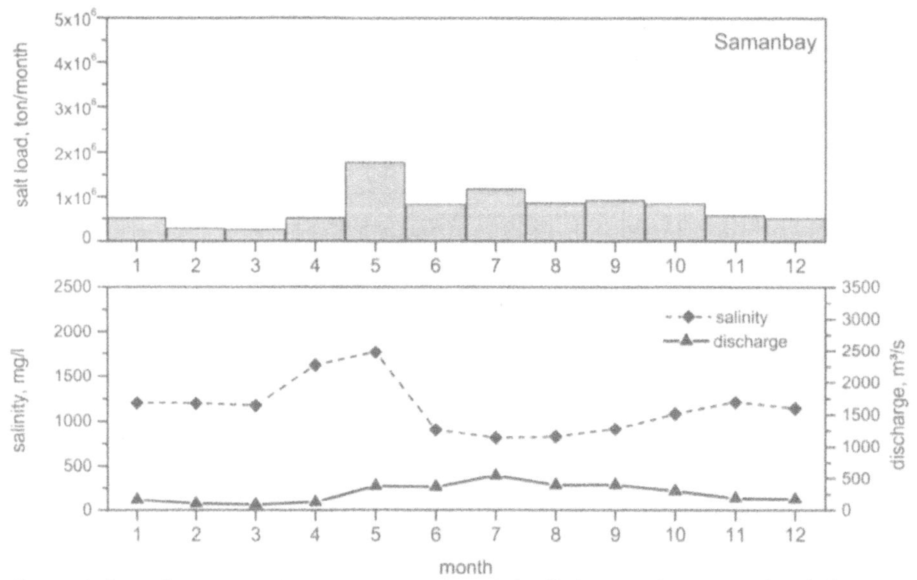

Figure 9. Samanbay; upper plot: average monthly salt loads; bottom plot: seasonal variation of averaged monthly flow and averaged monthly salinity (discharges data 1981-2000, salinity data 1991-2001) (FROEBRICH AND KAYUMOV, 2003)

The interconnection between the seasonal variation in flow volume and salinity reduction due to dilution effects is illustrated in Figures 8 and 9. Salt loads reach a peak in May and July at both Darganata and Samanbay stations.

In addition to seasonal water quality variations during one year, longer-term variability is also of interest. Water quality deteriorates in dry years, when the magnitude of the summer flood is only a fraction of the flow rate during average years. (Table 1)

Table 1. Salinity at Darganata in humid, average and arid years (KAYUMOV AND IKRAMOVA, 1997)

		more humid years	average	more arid years
Annual discharge	km³	45 - 65	25 - 45	20 – 25
Salinity *range (average)*	g/l	0.6 … 1.8 (0.95)	0.7 … 2.0 (1.1)	0.85 … 2.2 (1.5)

Figure 10 shows the salinity change in the main sections of the Amu Darya for the period from 1932 to 1999 (IFAS, 2001). Since the beginning

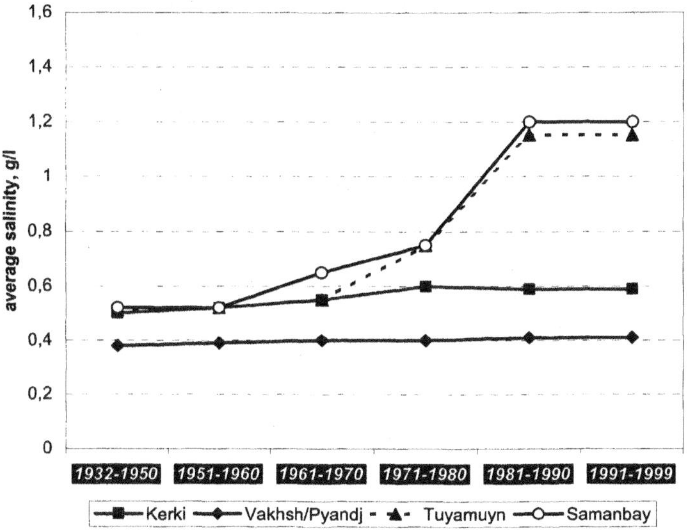

Figure 10. Long-term salinity development (IFAS, 2001)

of the 1960s, water salinity in the middle and downstream section of Amu Darya has increased sharply.

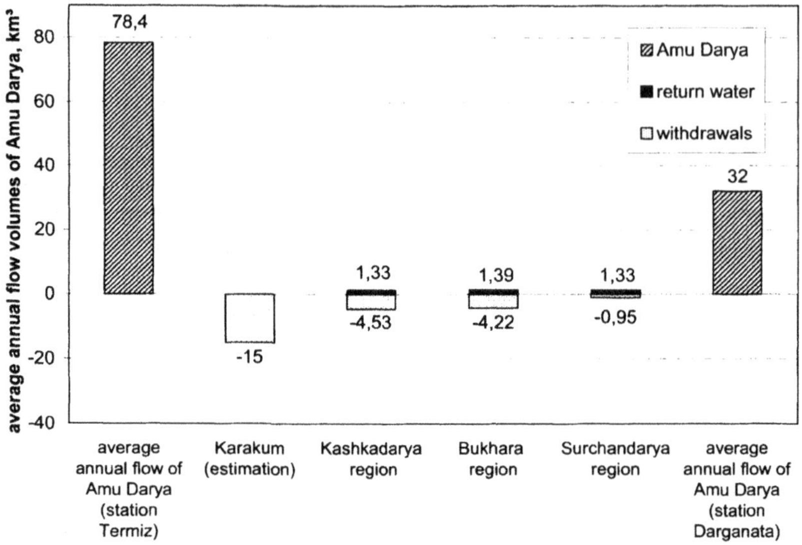

Figure 11. Flow volumes in Amu Darya basin (upper and middle section) (FROEBRICH AND KAYUMOV, 2003)

Table 2 summarizes the exceeding of concentration standards for dangerous substances in the Amu Darya. Different patterns of prior problems are visible for the different flow sections of the Amu Darya.

Table 2. Endangering substances in Amu Darya

	exceeded standards	
river section	**drinking water (GOST, 1983)**	**fishery**
Upstream (Termez)	phenols oil products suspended solids	phenols oil products
Middle section (Kerki, Chardgou, Dargangata, Tuya-muyn)	salts phenols oil products suspended solids	salts phenols oil products
Lower section (Samanbay, Kiziljar)	salts phenols oil products suspended solids water hardness pesticides	salts phenols oil products pesticides

Due to the high flow velocity of the river (high turbulence, high erosion capacity) and the large amount of eroded silt at the foot of the Pamir-mountains, the Amu Darya transports a huge amount of suspended matter. Owing to the high transport capacity and high suspended matter concentrations, the river bed changes rapidly in the flow section upstream of the THC.

Measured concentrations of suspended solids range between 1000 and 4000 mg/l. During the minimum flow period lower concentrations occur. With the beginning of the flood period concentrations of suspended matter reach a maximum. Nevertheless the suspended matter concentrations during the summer months are highly variable and this prevents the determination of a typical concentration/discharge relationship. The concentrations are much dependent on the transport capacity and conditions in the preceding year. Spatially, the highest suspended matter concentrations are present in the flow section before reaching THC. The settlement of particles in the channel reservoir leads to a significant reduction of suspended matter. The suspended matter concentration in the reach downstream of THC is therefore much lower.

The high concentrations of dissolved oxygen and the low concentrations of nitrogen compounds and BOD exclude the presence of significant organic pollution loads and intensive biological activities within the river. Samples from Termez and Tuyamuyn show comparable concentrations and stable water quality during all the sampled months. Data from the stations downstream of Tuyamuyn indicates a higher COD.

Pesticide use is difficult to estimate, so pesticide exposure risk is uncertain. Current data are available for DDT, alpha-HCH and gamma-HCH. Alpha-HCH and gamma-HCH are detected in the middle and upstream reaches, where concentration exceeds standards for fishery purpose, but not for potable water supply.

At the Amu Darya delta, an increase in the proportion of return water and reduced flow prevent a comparable dilution. Here alpha-HCH concentrations frequently exceed standards both for drinking water and fishery purposes.

The phenol concentrations measured in all the Amu Darya sections range from 0 to 0.008 mg/l and mostly exceed the critical threshold of 0.001 mg/l. In the upper and downstream sections the concentrations are at a comparable level. Major industrial pollution sources are located upstream of Termez. The stable concentrations along the flow course indicate that significant degradation processes do not occur.

The maximum permitted threshold of oil products (both for fishery use: 0.05 mg/l and for potable water supply: 0.10 mg/l) are exceeded by most samples at Samanbay and Kiziljar in the delta region. Concentrations for oil products exceed critical values also at the Termez and Tuyamuyn sampling points.

In the Amu Darya delta region not only the surface waters but also groundwater resources are affected by the extensive use of pesticides, defoliants and fertilizers. A study site in Khorezm district indicates maximum concentrations of 0.16 ppm (ground water) and 1.33 ppm (soil) - cumulative value for DDT and its metabolites (BODGADASAROV ET AL., 1998).

Summarizing the evaluation of data from 1990 to 2002 indicates the following characteristics of the Amu Darya:

- High salinity considerably affects the quality and the usability of the Amu Darya waters.
- Phenol concentration exceeds standard norms. There are both natural (foliage decomposition in upstream river sections) and industrial sources.
- Oil products particularly endanger some aquatic habitats.
- Pesticide concentrations declined slightly during the last decade, but remain a serious problem.
- Heavy metal concentrations generally do not exceed the critical values in the Amu Darya, but are a danger to some habitats owing to agglomeration processes.
- The impact of organic pollution sources within the Amu Darya catchment area (upper and middle river section) is considered low. The organic pollution of the Amu Darya is reduced by intensive degradation processes and considerable dilution.
- Traces of faecal contamination are below the limits indicative of severe pollution and do not greatly affect river and reservoir quality.

4. RESERVOIRS OF THE TUYAMUYN HYDROENGINEERING COMPLEX

Water quantity and water quality dynamics of the Amu Darya are largely affected by the Nurek dam in the upstream region and Tuyamuyn Hydroengineering Complex (THC) in the downstream region of the river.

The Tuyamuyn Hydroengineering Complex is the source of the irrigation water for the Khorezm, Darshauz, and Karakalpakstan regions. Operation of the THC greatly influences the downstream water quantity and quality of the Amu Darya. While the direct water abstraction from the reservoir is low, most of the water stored in THC is discharged to the Amu Darya during the irrigation season. Nevertheless the flow of the Amu Darya to the Aral Sea is at a very low level. The water, which passes the THC is almost completely consumed downstream by additional intake structures and irrigation channels. The total inflow to the Aral Sea was only 5 km^3/a in recent years, compared with 50 - 60 km^3/a before 1960 (UN, 2001).

Water quality in the THC mostly depends on the seasonal variation of the inflow water quality and its alteration by storage, mixing, and transformation within the reservoirs. Dam operation affects the suitability of water quality for irrigation and for potable water supply. Simultaneously, dam operation is determined by the seasonal demands for irrigation water, so the availability of an adequate supply of water of a given quality and dam operation cannot be considered independently.

As the salinity of the stored water increases, its suitability for irrigation and as potable water decreases. For a better understanding of the interdependencies and boundary conditions for water resource deterioration and for optimisation potentials, the main characteristics and dynamics of the THC are presented below.

4.1 General aspects

The THC was constructed to impound the Amu Darya to provide water for irrigation, industry and potable water supply for the lower Amu Darya region. The hydroengineering complex is situated at the beginning of the Amu Darya delta, 300 km south of the former Aral-Sea shore line. In 1981-1983 the THC construction was completed. It consists of four large dams with an initial total storage capacity of 7.8 km^3 (Figure 12). Due to siltation losses, the total storage was reduced to 6.8 km^3 in 2001 (Table 3).

Currently it is mainly used to provide irrigation water for the regions of Khorezm, Karakalpakstan and Daschauz. The entire irrigation area consists of 1,138 Mio ha. 0.731 Mio ha are located in Uzbek territory and 0.407 ha are part of Turkmenistan.

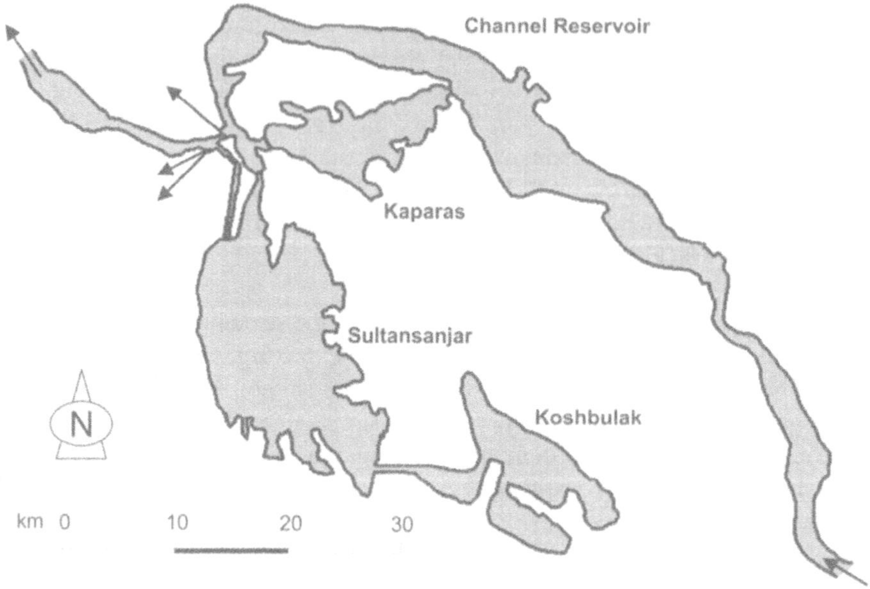

Figure 12. Sitemap of the Tuyamuyun Hydroengineering Complex (THC) and locations of sampling areas. Kaparas reservoir has only one in/outlflow structure at the western link to the channel reservoir. The eastern link is permanently closed by a dam.

The main crop is cotton; other frequent crops are rice and wheat. Furrow irrigation is the usual technology, using an extensive channel/collector system. Because main water use is irrigation, the highest water demand from THC is during the vegetation period from March until September. In spring (March and April) there is a large water demand for flooding fields and leaching salts from soils. In the remainder of the irrigation season, water use by plants is the main purpose of irrigation.

In general the irrigation period is well supported by the natural seasonal flow regime of Amu Darya (with a summer flood maximum in July), this enabled irrigation since ancient times. Because of the tremendous expansion of irrigation in the time of the Soviet Union, the dams are needed to satisfy increased water demand and changed seasonal water demands. The field flooding in spring coincides with the Amu Darya minimum flow.

Table 3: Hydrological characteristics of Tuyamuyn Hydroengineering Complex (THC) consisting of four interconnected reservoirs

Parameters			Channel reservoir	Kaparas	Sultansanjar	Koshbulak	Total
Full capacity (1981)		km^3	2.34	0.96	2.69	1.81	7.80
Useful capacity (1981)		km^3	2.07	0.55	1.63	1.02	5.27
Full capacity (2001)		km^3	1.40	0.96	2.63	1.81	6.80
Useful capacity (2001)		km^3	1.31	0.64	1.63	1.02	4.60
Area of surface at MOL'		km^2	303	70	149	128	650
MOL point		m a.s.l.	130	130	130	130	130
DLV'' point		m a.s.l.	120	120	116	120	
Length		km	102	15	24	26	
Width	Maximal	km	11	9	12	11	
	Average	km	4	4	8	6	
depth at MOL	Maximal	m	20	36	38	41	
	Average	m	7.7	13.7	18	14.2	
depth at DLV	Maximal	m	10	26	28	31	
	Average	m	208	9.3	10.8	12.7	
Shallows area of 2m depth	At MOL	km^2	93	6	10	7	116
	At DLV	km^2	59	4	2	9	74

'MOL – Maximum Operating Level
'' DLV – Dead Level Volume

Because of the expansion of irrigation annual variation of water stored by THC becomes a problem. The THC was not designed to store enough water to meet multiple year water needs. In some years the total flow of the summer flood is less than needed to meet water demand, and since available water supply under low flow conditions depends largely on the previous years flow, agricultural production is very much affected by water shortages when prior year flows are inadequate. This occurred in 2001.

The large dependency of irrigation on the THC highlights the strategic importance of the THC. In comparison to the past Soviet system, the independency of Uzbekistan and Turkmenistan results in a much more complicated situation. As indicated in Figure 12 the border between Uzbekistan and Turkmenistan runs along the Channel reservoir. The river dam is situated both on Uzbek and Turkmen territory. All bypass reservoirs Koshbulak, Sultansanjar and Kaparas are located in Turkmenistan. Uzbekistan has a large irrigated area, the question of controlling the dam operation is quite critical. There are a number of agreements, which secures water supply for Uzbekistan, but the overall coordinated operation of the THC mainly depends on actual intergovernmental relationships.

Irrigation water, discharged to the Amu Darya downstream, is also used by a hydroelectric power plant. The plant is located at the main dam of the Channel reservoir and used for regional and local electric power supply.

While THC is used for irrigation in the first place, it also provides water for some municipalities in the lower Amu Darya delta. There are two pumping stations located on the both sides of the river. There is an operating purification plant for domestic and industrial use.

4.2 Reservoir Management

In general, water management of the THC is based on the concept that water can be directed to Kaparas, Koshbulak and Sultansanjar reservoir by free surface flow, if the water at the Channel Reservoir reaches a level higher than the three bypass reservoirs. Filling of Kaparas reservoir might be independent of filling Sultansanjar and Koshbulak reservoirs. All water to Kaparas reservoir until recently has passed the channel reservoir. Water stored in Koshbulak reservoir first must pass the Channel Reservoir and then the Sultansanjar reservoir. All side dams can be closed by an intake structure, so the outflow control is potentially independent as long as the water level of reservoirs located in front of it remains at a lower elevation.

Due to the dependency of water elevation in the Channel reservoir, the entire THC control depends largely on the overall hydrological conditions of the Amu Darya region. In short water years, both total water availability and operation schemes are limited.

Figure 13 demonstrates the over annual variability of the THC operation. At the beginning of the year depletion prevails (leaching the agricultural fields). Filling takes place during the summer flood of the Amu Darya and the subsequent months until the next spring.

The inflow of Amu Darya is about 30-35 km^3/a under average conditions. The time series in Figure 14 indicates that most of the outflow is discharged to the Amu Darya downstream (80%). An average of 20% is abstracted directly to irrigation channels from the reservoir, with a maximum flow of 500 m^3/s. The time series of reservoir outflow indicates also that the largest volume of water is discharged during the vegetation period. The total volume for irrigation is greater than the amount discharged for leaching the fields.

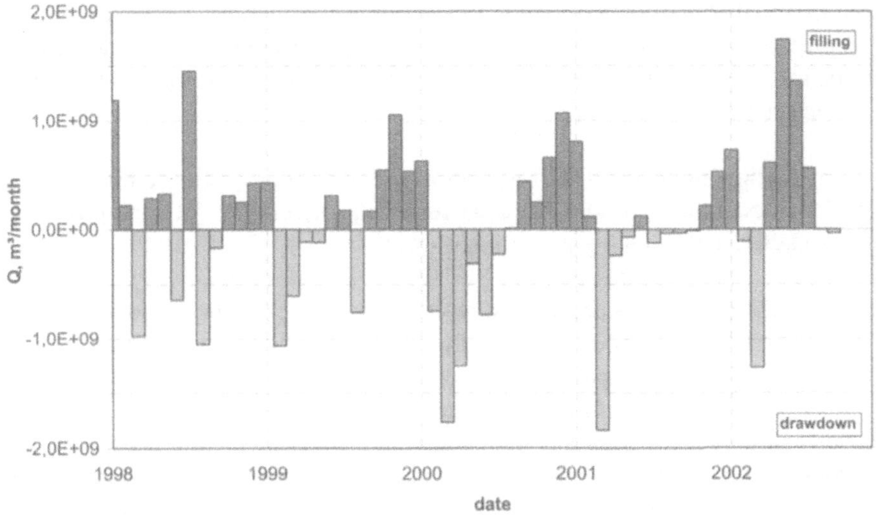

Figure 13. THC flow regime; difference of total outflow and inflow (GLAVIDROMET)

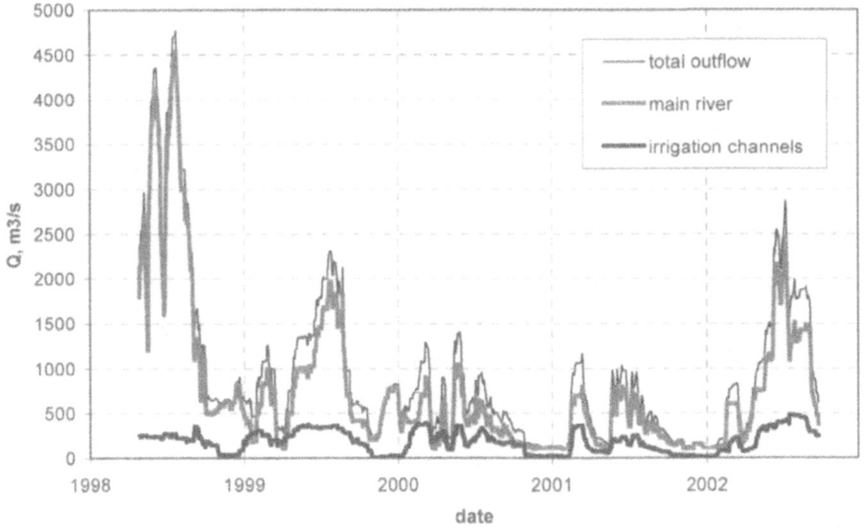

Figure 14. THC flow regime; distribution of total outflow, discharge to the Amu Darya downstream and, direct abstraction for irrigation channels (GLAVIDROMET)

The current THC reservoir management strategy indicates an extended water exchange in the entire complex. Due to the present mass exchange, water quality dynamics are very much dependent on the temporal variations of inflow water quality. In an average water year approximately 25-time of the Channel Reservoir volume is flowing through it. At Kaparas reservoir a volume of 320 million m^3 is not useable by the design of the water intake structure. Nevertheless the complete usable volume of 640 million m^3 is being emptied during the year.

Interaction between THC operation and resulting water quality can be well demonstrated for the Kaparas reservoir. Salinity is of major interest amongst the water quality parameters.

The salinity of the Kaparas reservoir is depending on the seasonal salinity of the Amu Darya inflow and on the water already stored in the Channel reservoir. The falling part of the summer flood wave is characterised by higher salinity. Storing this part of the flood wave in the Channel reservoir results in a salinity increase. The impact of water with low salinity at the beginning of the summer flood by mixing with a high salinity in the stored water depends also on the amount of water discharged from Channel Reservoir downstream in spring. Figure 15 demonstrates the bi-annual variability of the salinity at the Kaparas reservoir as measured in 1990 and 1991 (NIKOLAYENKO, 1991). The beginning period is determined by high salt concentrations up to 1500 mg/l. During the summer stratification occurs and salinity in top layers is lower than in deeper ones. After lowering the reservoir in the early summer, the increasing volume after August is related to a reduction of salinity. Nevertheless the salinity remains high (about 1000 mg/l) compared to the minimum 692 mg/l on July 13th, 1990 and 697 mg/l on July 30, 1991. During the following year, salinity increases again after March, which is directly related to higher salinity of the inflow during the Amu Darya low flow period.

The impact of salt release was presented by SAVITSKY (2003). He states that the release rate reached an equilibrium level of 5300 ton/month. The intensity of the salt release from the bottom depends on the concentration gradient, the intrusion in upper layers by wind mixing and by the residence time of the water above. Modelling results by REDER (2003) indicates a stable stratification during the summer month and wind mixing in spring and autumn.

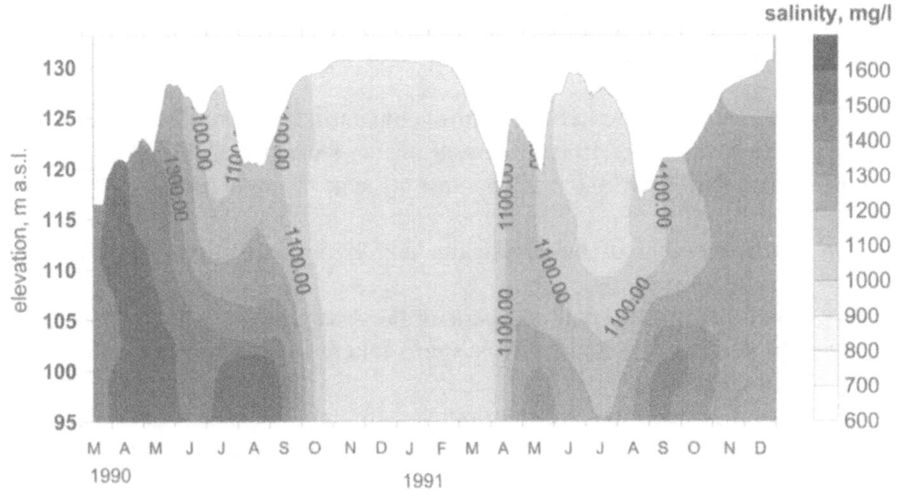

Figure 15. Time-depth plot of salinity dynamics in Kaparas reservoir, period from March 1990 until December 1991

Kaparas reservoir could be considered as a non-eutrophicated system. Oxygen profiles refer to frequent absence of anaerobic conditions in the Hypolimnia. There is some oversaturation present in the surface layers, but both algae biomasses (by chlorophyll-a and cell numbers) and nutrient availability remains low (CROSA AND CALAMARI, 2003)

Due to sedimentation during the water passes the Channel Reservoir, the major part of suspended matter has already deposited when the water is inflowing to Kaparas. Nevertheless there is an annual variability of water transparency. Small values of secchi-depth are 0,5 m (e.g. September 5th, 1990); high secchi-depth are about 2,5 m (e.g. June 30, 1990). Sometimes secchi-depths of only 5 cm have been measured.

Low secchi-depths and high concentration of suspended matte occur during the weeks after inflow to Kaparas. High secchi-depths can be observed after the period of sediment settling which can last up to one month. The comparable low amount of suspended matter makes the Kaparas reservoir water quite suitable as a drinking water supply with reduced costs for filtration.

The EU-INTAS project IWMT (INTAS-1043) is committed to test and recommend enhanced water management strategies for the THC aiming to document the feasibility of increased use of the Kaparas reservoir for drinking water supply.

5. FUTURE PERSPECTIVE, MANAGEMENT OPTIONS

Despite a tremendous effort, improvements in the Amu Darya basin are slow. In the past, most efforts focused on the shrinking Aral Sea, more recently, there has been a focus on other aspects. Present project objectives are:

- improvement of human health and living conditions for the local population
- stabilizing the remaining parts of the Aral Sea
- protection of sensitive ecosystems-lakes, wetlands, rivers and Tugay forests

The specific activities are interlinked with each other and have different time scales in planning and completion. All activities are urgently needed but they compete for the remaining water resources and suffer additionally from reduced water quality. The present water stress points up the impact of water quality deterioration. As safe water resources are reduced by pollution the usable fresh water is reduced. In particular in minimum run-off years the conditions are critical.

As return water in the Amu Darya increases along the course of the Amu Darya water quality problems also increase. Return water reaches almost 100% of the total flow in the delta to the Aral Sea.

In UNESCO (2001) it is stated that a two or three fold increase in productivity per cubic meter of water is required. The productivity increases would come about through land improvements that diminished soil salinity, lower conveyance losses, improved water distribution, and a more economic use of water at the farm through technical and economic measures. Better-adapted varieties are needed. More knowledgeable scientists, engineers and farmers are needed.

Due to an extremely large cost, up to 16 billion dollars for a substantial improvement (MICKLIN, 2003), it is not feasible to simultaneously apply an improvement program for the entire basin.

Spatial and temporal pollution variability

For long term safe fresh water resources, highest priority must be given to combating pollution sources. Past attempts in Central Asia demonstrates the risk that the financial expenses will result in a negligible effect, unless treatments are limited in spatial extent.

Consideration of spatial and temporal variations in pollution and transport processes may be important. If low and high quality water can be separated, there will be a reduction in treatment volume, an increase in efficiency and reduced costs. For example, a treatment of pesticides in constructed wetlands is feasible (MOORE et al., 1999). The feasibility depends on the flow and total treatment volume. Same is valid for the disposal of high saline waters. A collection, pumping and disposal in natural depressions will only be economic, if the volume (with highest concentrations) will be kept as small as possible.

Another priority should be the detection of areas with high concentrations of pollution, and to reduce pollution from sources that discharge to these areas. Adsorption capacity, exchange rates and transmissibility/drainage of soils may be used to identify sources discharging high volumes of pollutants. Treatment of specific sub areas may lead to reduce downstream loads. An example would be the treatment of the Bukhara or Karshi region to improve the inflowing water quality to THC and hence the water supply downstream.

Enhanced reservoir operation

Basic principle of an enhanced reservoir operation is to alter the time and volume of filling, storage and withdrawal in a way to achieve both highest available water quality and water quantity. When inflow to a reservoir is polluted, the amount of high quality water is reduced.

In general the options have to be distinguished for two basic cases:

- single reservoirs: prevention of high polluted inflow by bypassing or disposal upstream
- multi-reservoir systems: storing of water with different quality in different reservoirs to protect the best water quality for the highest demand (drinking water supply and irrigation of sensitive crops)

The improvement of water quality by modification of operation modes has the advantage that it uses mostly existing infrastructure and is cost effective.

The Tuyamuyn Hydroengineering Complex demonstrates in an exceptional way the potential for improvement of regional water supply by changing operation strategies. This is mainly feasible because the Kaparas reservoir is located as a side reservoir of the channel reservoir and is also independent of Sultansanjar and Koshbulak reservoirs.

Current THC reservoir operation schemes are largely based on the rule to store all inflowing water dependent only on the aspects of temporal variations in demand and availability and not considering different inflow water quality. Filling of Kaparas takes place predominantly in the period of October to January. Results of Reder (2003) and the IWMT project (FROEBRICH AND KAYUMOV, 2003) indicates, that by changing the operation modes, salinity reductions of up to 40% might be feasible. Future activities should therefore focus on facilities enabling a more independent flow control and operation.

Implementing an enhanced reservoir operation scheme for the Kaparas reservoir would have a significant impact on the future potable water supply of the Amu Darya delta region. It is recommended to store as much water as possible from the less mineralised summer flood (June – July) since higher dilution during this period also indicates lower concentrations of toxic substances. This would increase the potable water supply, even a 4 times increase of potable water would not be limited by reservoir capacity.

The uncontrolled use of highly polluted irrigation water in the downstream regions is a problem. With an increasingly high proportion of return water in the irrigation channels, a feasible pollution reduction up to potable water standards is questionable. Therefore the extension of the water supply network from the Kaparas reservoir is needed to provide potable water for human needs.

Enhanced reservoir operation depends on the availability of data that, describe the temporal fluctuations of water quality and the movement of pollution waves across the catchment. The water quality monitoring system has declined within the Amu Darya and Syr Darya. Many monitoring stations fail to provide systematic data; others simply are not operated. The equipment and methods used to sample and analyse water are not adequate. There are no automatic monitoring devices to check water quality 24 hours a day. Consequently there is a lack of detailed and reliable monitoring information on surface water, groundwater and drinking water (UN, 2001)

Additional problems are caused by differences in monitoring methodologies and equipment in the riparian countries of Uzbekistan (Kazakhstan, Kyrgyzstan, Tajikistan, and Turkmenistan), this casts doubt on the reliability of the data and comparison of data between countries (UN, 2001).

An improvement at this sector could be achieved simply. A more dense estimation of salinity time series by daily manual measures will provide sufficient information to adapt the reservoir operation to various inflow conditions.

Reduction of water losses

The main limitation of enhanced reservoir operation is that it is only applicable within the range of available water resources. There is no way to increase the total amount of water to combat problems resulting from minimum-flow years.

To cover additional water demands, and also for ecological needs or stabilisation of Aral Sea level, the best option is to reduce the very large water losses. Future activities should not only focus on the irrigation system, but also on the losses from the dam operation. In some cases (but most evident in the Syr Darya basin) the demands of the upstream regions for Hydropower generation may lead to a seasonal shift of flow in a way that exceeds the total storage capacity downstream. Such excess water may be used to support ecological needs. To obtain an extended use of such waters, future optimisation strategies should include aspects of transboundary water management and seasonal occurrence of different types of losses. Also the building of specific additional reservoir capacity or the use of ponds should be considered.

Site specific farming

As irrigation is the major water consumer of Amu Darya, new farming strategies will provide additional water stress mitigation options. From the viewpoint of reservoir operation, the current practice of leaching the fields should be accomplished by other measures to reduce soil salinity. Highest priority must be given to high groundwater levels and adequate groundwater management. The drainage system, groundwater management and irrigation practice should be dependent on local soil conditions and focus first on the highest polluted areas (in terms of salinity and maybe toxic chemicals). There are numerous past and current projects which describe the potential for improved farming techniques.

The use of pesticides and fertilizers should be controlled. But, it is difficult to obtain sound data about the real application of agro-chemicals. New indices for a risk assessment (such as those. developed by CROSA AND

CALAMARI (2003) should be used to encourage a shift towards less toxic chemicals.)

Outlook

Independent of the Aral Sea shrinking, first priority must be given to enable the long term access to save water and soil resources for the inhabitants in the Amu Darya Basin.

Related to the catastrophic dimension of the problem the question must be answered, how much industrialisation, agriculture and population might be kept in the Delta or could be concentrated in other regions. Obviously a short term change is not feasible. But there is a need to develop precise planning indicators on the scale of activities that are tolerable under the present and future water supplies. The existing problems in minimum-flow years indicate the severe consequences of water scarcity due to climate or global change. Such aspects are not well represented in ongoing studies and are difficult to quantify. Other global change aspects as the expected increased consumption of Amu Darya water by Afghanistan and the Karakum channel in Turkmenistan (GLANTZ, 2002) must be stated more precisely in future studies.

Acknowledgements

Odil Kayumov initiated the project with his idea of focusing on the Kaparas reservoir. This generated a stimulating period of activity, providing important insights and understanding. Viktor Nikolayenko provided an abundance of original data for the ongoing investigations. Special thanks also go to Christian Reder, Robert Gorbatchev, Lev Poberejsky, Shavkat Rakhimov, Giuseppe Crosa, Davide Calamari, and Herwig Lehman for their important personal support. We also thank all the other team members who cannot be listed here for lack of space. The authors are much obliged to the Deutsche Forschungsgemeinschaft (DFG) and INTAS who provide the financial support for large parts of the work presented.

6. REFERENCES

Alekin, O. A., Moricheva, N.P. (1955) : To the problem of the origin and the future of the Aral Sea water ionic composition. Gidrokhim. Materialy, Leningrad: AN SSSR, V.25, p. 3-15 (in Russian)

Bogdasarov,V.M., Bolshakova,L., Mollaev,T., Kuturina,N. and I. Sergeeva. (1998): Migration of pesticides in time and their distribution in the soil-water-vegetaion system. In: Ecological Research and Monitoring of the Aral Sea Deltas, UNESCO, Paris, pp. 81-91.

Crosa, G., Calamari, D. (2003): Assessment of the quality status of the Amu Darya and Kaparas Reservoir. Presentation at Aral Sea Basin Conference in Bukhara – Uzbekistan; 4-8 April 2003, Bukhara, Uzbekistan.

Dukhovny V.A., Sokolov V.I (2000): Integrated water resources management in the Aral Sea Basin. Scientific-Information Center of the Interstate Comission for Water Coordination in the Aral Sea Basin. Tashkent, Uzbekistan.

Elpiner, L.: Public health in the Aral Sea coastal region and the dynamics of changes in the ecological situation. in: Glantz, Michael H. (1999): *Creeping environmental Problems and Sustainable Development in the Aral Sea Basin.* Cambridge University Press. United Kingdom.

Froebrich, J.; Kayumov, O. (2003) (eds.): Annual project report February 2003 of EU-INTAS project 1043 (IWMT).

Glantz, M. (2002): Water, Climate and Development Issues in the Amu Darya Basin. Workshop report of the Informal Planning Meeting 18-19 June 2002, Philadephia, USA.

Glavgidromet (Main Administration on Hydrometeorology at the Cabinet of Ministers of the Republic of Uzbekistan): State water cadastre. Tashkent, Uzbekistan.

GOST - Gosudarstevenniy Standart (1983): Drinking water standards. Moscow. USSR.

GRDC / UHC: Global Runoff Data Centre; www.grdc.sr.unh.edu

Hannan-T.; O'-Hara-S.L. (1998): Managing Turkmenistan's Kara Kum Canal: problems and prospects. Post-Soviet-Geography-and-Economics. 1998; 39(4): 225-235.

IFAS Project report (2001): Water Resources and Environmental Management in the Aral Sea Basin, Component A1 "Water Resources and Salts Management". Agency GEF of International fund for Saving the Aral Sea, volume 2, Tashkent.

Kayumov O. A., Ikramova M. R. (1997): Influence of Tuyamuyn Hydrosystem on bed processes and water intake in Amu-Darya lowers. Agriculture of Uzbekistan, 2.

Liebmann, C. Chr. (1990): Die Austrocknung des Aralsees. Praxis Geographie 20 (1990), H.3, p. 18-20

Micklin, P.M. (1991): The water management crisis in Soviet Central Asia. Carl Beck Papers in Russian and East European Studies, No.905, p.120. in Ressl, R. (1999): *Fernerkundungs- und GIS-gestützte Optimierung des Bewässerungsfeldbaus am Amu-Darja-Unterlauf und in seinem Delta.* DLR Forschungsbericht. Oberpfaffenhofen. Germany.

Micklin, P.M. (2003): The Aral Sea crisis. At NATO Advanced Research Workshop (Dying and Dead Seas) May 8-10 2003, Liege, Belgium.

Moore, M.T., Rodgers, J.H., Cooper, C.M., Smith, S. (1999): Constructed wetlands for mitigation of atrazine-associated agricultural runoff. Environmental Pollution 110 (2000), p. 393-399., Elsevier Science.

Nikolayenko, V. (1991): Internal publication SANIIRI. Tashkent, Uzbekistan.

Orlovsky, Nikolai S. (1999): Creeping environmental changes in the Karakum Canal's zone of impact, in: Glantz, M.: *Creeping environmental Problems and Sustainable Development in the Aral Sea Basin.* Cambridge University Press. United Kingdom. Pp. 225-245.

Reder, C.; Froebrich, J. (in preparation): Water quality improvement potentials by optimising Tuyamuyn reservoir management.

Ressel, R. (1999): Satellite remote sensing and GIS based optimizatzion of irrigated lands in the Amu Darya lower course and in its delta area, DLR, rep. 1999-30, Oberpfaffenhofen.

Rubinova F.E., Gorelkin N.E. (1998): Salt drain of Aral Sea basin and its transformation under economic activities influence. Theses of participants reports in 1st Regional conference on salts migration in the Central Asian territory. Tashkent, 1998, p. 65-66.

Savitsky, A. (2003): Presentation at Aral Sea Basin Conference in Bukhara – Uzbekistan; 4-8 April 2003, Bukhara, Uzbekistan.
UNESCO (2000): Water-related vision for the Aral Sea basin for the year 2025. UN Educational, Scientific and Cultural Organisation. Paris, France.
UN (2001): Environmental performance review of Uzbekistan. http://www.unece.org/env/epr/studies/uzbekistan
ZEF/UNESCO (2002): Economic and Ecological restructuring of Land- and Water use in the region Khorezm (Uzbekistan): a pilot project on development research. – project proposal.

Chapter 3

Morphological changes in the Aral Sea: Satellite imagery and water balance model.

Rainer Ressl
German Aerospace Center (DLR), German Remote Sensing Data Center (DFD) Environment and Geoinformation Division, Kalkhorstweg 53, 17235 Neustrelitz

Philip Micklin
Western Michigan University (emeritus), Department of Geography, 3219 Wood Hall Kalamazoo, Michigan 49008-5424

1 INTRODUCTION

The Aral Sea, once the world's fourth largest lake, has suffered a dramatic loss of both area and volume, as a result of greatly reduced river inflow brought about by increasing irrigation (mainly of cotton and rice) along the two rivers, Amu Darja and Syr Darja, that flow into it. The ecological consequences are severe and have been discussed in a variety of scientific publications (Micklin, 1991 and 2000, Giese, 1998, Letolle, & Mainguet, 1996).

Satellite images have helped to monitor and document the desiccation process and to describe the associated morphological changes. This includes changes of the water body itself as well as the related changes

J.C.J. Nihoul et al. (eds.), Dying and Dead Seas, 77–89.

of the surrounding environment and plant communities, such as the newly formed desert areas like the so-called "Aral Kum". Low spatial resolution data such as NOAA-AVHRR has proven to be valuable to monitor the desiccation on a frequent basis. The use of high resolution satellite data has been limited for morphological change studies due to higher costs (Resurs-MSU-SK/E, Landsat-TM) and only partial coverage of the lake surface. New satellite systems such as Terra-MODIS are useful because of full coverage of the lake, easy access on a daily basis and provision to researchers at no cost. Besides remote sensing data, GIS modelling techniques have proven to be a useful tool for forecasting the desiccation process of the Aral Sea under specific circumstances. Several probable water inflow scenarios have been assumed to forecast the future desiccation of the lake. Further investigations will include automatic change detection and automated monitoring of the desiccation process.

2 DESICCATION MONITORING USING MULTISENSORAL SATELLITE DATA

For the years 1985 to 1992 AVHRR satellite imagery of NOAA-11 and NOAA-9 were used with spatial resolution between 1.1 km (LAC) and 4.4 km (GAC). For the year 1984 an old Challenger Hasselblad photograph was scanned, geocoded and resampled to 500m ground resolution for comparison. Starting from 1996 Russian Resurs-MSU-SK data were used to calculate the water surface, as this sensor with 170 m ground resolution provides a more accurate source than NOAA-AVHRR (Ressl, 1996). Starting from the year 2002 TERRA-Modis data were used in addition as these data are free of charge and provide a ground resolution of 250 m (channel 1 and 2). These data are also an excellent source for the area estimation of the lake.

For each year so called "land-water masks" were calculated, which make use of the different spectral behavior of water in the electromagnetic spectrum. Figure 1 shows the characteristics of water within the spectral range of 0.5 μm to 1.0 μm. It can be clearly seen that the reflectance of clear water differs from turbid water. Because of the suspended material in the water the reflection increases in the red portion as well as in the infrared portion of the spectrum. On the other hand there is a significant absorption of radiation of clear water in the infrared. This fact can be used to calculate a simple ratio between channel 1 (red) and channel 2 (NIR) to enhance these differences. In the following step a threshold value has to be defined, which separates the water pixels from land pixels. This "borderline" was extracted by a visual interpretation and by a statistical analysis of the associated pixels. Finally, a reclassification of all pixels below and above this threshold was undertaken to receive a "binary mask" of 0 (water) and 1 (no water). Summarizing the number of water pixels and multiplying the pixel resolution gives the water area of the Aral Sea. Figure 2 illustrates the water masks derived for the observation period.

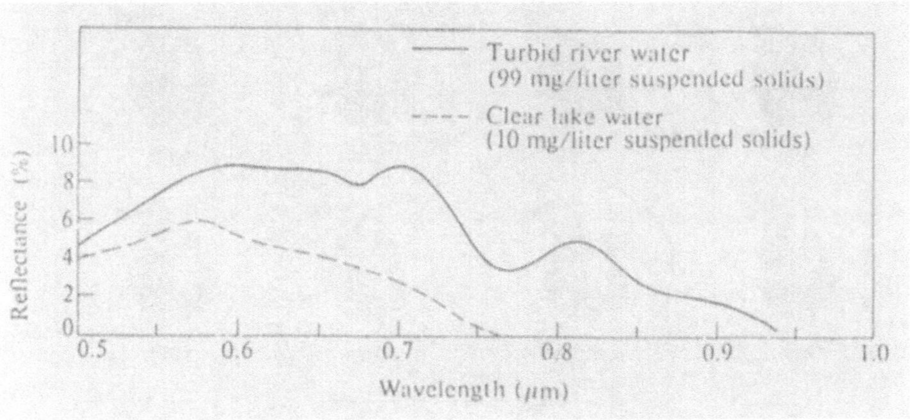

Figure 1. Spectral characteristics of clear lake and turbid river water

The changing morphology of the Aral Sea shown in Fig. 2, which was derived from satellite imagery for the different years, clearly demonstrates the rapid desiccation process of the lake since 1960. This development is especially obvious if the "growth" of the islands within the lake is considered. The small island in the northern part of the lake, Barsakelmes, connected to the mainland in 1996 and since then has literally been "swallowed" by the expanding eastern shoreline of the sea. Even more striking has been the growth of the island Vozrozhdeniya. Originally a tiny island in the middle of the Aral, which was used for biological and chemical weapon testing during the Soviet era, the island has enormously increased in size owing to the falling sea level. For the first time in the year 2002, this island also connected to the mainland to the south.

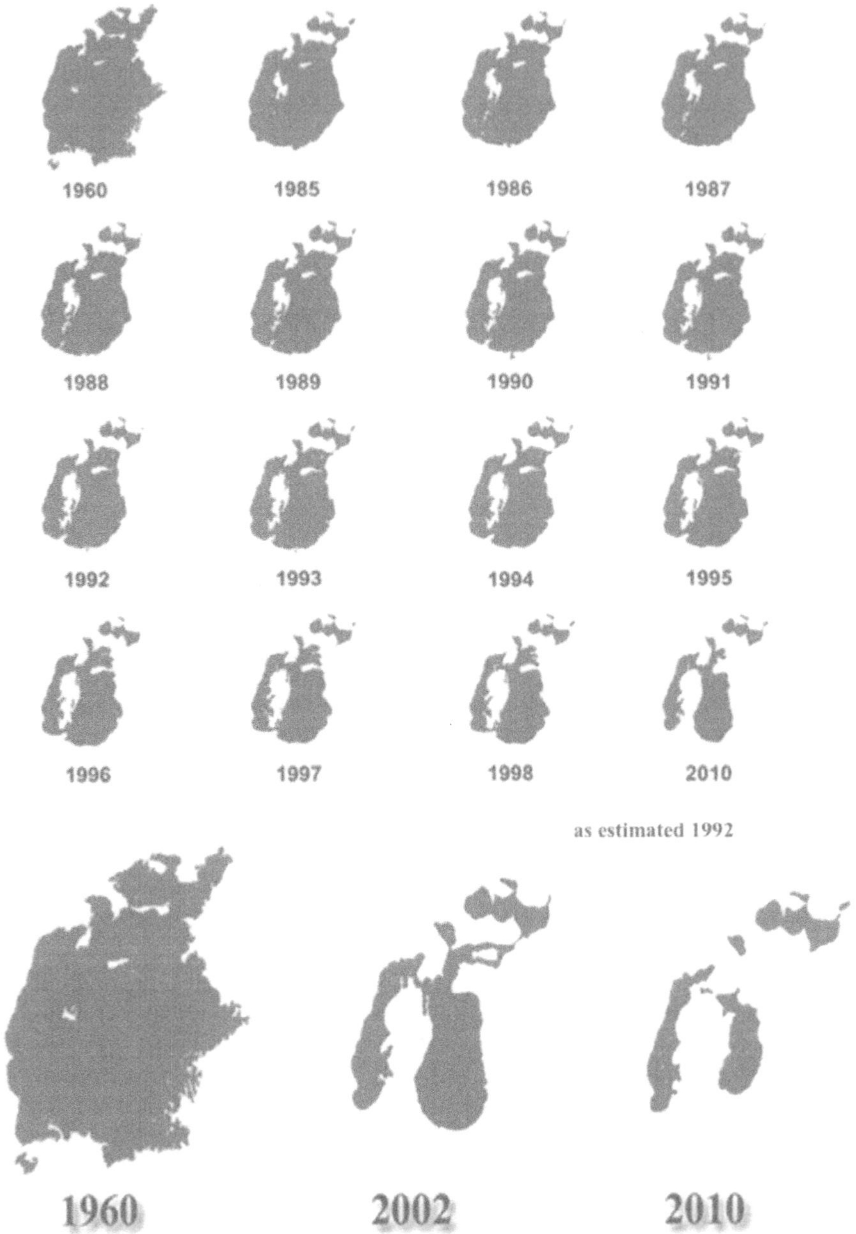

Figure 2. Land water masks derived by satellite imagery

The eastern shallow part of the large (southern) Aral Sea is suffering the severest effects of the desiccation process, whereas the western part of this waterbody, with much greater depths and steeper shoreline has lost relatively much less surface area. The reason for this can be understood if the lake's bathymetry is analysed. For this purpose a bathymetric map, scale 1:500000 was digitized and analysed within a Geographical Information System (GIS). Figure 3 shows the bathymetric map and a calculated 3-dimensional view of the elevations derived from the map's isolines. It can clearly be seen, that the shallow eastern part, with a maximum depth of 28 m, is especially sensitive to the desiccation process. The determination of the shoreline for different years as the sea dried and shrank is especially difficult in these parts of the lake as the high content of suspended material doesn't allow a clear discrimination of clear water pixels from land pixels. Especially when using NOAA-AVHRR data for area estimations this may result in significant differences due to the large size of the water body and the coarse pixel resolution. Up to the year 2002 the water surface of the whole sea decreased to approx. 21,000 km^2 from originally around 69,000 km^2 in 1960. Over the same time period time the level of the large Sea declined by more than 20 meters. The maximum water depth of the shallow eastern part of the large Aral Sea remains only around 7-8 m.

In addition in Figure 2 it can be seen that the separation of the large water body to the south from the smaller portion to the north started around 1989. Since then the "Small Aral Sea" and the "Large Aral Sea" have undergone different hydrological conditions. Originally the two tributaries, Syr Darja and Amu Darja, discharged into a single water body. The Syr Darja discharges now only into the small Aral sea. This discharge has been sufficient to stabilize the size of the small water body since the early 1990s. The discharge of the Amu Darja River to the Aral Sea has been greatly diminished owing to heavy irrigation in its drainage basin. since the 1960s. Some years in the 1980s and more recently in 2000-2001 it was zero or near zero. Due to the high evaporation of approx. 1000 mm/yr. from the sea's surface and insignificant precipitation of about 130 mm/yr. on the sea's surface, along with a small groundwater contribution, the greatly reduced inflow has led to a rapid desiccation of the large Aral Sea.

Satellite imagery modelling was also used to derive the actual sea level associated with each year's water surface area. For this purpose a high pass filter was calculated over each binary land sea mask for every year. The resulting "borderlines" describe the maximum extension of the water surface for the individual years. These borderlines were interpreted as shorelines and were overlaid onto the digital elevation model derived from the bathymetry map. The pixels below the borderline were extracted in the following step and statistically interpreted. The highest frequency was taken as the representative sea level. The derived values were compared to known actual measurements of the sea level from the literature. The comparison showed some differences between measured sea level and the values derived by the bathymetric map. This is due to the fact that the map scale of 1:500.000 is relatively crude and errors occur due to generalization, digitising and interpolation routines. Available bathymetric maps with a scale of 1:200.000 should be able to minimize these errors. The values for sea level in this study should therefore be seen as indications only and as a demonstration of a method for deriving sea level using bathymetric map information (Table1). A Soviet era bathymetric map of the Aral Sea in the scale of 1:200,000 is in possession of the authors. If efforts to digitise this map are successful, then more accurate modelling of sea level will be possible. However, it should be mentioned that new altimeter sensors with advanced accuracy soon to be available on earth orbiting satellites will be able to measure sea levels directly without additional information sources such as bathymetric maps.

The results of the sea surface and sea level calculations are shown in Table 1.

Table 1. Area and sea level estimations using satellite imagery

Year	**Area (in km²)**	**Area (Zakovu) (in km²)**	**Sea Level (in m)**	**Sea level (Zakovu) (in m)**
1960	69,384	70,000	53.0	53.4
1985	44,468	44,257	41.5	41.6
1986	43,278	42,517	40.5	40.9
1987	42,517	41,647	40.0	40.5
1988	41,470	40,777	39.5	39.9
1989	39,543	39,907	39.0	38.8
1990	38,163	38,167	38.0	38.0
1991	35,412	35,732	37.0	36.9
1992	33,635	34,460	36.5	36.2
1996	31,427	-------	36.0	-----
2000	22,500	-------	32.9	-----
2002	21,200	-------	32.5	-----

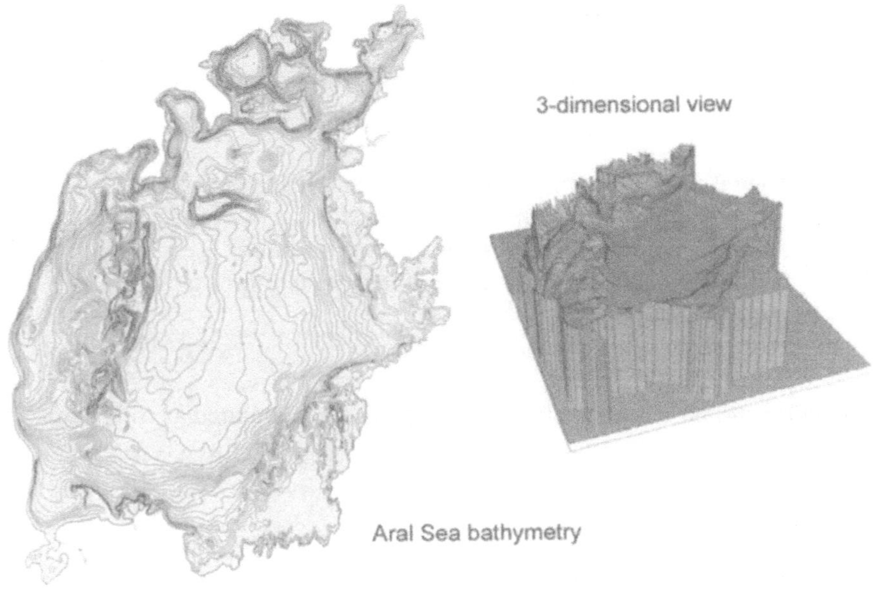

Figure 3. Bathymetry map of the Aral Sea and 3-dimensional view

3 FORECASTING THE DESICCATION OF THE ARAL SEA USING GIS METHODS AND WATER BALANCE MODELS

The more significant indicator for the desiccation of the Aral Sea is the yearly water volume loss. The desiccation from one year to another can appear dramatic due to large area losses although the actual volume loss may be small. This is especially true in the very shallow parts of the eastern basin of the Aral Sea. For calculating actual volume losses, water budget models can be used as well as GIS techniques using the bathymetric information. The latter seems to be promising as there is fewer generalisation involved. Besides of monitoring the actual desiccation of the Aral Sea and its consequences, it is more interesting from the ecological point of view to forecast the future state. For this purpose several scenarios were calculated under different hydrological conditions (RESSL, 1993).

3.1 Water balance components

The water balance of the Aral Sea consists of the following components (Micklin, 1991):

INPUT OUTPUT

$$Qr + Qu + (P*F)/10^6 = (E*F)/10^6 \pm (dh*F)/10^6$$

Qr = annual river inflow by Amu Darja and Syr Darja (in km^3)
Qu = annual net groundwater inflow (in km^3)
P = annual precipitation on the sea surface (in mm)
E = annual evaporation from the sea surface (in mm)
F = average annual area (in km^2)
dh = Net annual sea level change (in mm)

The annual volume losses of the Aral Sea can be calculated as following:

$$Vl = F*(En*10^{-3})-Z-G$$

Vl = Volume loss in km^3
F = average area for year $_I$ in km^2
En = net evaporation (evaporation from sea surface minus precipitation on sea surface) in meters
Z = water inflow by Amu Darja and Syr Darja (in km^3)
G = groundwater inflow (in km^3)

The resulting net annual sea level change (dh) in meters can be calculated from:

$$dh = (Vl / F) * 10^3$$

3.2 Desiccation prognosis for the Aral Sea

Several prognoses up to the years 2000 and 2010 were calculated on the basis of the values of the Aral Sea of the year 1992. This year was taken as “Status Quo” at that time. Crucial determining input in the prognosis was the water input by the two rivers besides of groundwater input and precipitation which was assumed to constant. The estimated average runoff into the Aral Sea was calculated with 10 km^3. This value is probably still to optimistic as runoff at the hydrological stations closest to the Aral Sea in 1991 was only at 5 km^3. This probably too optimistic view gave somewhat higher values for the key hydrologic parameters of the sea (area, level, volume) than were later determined by direct observations and analysis of satellite imagery.

Table. **2** shows retrospective the predicted values for the Aral Sea calculated in 1992 for the period 1992 – 2010

Table. 2. Desiccation prognosis for the Aral Sea

Year	Sl	S	I	E-P	Net-E	G	Slc	V	Al
1992	36.5	32,985	10.0	0.887	29.3	2.0	0.524	234	1369
1993	36.0	31,616	10.0	0.887	28.0	2.0	0.507	217	1262
1994	35.5	30,354	10.0	0.887	26.9	2.0	0.491	201	1222
1995	35.0	29,132	10.0	0.887	25.8	2.0	0.475	186	1496
1996	34.5	27,635	10.0	0.887	24.5	2.0	0.452	172	1426
1997	34.0	26,209	10.0	0.887	23.2	2.0	0.429	160	1548
1998	33.6	24,661	10.0	0.887	21.9	2.0	0.400	149	1444
1999	33.2	23,216	10.0	0.887	20.6	2.0	0.370	140	1183
2000	32.8	22,033	10.0	0.887	19.5	2.0	0.342	132	931
2001	32.5	21,101	10.0	0.887	18.7	2.0	0.318	124	866
2002	32.2	20,235	10.0	0.887	17.9	2.0	0.293	117	808
2003	31.9	19,427	10.0	0.887	17.2	2.0	0.269	111	758
2004	31.6	18,668	10.0	0.887	16.5	2.0	0.244	105	688
2005	31.4	17,980	10.0	0.887	15.9	2.0	0.220	101	618
2006	31.2	17,361	10.0	0.887	15.4	2.0	0.196	97	551
2007	31.0	16,810	10.0	0.887	14.9	2.0	0.173	94	582
2008	30.8	16,228	10.0	0.887	14.4	2.0	0.148	91	496
2009	30.7	15,732	10.0	0.887	13.9	2.0	0.124	88	417
2010	30.6	15,314	10.0	0.887	13.6	2.0	0.103	86	347

Sl = Aral Sea level in meters above mean sea level (msl)
S = Surface area in km^2
I = Inflow in km^3
E-P = Net evaporation in meters
Net-E= annual evaporation from the water surface in km^3
G = annual net groundwater inflow in km^3
Slc = Sea level change in m
V = remaining volume in km^3
Al = annual area loss in km^2

For the year 2000 the remaining water surface was calculated at 22,033 km² with an approximate water volume of 132 km³. Sea level was estimated at 32.8 m above msl. In the year 2010 the water surface was estimated at around 15,300 km², with a volume of around 86 km³ and a sea level of 30.6 m. If this scenario proves accurate, there could be an end of the desiccation process of the Aral Sea in the year 2017 as the diminished evaporation due to the smaller water body would be equal to the annual water input. With a level of 30 m above msl and an approx. volume of 80 km³ and a sea surface area of 15,300 km², the Aral will only be a fraction of its original size. The surface will be diminished about 78 %, the volume by more than 90% with a sea level drop of about 23 m (Ressl, 1996). Following that scenario the "Large Aral Sea" will divide again into two bodies as there will form another land bridge at the northern tip of Vozrozhdenya island. Both parts of the large Sea will have become so saline that all fish species, even the most salinity tolerant introduced varieties, will have disappeared.

Figure 4 depicts the desiccation of the Aral sea for the years 2001 and 2002 on basis of Terra-MODIS satellite data. Morphological changes can clearly be seen in the northern part of the large Sea from one year to the next.. Although the overall size of the large Sea diminished over those two years, in the northern part of the large Sea amount of water covered surface is decidedly larger in 2002. This effect is due to the fact that in 2002 the outflow from the small Aral Sea to the large Aral Sea was much greater than in 2001, which for the period from late spring to late summer created an extremely shallow, but extensive "wetland zone" here. The morphological consequences because of this can be significant and the actual size of the lake, especially in the northern part of the eastern portion of the large Sea can vary considerable. In order to calculate a representative size of the large Sea these effects have to be considered.

Overall the desiccation process for the large Sea will slow down in the future as the absolute evaporation diminishes as the water surface of the shrinks (illustrating the operation of the classic negative feedback process). The Small Aral Sea in the north most likely remains the same, as its level has been stabilized by the inflow of the Syr Darja since the early 1990s.

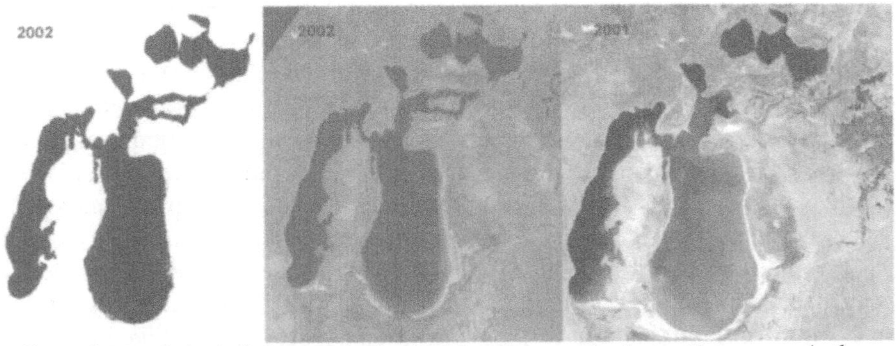

Figure 4. Morphological changes for the Aral Sea owing to the desiccation process in the years 2001 / 2002

4 FUTURE AUTOMATIC MODELLING OF MORPHOLOGICAL CHANGES OF THE ARAL SEA AND ITS DRY BOTTOM

Future focus will lie on the automated detection of changes of the Aral Sea using remote sensing information in combination with GIS techniques. Up to now there is a high portion of interaction in the process, starting from the calibration of the data, the accurate rectification, the calculation of the channel ratios, the determination of threshold values, the extraction of water pixels to the calculation of the water surface, volume and sea level. A major goal is the automation of these processing steps in one operational processing chain without the need for human intervention.

Each step requires a high degree of accuracy to receive comparable results. Automation doesn't appear to be so necessary in monitoring some of the morphological changes of the Aral Sea, such as the annual sea surface area, but for other related applications it is quite useful. For comparison purposes, the water budget of the Aral Sea is usually calculated at the same time of the year as seasonal variations can be significant. On the other hand, there may be occasions where exactly these variations are of interest and a weekly or even daily observation of the water budget of the lake is desired. In this case automation is highly desirable. This can also be true when it comes to the investigation of dynamic geophysical properties such as sea surface temperature (SST), major water circulation patterns, or even the extent of the ice cover during the winter months.

Automation is especially important when it comes to studying dynamic processes such as those related to the evolution of the dry bottom and the remaining ecosystems of the Aral Sea. Seasonal dynamics within the plant communities can be indicators of major ecological changes (Ptichnikov, 1991). Phenological (i.e., stages of development) anomalies can rapidly be investigated and also analysed within thematic processors as part of the automated processing chain. Depending on the investigation subject, the type of remote sensing input data is crucial. A return period of 16 days, such as is characteristic of Landsat TM, may not be sufficiently repetitive to accurately follow the phenology of certain plant communities, especially taking into consideration that not all satellite images will be cloud-free. On the other hand satellite systems with high repetition such as Terra-MODIS may be of limited use due to their lower spatial resolution. The automation process can be extended to GIS applications for further data analysis, modelling and data dissemination as well as product generation (e.g. maps). Also from the point of view of product production costs automation is an important point to consider.

5 CONCLUSION

Morphological changes in the Aral Sea have been dramatic since 1960. Satellite images have helped to monitor, document and to investigate the desiccation of the lake. Modern modelling techniques help to understand the complexity of the problem and to forecast the desiccation process as well as associated ecological developments. The combination of satellite remote sensing data and GIS techniques has proven to be an especially useful combination for these investigations. In retrospect the scenarios provided for the Aral Sea in 1992 for the year 2000 were quite accurate. It can be assumed with reasonable assurance, therefore, that the extrapolation of the desiccation trend up to the year 2010 has similar accuracy. It is also apparent that remote sensing data from the earth orbiting satellites such as NOAA-AVHRR, Terra-MODIS, Landsat TM and Resurs-MSU-SK have been extremely valuable data sources for the desiccation monitoring.

The morphological changes of the Aral Sea will continue to be significant in coming years. More important, nevertheless, will be the continuous changes of the environment. These changes are well documented though not understood in every detail. Satellite imagery from the new generation satellite systems will help to enhance this knowledge.

REFERENCES

Giese, E. (1998). „Die Umweltzerstörung in den Trockengebieten Zentralasiens (West- und Ost-Turkestan).-Ursachen, Auswirkungen, Maßnahmen". -In: *Erderkundliches Wissen*, Band 125, pp. 55-119.

Letolle, R and M. Mainguet (1996). *Der Aralsee – Eine ökologische Katastrophe*. Springer Verlag, Berlin, 517 pages.

Micklin, Philip P. (2000) *Managing Water in Central Asia* (London: The Royal Institute of International Affairs, Central Asian and Caucasian Prospects).

Micklin, Philip P. (1991). *The Water Management Crisis in Soviet Central Asia*, The Carl Beck Papers in Russian and East European Studies, No. 905 (Pittsburgh: The Center for Russian and East European Studies).

Ptichnikov, A.V. (1991). "Physical geographical consequences of anthropogenic desertification in the Aral region." -In: *Aral'skij krizis*, Academy of Sciences, Moskau, pp. 28-48 (in Russian).

Ressl, R.A. (1993). Multitemporale Flächen- und Volumenbilanzen des Aralsees mit NOAA-AVHRR und GIS-Verfahren. Diplomarbeit am Geographischen Institut der LMU München, unveröffentlicht.

Ressl, R.A. (1996). "Monitoring of recent area and volume changes of the Aral Sea and development of an optimized land and water use model for the Amu Darja-Delta." In Micklin, Philip P. and William D. Williams (eds.). *The Aral Sea Basin* (Proceedings of an Advanced Research Workshop, May 2-5, 1994, Tashkent, Uzbekistan), NATO ASI series, Vol. 12 (Springer-Verlag: Heidelberg, 1996), pp. 149-160.

Chapter 4

ARAL SEA BASIN EVOLUTION : GEODYNAMIC ASPECT

Bakhtiar Nurtaev
Institute of Geology and Geophysics, Academy of Sciences of Uzbekistan

INTRODUCTION

The Aral Sea lies in the Aral-Sarykamysh depression, which is bordered by the low plains of Central Asia. The climate is continental and extremely dry, and surface runoff is virtually zero. Since direct precipitation over the lake comprises only 10% of the water budget, lake-level fluctuations are largely determined by changes in inflow from the Amu Darya and Syr Darya rivers. These rivers originate in the highest part of the Pamir and Tien-Shan mountains 1500 km southeast of the Aral Sea. Since 1960 the lake level has been affected by irrigation activity in the basins of the Amu Darya and Syr Darya. The evolution of continental erosion and landforms is controlled by several parameters: tectonic, climatic and anthropogenic factors. The arid climate developing in the region also lead to a decrease of the Aral sea surface and an increase of desertification (Kovda,1980). The important development of the agriculture (cotton cultivation) since 50 years ago has led to a significant modification of the geomorphic system of the Aral basin: (1) the natural drainage network has been diverted for the irrigation, (2) the Aral sea surface has strongly decreased, (3) the desertification has increased.

J.C.J. Nihoul et al. (eds.), Dying and Dead Seas, 91–97.

Caution is important when identifying these observations as manifestation of only antropogenic impacts, because sea level changes may be complicated by strong natural variability. For example, in the Arctic, this variability is dominated by multi-decadal fluctuations with a time scale of 50-80 years (Polyakov et al, 2002). But at the geological scale during the last 6000 yr, the Aral Sea experienced many times large regressions and transgressions, and we can't explain it by anthropogenic factors or climate changes only. Over recent years, a number of provoking studies have explored the connections between sea level decline, climate change and anthropogenic factors. So, the Aral basin is an exceptional natural laboratory for studying these parameters, including the links between orogenic processes and landscape modification. Geological processes operate at rapid rates in this region, which allows scientists to concurrently collect data on tectonic deformation uplift, erosion and sedimentation, and develop comprehensive models that connect these diverse problems. In such a context, the Aral basin is a preferential study object of interaction between the respective parts of climate, tectonic and anthropogenic impact on the evolution of continental erosion and landform.

DATING OF ARAL SEA LEVEL IN THE LATE PLEISTOCENE AND HOLOCENE

Turan depression is the large depressional part of inner Asia which has no connection with oceans. It is a part of the large belt of Mezozoic-Cenozoic oil and gas bearing basins, oriented north-north-east extending from the Western Siberian depression through Turan, Kaspian, Western Iran and Mesopotamia to the Persian Gulf and South East Arabia. The most important cause of landforms processes in the Turan plate is : recent tectonic movements and sediment transport by rivers from mountains.The Aral Sea lies in the Aral-Sarykamysh depression, which was formed by tectonic processes and wind erosion during the late Neogene (Pinhasov,2000). Dating of Aral sea level in the Late Pleistocene and Holocene has been unusually difficult owing to a variety of reasons, and large discrepancies exist between results from different research teams. The time of the last flooding of the Aral depression is debated: estimates vary from some time after 24,000 yr BP (Rubanov et al, 1987, Pshenin *et al.*, 1984) to the beginning of Holocene, 10,000 yr BP (Kvasov, 1991). A more widely-held hypothesis is that the modern configuration of the Aral Sea resulted from the Amu Darya and Syr Darya flowing in the Aral Depression 15,000 –17 000 yr BP. Both higher and lower lake levels are indicated by terraces at 73, 65-68, 57-58, 54.5, 53, 43-44, 40-41, 35-36, and 31 m above sea level. According to Gorodetskaya (1978) the older they are, the higher the elevation. In the Middle Ages, sea level decreased up to 31-35 m and a gypsum layer was sedimented. More severe decline took place nearly 1500 years ago when during regression, the level of the sea was minimal – 27-28 m (mirabilit layers sedimented) and we may assume that the situation was similar with the one existing now. We made an attempt to compile all available data of higher and lower lake levels, obtained by previous researchers (Fig.1).

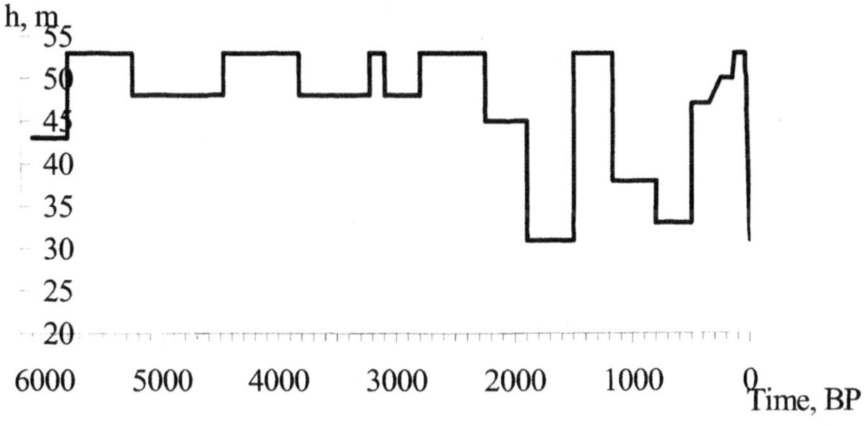

Figure I : Aral Sea level changes during the last 6000 years.

It shows distinct periodicity in sea level variations. The mechanisms for such major shifts in sea level variability in the past are not wellfor such major shifts in sea level variability in the past are not well understood, and currently, it seems there is no explanation of a climatic process that could lead to a mode change. Climate itself is formed as a result of the combination of different factors: external or astronomic factors, geographical and geophysical parameters, atmospheric and only recently antropogenic factors. If we analyze long term data, we can find evidence that all these factors were characterized by strong variability and periodicity.

RECENT VERTICAL DISPLACEMENTS DATA

Regional data of present day vertical displacements are available for Western Central Asia. At a regional scale two domains can be distinguished: the northern domain, north of the 44 parallel with little or no vertical displacement and the southern domain with significant uplift >5 mm/year (Fig.2). This domain undergoes regional uplift except in the central part corresponding to the Amu Darya basin. Our observations attest of an active propagation of the deformation within the Asian platform several hundreds kilometers from the collision zones of the Pamir and the Kopet Dagh (Thomas et all,1999). In the Central Kyzylkum desert, characterized by uplift rates of 10 –12mm, very active seismicity appeared since the late 60th as a manifestation of tectonic processes development. Large scale cracks and fissures, Kyzyk-Kum earthquakes of 1968, Gazli 1976 and 1984 showed activization of geodynamic processes in the region. Previous seismic activity period in this region has been known, from historical data, in the period 820-1200th.

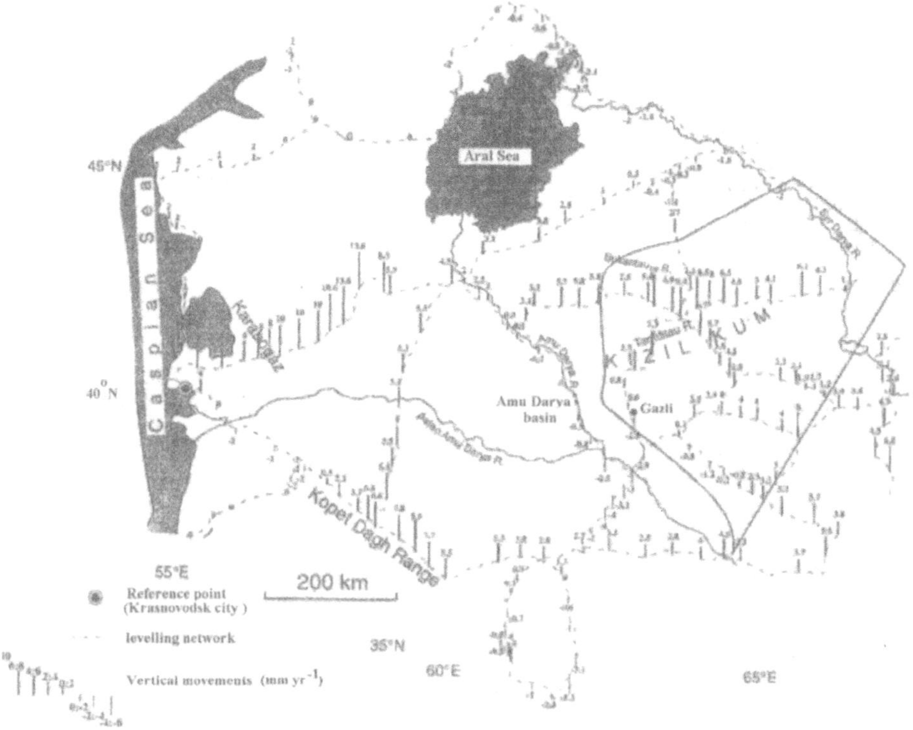

Figure 2 : Map of present-day vertical displacements for western Central Asia.

Paleo Amu Darya river runs westward from the present day Amu Darya river, to the Southern Caspian sea and was active most probably until the Pleistocen. Such major change of the Amu Darya flow from E-W to SE-NW is evidence for a regional uplift and in agreement with recent instrumental data. The EW paleo-Amu Darya river bed stands within a region showing about a 6 mm year uplift rate whereas it now flows to an area where uplift almost vanishes relative to the surroundings. Furthermore, before the last 7000 years the Aral sea was connected with the Caspian sea through a fluvial network running to the south-west and joining with the Paleo-Amu Darya south-east of the Karabogaz. The closure of this link is attributed to active tectonic uplift of the Karabogaz relative to the Aral sea. 6000 years before the present history of the Aral Sea may be explained as periodical fluctuations of the Amu Darya river from the Aral Sea depression to the Sarykamysh one and back. Another factor, never taken into account before is the intensive exploration of oil and gas in the Bukhara-Khiva region, beginning in 1957. Our observations showed that billions of cubic meters of water appeared in largest gas-oil fields in the process of exploration.

CONCLUSIONS

Over recent years, a number of provoking studies have explored the connections between sea level decline, climate change and anthropogenic factors. But anthropogenic factors played a key role only in the last 40 years. At the geological scale during the last 6000 yr, the Aral Sea has experienced large regressions and transgressions, and we can't explain them by anthropogenic factors or climate changes only. Geodynamic factors and river sediment transport played a significant role in the Aral Sea level changes in the past.

These research results suggest that the current mode of sea level variability encompassing the modern instrumental record is not representative of the full range of past sea level variability.

ACKNOWLEDGMENTS

This research is made possible partly due to the NATO grant CLG - EST 977831 "Climate change in the Aral Region, past Variability, Present Challenges, and future Scenarios".

REFERENCES

Akulov V.V. Delta sediments of Amu Darya In: Izvestia Uzb. Geogr.Obshestva, *v.4, 1960*

Aleshinskaya, Z.V., 1991a. Diatom analyses of continental deposits. Aral. Itogi nauki i tekhniki. Ser. Paleogeog., vol. 7., pp. 114-120.

Aleshinskaya, Z.V., 1991b. Diatomic and palynologic characteristic of bottom sediments. In: Istoriya ozer Sevan, Issyk-Kul', Balkhash, Zaisan i Aral (The history of lakes Sevan, Issyk-Kul, Balkhash, Zaisan and Aral). Leningrad, Nauka, pp. 243-246.

Gorodetskaya M.E. (1978) Terraces of Aral sea. Geomorphology, N1, p.46-55.

Kes', A.M., 1983. Upper Pleistocene and Holocene palaeogeography of the Aral Sea. In: Paleogeografiya Kaspiiskogo i Aral'skogo morei v Kainozoe (Cainozoic palaeogeography of the Kaspian and Aral Seas). Moscow, Izd. Mosk. Univ., vol. 2, pp. 97-106.

Kovda, V.A., 1980: *Land Aridization and Drought Control.* Westview Special Study in Natural Resources and Energy Management. Boulder, CO: Westview Press

Kvasov, D.D., 1991. Principal information on the Aral Sea. In: Istoriya ozer Sevan, Issyk-Kul', Balkhash, Zaisan i Aral (The history of lakes Sevan, Issyk-Kul, Balkhash, Zaisan and Aral). Leningrad, Nauka, pp. 216-219.

Lymarev V.I. 1967. Shores of the Aral Sea – the enclosed water body of the arid zone. Leningrad: Nauka, 252pp.

Maev, E.G. and Maeva, S.A., 1991. Bottom sediments of the Aral. In: Istoriya ozer Sevan, Issyk-Kul', Balkhash, Zaisan i Aral (The history of lakes Sevan, Issyk-Kul, Balkhash, Zaisan and Aral). Leningrad, Nauka, pp. 239-243.

Maev, E.G., Maeva, S.A., Nikolaev, S.D. and Parunin, O.B., 1983a. New data on the Aral Sea Holocene history. In: Paleogeografiya Kaspiiskogo i Aral'skogo morei v Kainozoe (Cainozoic palaeogeography of the Kaspian and Aral Seas). Moscow, Izd. Mosk. Univ., vol. 2, Cainozoic palaeogeography of the Kaspian and Aral Seas. Moskva, Izd. Mosk. Univ., part 2, pp. 133-143.

Maev, E.G., Maeva, S.A., Nikolaev, S.D., Vronskii, V.A., Il'ina, L.B., Maier, E.M., Naidina, N.N. and Sorokin, V.M., 1983b. Sequence of bottom deposits from the central part of the Aral Sea. In: Paleogeografiya Kaspiiskogo i Aral'skogo morei v Kainozoe (Cainozoic palaeogeography of the Kaspian and Aral Seas). Moscow, Izd. Mosk. Univ., vol. 2, pp. 119-132.

Mamedov, E.D., 1991. Paleogeographic conception. In: Istoriya ozer Sevan, Issyk-Kul', Balkhash, Zaisan i Aral (The history of lakes Sevan, Issyk-Kul, Balkhash, Zaisan and Aral). Leningrad, Nauka, pp. 219-222.

Nikolaev, S.D., 1991. The Aral development according to isotope-oxygenic data. In: Istoriya ozer Sevan, Issyk-Kul', Balkhash, Zaisan i Aral (The history of lakes Sevan, Issyk-Kul, Balkhash, Zaisan and Aral). Leningrad, Nauka, pp. 246-250.

Polyakov I.V. et al, 2002. Trends and variations in Arctic climate system. EOS Trans., AGU, V.83, N47, pp.547-548

Pinhasov's B.I., 2000, Paleography of Turan and Tien Shan plains in Neogene. Geoloqiya va mineral resurslar. № 6. pp.3 - 8.

Pshenin, G.N., Steklenkov, A.P. and Cherkinskii, A.E., 1984. The origin and age of pre-Holocene terraces of the Aral Sea. *Doklady AN SSSR*, vol. 276, no. 3, pp. 675-677.

Rubanov, I.V., 1991. Salt accumulation in the Aral hollow. In: Istoriya ozer Sevan, Issyk-Kul', Balkhash, Zaisan i Aral (The history of lakes Sevan, Issyk-Kul, Balkhash, Zaisan and Aral). Leningrad, Nauka, pp. 231-235.

Rubanov, I.V., Ishniyazov D.P., Baskakova M.P., Chistyakov P.A.,1987. Geology of Aral sea. Tashkent, FAN, 247 p.

Rubanova, F.E. and Gorelkin, N.E., 1991. Present status and the Aral development tendencies. In: Istoriya ozer Sevan, Issyk-Kul', Balkhash, Zaisan i Aral (The history of lakes Sevan, Issyk-Kul, Balkhash, Zaisan and Aral). Leningrad, Nauka, pp. 261-267.

Chapter 5

THE ARAL SEA CRISIS

Dr. Philip Micklin

Department of Geography, Western Michigan University (emeritus), Kalamazoo, Michigan USA, 49006

1. INTRODUCTION

The Aral Sea is a terminal lake amidst the deserts of Central Asia. Its size and water balance are fundamentally determined by river inflow and evaporation from its surface. Until the 1960s, the Aral was the world's fourth largest lake in surface area. Over the past four decades, this water body has rapidly and steadily shrunk as countries in the Aral Sea Basin have increasingly taken inflow from its two influents, the Syr Dar'ya and Amu Dar'ya, for expansion of irrigation. The Aral's diminution has directly and indirectly led to an array of severe problems in the surrounding region, ranging from degradation of major terrestrial and aquatic ecosystems to deterioration of human health and welfare. National, regional, and international efforts are underway to cope with these, but even their partial alleviation will be enormously costly and require many years. Full restoration of the Aral Sea to its former state is, at best, a remote possibility for the more distant future. However, rehabilitation of portions of it and adjacent areas that would partially restore former ecological functions and economic uses are feasible.

J.C.J. Nihoul et al. (eds.), Dying and Dead Seas, 99–123.

2. THE ARAL SEA: LOCATION AND CHARACTER

The Aral Sea resides amidst the great deserts (Kara-Kum, Kyzyl-Kum, Betpakdala) of Central Asia (Fig. 1). A terminal lake, it has surface

Figure 1. The Aral Sea Basin

inflow but no surface outflow. Therefore, the balance between inflows from two rivers, the Amu Dar'ya and Syr Dar'ya, and net evaporation (evaporation from its surface minus precipitation on it) fundamentally determine its level. Net groundwater exchange, which is difficult to measure, played an insignificant role in the sea's water balance (Bortnik and Chistyaevaya, 1990, p. 38). In the recent geologic past (last 10,000-15,000 years), the sea has endured significant level fluctuations, perhaps as much as 40 meters (Micklin, 1991, p. 42-43; Kes' 1978; Kes', Andrianov and Itina, 1980; Kes and Klyukanova, 1990; Kosarev, Kostyanoy and Mikhailov, 2003). The major level changes prior to 1960 resulted from diversion of the Amu Dar'ya westward so that it flowed into the Sarykamysh hollow (and sometimes farther through the Uzboy channel to the Caspian Sea after it overtopped Sarykamysh) rather than the Aral Sea. These diversions resulted from natural events (sedimentation of the bed and subsequent breaching of the rivers left bank during spring floods) and from advertant human actions (destruction of dikes and levees, built to keep the river flowing to the, during times of conflict). However, from the mid-18th century until the 1960s, sea level changes were less than 4.5 meters (Bortnik, 1996). For the period of instrumental observation (1911-1960), the sea was unusually stable, with annual inflow and net evaporation never far apart (the average of each of these water balance components was around 55 km^3 for the period) (Bortnik and Chistyaevaya, 1990, p. 36; Micklin, 1994). Hence, the water balance was in long-term equilibrium with a maximum lake level variation of less than one meter.

The Aral Sea's drainage basin encompasses 1.8 million km^2 falling within seven nations: Uzbekistan, Turkmenistan, Kazakhstan, Afghanistan, Tajikistan, and Iran. However, only Kazahstan and Uzbekistan are riparian on the Sea proper, with each possessing an approximately equal length of shoreline. The entire Aral coastline within Uzbekistan lies within that nation's Karakalpakstan Republic.

At slightly more than 67,000 km^2, the Aral Sea, according to area, was the world's fourth largest inland water body in 1960 (Micklin, 1991, pp. 42-54). As a brackish lake with salinity averaging near 10 g/l, less than a third of the ocean, it was inhabited chiefly by fresh water species. The sea supported a major fishery and functioned as a key regional transportation route. The extensive deltas of the Syr Dar'ya and Amu Dar'ya sustained a diversity of flora and fauna. They also supported irrigated agriculture, animal husbandry, hunting and trapping, fishing, and harvesting of reeds, which served as fodder for livestock as well as building materials.

Over the past four decades the sea has steadily shrunk and salinized (Figures 2, 3, 4 and Table 1). The main cause has been expanding irrigation that diminished discharge from the two tributary rivers to a fraction of earlier volumes. The Aral separated into two water bodies in 1987 - a small Aral Sea in the north and a large Aral Sea in the south. The Syr Dar'ya flows into the former, and the Amu Dar'ya into the latter. Between 1960 and January 2003, the level of the small Aral fell by 13 meters and the large Aral by 23 meters. A channel (river) has intermittently connected the two lakes, with the flow from the large sea to the small. The area of both seas taken together diminished by 75 % and the volume by 90%. Salinity in the small sea is estimated to have doubled whereas in the western part of the large sea it has increased by more than 6 fold. Within a year or two, the large Aral is likely to divide into two parts - a "deep" western lake and a "shallow" eastern lake.

3. ECOLOGICAL, ECONOMIC, HEALTH CONSEQUENCES

The mainly human-induced desiccation of the Aral Sea and flow reduction, salinization, and pollution of its influent rivers has had severe negative impacts (Micklin, 2000, pp. 13-23). Besides the consequences for the sea proper, a zone around the waterbody of several hundred thousand

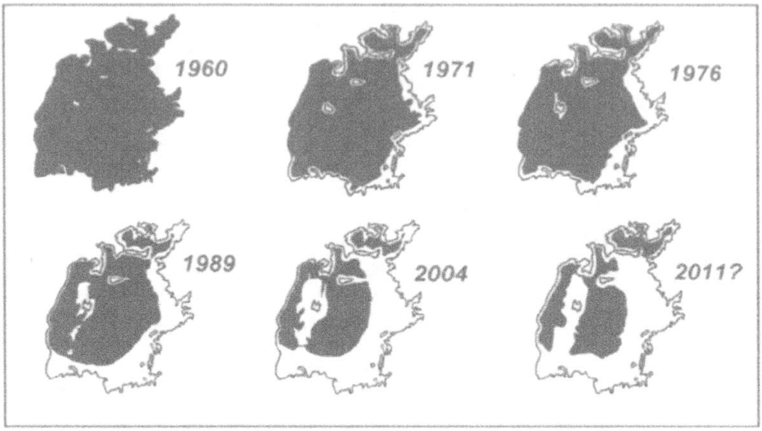

Figure 2. The Changing Profile of the Aral Sea

Table 1. Hydrologic and Salinity Characteristics of the Aral Sea, 1960-2011[a]

Year	Level (meters)	Area km^2	% of 1960	Volume km^3	% of 1960	Avg. depth (meters)	Average salinity g/l	% of 1960
1960(whole sea)[b]	53.4	67,499	100	1089	100	16.1	10	100
Large Sea	53.4	61,381	100	1007	100	16.4	10	100
Small Sea	53.4	6,118	100	82	100	13.4	10	100
1971 (whole sea)[b]	51.1	60,200	89	940	86	15.6	11	110
1976(whole sea)[b]	48.3	55,700	83	763	70	13.7	14	141
1989 (whole sea)[c]		39,734	59	332	33	9.1		
Large Sea	39.32	36,307	60	332	34	9.1	28	280
Small Sea	40.2	2,804	46	23	28	8.2	28	280
2004 (whole sea)[c]		17,158	25	108	10	6.3		
Large Sea	30.4	14,293	23	85	8	5.9	80 to over 100	480 600
Small Sea	40.4	2865	47	23	28	8.0	~20	200
2011 (whole sea)		10,033	15	92	8	9.1		
Large sea[d]	27.3	6,113	10	53	5	8.7	>150	1500
Small Sea[e]	44.5	3,807	62	37	37	9.7	< 15	150

[a]Values derived from Soviet data, data from Glavgidromet of Uzbekistan, data from Shivareva, et al., *1998,* from Mirabdullayev et al., 2003, from Friedich, Jani and Heidi Oberhansli, 2003, from annualized water balance and salt balance models developed by the author as well as from 1:500,000and 1:200,000 bathymetric maps and satellite imagery. [b] annual average [c] on January 1 [d] the sea will have divided into a western and eastern part [e] assumes implementation of north Aral project in 2004 [f] based on few measurements taken in very shallow, coastal water and may significantly exaggerate average salinity

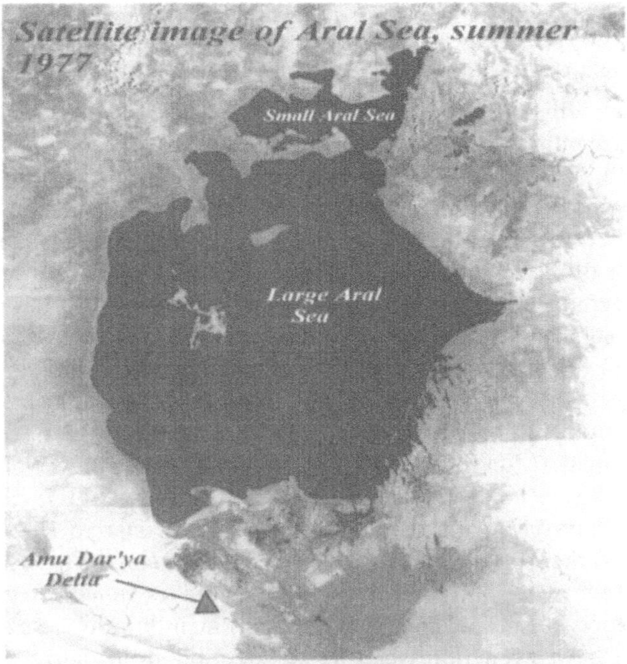

Figure 3. Landsat image of Aral Sea Summer 1977

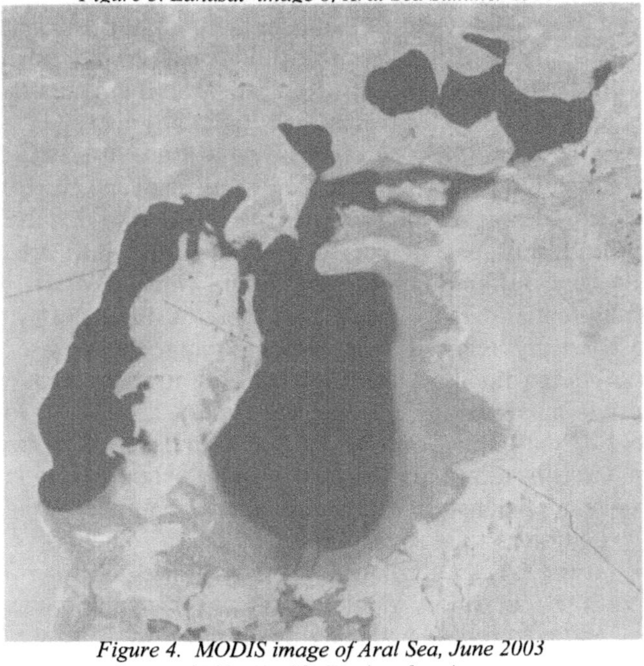

Figure 4. MODIS image of Aral Sea, June 2003 (NASA Visible Earth website)

square kilometers with a population of several million has also been damaged. (Khvorog, 1992). The Republic of Karakalpakstan in Uzbekistan and portions of Kzyl-Orda Oblast in Kazakhstan, have suffered the most harm. Turkmenistan, although not abutting on the sea, has one Oblast, Dashauz, that has been substantially impacted. Even within Kazakhstan, Uzbekistan, and Turkmenistan, the territory suffering significant impacts is a small part of each country's area and contains a minor portion of its population. The other states of the Aral Sea Basin (Kyrgyzstan, Tajikistan, Afghanistan, and Iran) are so distant from the zone where intense effects are apparent that they have suffered no demonstrable harm from the drying of the sea.

The substantial Aral fishing industries developed by Kazakhstan and Uzbekistan in the first half of the 20th century ended in 1983 as indigenous fish (20 species), which provided the basis for the commercial fishery, disappeared from the effects of rising salinity and loss of shallow spawning and feeding areas (Micklin, 2000, p. 16; Micklin, 1991, pp. 49-50; Williams and Aladin, 1991; Zholdasova *et. al.*, 1998; Ptichnikov, 2002). However, all of these still survive in the deltaic lakes and Amu Dar'ya and Syr Dar'ya rivers, except the Aral salmon (*Salmo trutta Aralensis)* that has become extinct. A few of the introduced salt-tolerant fishes remain (e.g., flounder, known as plaice in Europe, sprat, and herring) mainly in the less saline small Aral. Some have even flourished as competition from native varieties for food disappeared. But all of these will vanish from the large Aral Sea by the early 21st century as salinity rises. Flounder are both plentiful and edible but are not caught commercially. Because of the loss of the fishery tens-of-thousands were thrown out of work. Navigation on the Aral also ceased by the 1980s as efforts to keep the increasingly long channels open to the major ports of Aral'sk at the northern end of the sea in Kazakhstan and Muynak at the southern end in Karakalpakstan became too difficult and costly.

The rich ecosystems of the extensive Amu Dar'ya delta, primarily located in the Karakalpak Republic of Uzbekistan but stretching into Turkmenistan have suffered considerable harm (Micklin, 1991, pp. 50-52). Lesser, but substantial damage, has also accrued to the Syr Dar'ya delta in Kazakhstan. Greatly reduced river flows through the deltas, the virtual elimination of spring floods in them (owing both to reduced river flow and construction of upstream storage reservoirs) and declining ground water levels, caused by the falling level of the Aral Sea, have led to spreading and intensifying desertification. Halophytes, plants tolerant of saline soils, and xerophytes, plants tolerant of dry conditions, are rapidly replacing endemic vegetation communities (Novikova, 1996, 1997). In some places, salts have accumulated on the surface forming *solonchak* (salt pans) where practically nothing will grow. Expanses of unique tugay (vegetation communities of trees, bushes, and tall grasses, including popular, willow, oleaster, salt cedar, and reeds) that formerly stretched along all the main rivers and distributary channels here have been particularly hard hit. According to Dr. Novikova (1996), a Russian expert, whereas tugay covered 100,000 ha in the Amu Dar'ya delta in 1950, it shrank to 52,000 ha by the 1970s and to only 15-20,000 ha by the mid-1990s. Tugay complexes around the Aral Sea are habitats for a diversity of animals, including 60 species of mammals, more than 300 types of birds and 20 varieties of amphibians.

Desiccation of the deltas has significantly diminished the area of lakes, wetlands, and their associated reed communities. Between 1960 and 1980, the area of lakes in the Amu Dar'ya delta is estimated to have decreased from 49,000 to 8,000 km2 whereas the area of reeds may have diminished from 500,000 ha to as little as 1,000 ha from 1965-1986 (Chub, 2000, Fig. 3.3, p. 125; Palvaniyazov, 1989). This has resulted in serious ecological consequences as these areas provide prime habitat for a variety of permanent and migratory waterfowl, a number of which are endangered (Micklin, 1991, p. 116). Diminution of the aggregate water surface area coupled with increasing pollution of the remaining water bodies (primarily from irrigation runoff containing salts, fertilizers, pesticides, herbicides, and cotton defoliants) has decimated aquatic bird populations. However, in the late 1980s and in the 1990s, significant efforts were made to restore wetlands and their reed communities and to improve habitat conditions (Chub, 2000, p. 125). A 1999 survey, for example, indicated that the area of reeds for the key lake/wetland in the lower delta, Sudochye, was 12,000 ha. (Dukhovnyy, 2003).

Irrigated agriculture in the deltas of the Amu Dar'ya and Syr Dar'ya has suffered from an inadequacy of water as inflow to the deltas has decreased owing to heavy upstream consumptive use for irrigation. Additionally, water that does reach the deltas has elevated salinity from the leaching of salts caused by repeated usage in the middle and upper courses of the rivers (World Bank, 1998, pp. 3-5). At times over 2 grams/liter, these saline flows have lowered crop yields and, in conjunction with inadequate drainage of irrigated fields, promoted secondary soil salinization. Animal husbandry, both in the deltas and desert regions adjacent to the Aral Sea, has been damaged by reduction of the area and declining productivity of pastures resulting from desertification, dropping groundwater levels, and replacement of natural vegetation suitable for grazing by inedible species.

Strong winds blow sand, salt and dust from the dried bottom of the Aral Sea, now largely a barren, salt covered desert with an area near 50,000 km^2, onto adjacent lands. Since the mid-1970s, satellite images have revealed major salt/dust plumes extending 200 to more than 500 km, allowing dust and salt to settle over a considerable area adjacent to the sea in Uzbekistan, Kazakhstan, and to a lesser degree, in Turkmenistan (Micklin, 1991, pp. 48-49; Glazovskiy, 1990, pp. 20-23; Ptichnikov, 2002). One investigator, however, has asserted that very light, fine particles of dust and toxic salts are lifted several kilometers and travel 5-10 thousand kilometers before settling back to earth (Tursunov, 1989). However, there is no scientific proof to date that would validate such a claim. Satellite imagery for 2002 and 2003 (MODIS, SeaWiffs) does show salt and dust plumes extending for more than 500 km, and it is possible fine aerosols that would not show up on satellite imagery could be carried farther.

Although dust/salt storms affect the entire zone surrounding the Aral, most of the major storms occur with north and northeast winds, which most seriously impact the Ust-Urt Plateau to the sea's west and the Amu Dar'ya delta at the south end of the water body (Bortnik and Chistyaevaya, 1990, p. 27, Fig. 2.7). The latter is the most densely settled as well as economically and ecologically important region around the sea. N. Glazovskiy (1990, pp. 21-22) reviewed the various estimates of the total deflated material, ranging from 13 million to as high as 231 million metric tonnes/year, which were made in the 1980s. He concluded that the most probable figure was from 40 to 150 million tons. A careful study, completed in the mid-1980s, by well-known geologists and experts on the Aral concluded annual aeolian transport of salt alone from the dried bottom was around 43 million metric tons but would decrease slightly to 39 million metric tons by 2000.[1]

Salts in dry and aerosol forms, the most harmful of which include sodium bicarbonate, sodium chloride, and sodium sulfate, are settling on natural vegetation and crops, particularly in the Amu Dar'ya delta (Bel'gibayev, 1984). In some cases, plants are killed outright but more commonly their growth (and for crops, yields) is substantially reduced. The salt and dust also has ill effects on wild and domestic animals by directly harming them and by reducing their food supply (Palvaniyazov, 1989). Local health experts also consider airborne salt and dust as a factor contributing to high levels of respiratory illnesses and impairments, eye problems, and possibly even throat and esophageal cancer in the near Aral region (Abdirov et. al, 1993; Tursunov, 1989). More recent field work by a British led group indicates that salt and dust blowing from the dried bottom (and likely from irrigated farmland in regions adjacent to the Aral Sea) is laced with pesticides and heavy metals, which, of course would enhance the negative impacts on humans and other animals (O'Hara et. al., 2000).

Owing to the sea's shrinkage, climate has changed in a band up to 100 km wide along the former shoreline in Kazakhstan and Uzbekistan (Micklin, 1991, pp. 52-53; Glazovskiy, 1990, pp. 19-21). Maritime conditions have been replaced by more continental and deseritic regimes. Summers have warmed and winters cooled, spring frosts are later and fall frosts earlier, humidity is lower, and the growing season shorter. Uzbekistani climatological experts also believe that the increase in the levels of salt and dust in the atmosphere are reducing levels of surface radiation and thereby photosynthetic activity as will as increasing the acidity of precipitation (Chub, 1998).

The population living in the "ecological disaster zone" suffers acute health problems (Micklin, 1992; Medicins sans Frontieres, 2000). Some of these are direct consequences of the sea's recession (e.g. respiratory and digestive afflictions and possibly cancer from inhalation and ingestion of blowing salt and dust and poorer diets from the loss of Aral fish as a major food source). Other serious health related problems owe to environmental pollution associated with the heavy use of toxic chemicals (e.g., pesticides and defoliants for cotton) in irrigated agriculture, mainly during the Soviet era. However, the most serious health issues are directly related to 'Third World' medical, health, nutrition and hygienic conditions and practices. Bacterial contamination of drinking water is pervasive and has led to very high rates of typhoid, paratyphoid, viral hepatitis, and dysentery. Tuberculosis is prevalent as is anemia, particularly in pregnant woman. Liver and kidney ailments are widespread; the latter is probably closely related to the excessively high salt content of much of the drinking water. Medical care is very poor, diets lack variety, and adequate sewage systems are rare.

Health conditions in the Karakalpak Republic in Uzbekistan, with the possible exception of places in the formerly civil war torn Tajikistan, are likely the worst in the Aral Sea Basin. Surveys conducted in the mid to late 1980s showed the average infant mortality rate at more than 70/1000 live births whereas several districts adjacent to the former seashore ranged from 80 to over 100/1000 live births (Micklin, 1992). These rates are three to four times the national level in the former USSR and 7-10 times that of the U.S. Although efforts have been made in the post-Soviet period to improve health conditions here, it is doubtful these rates have declined in any substantial way.

Perhaps the most ironic and dark consequence of the Aral's shrinkage is the story of Vozrozhdeniya (Resurrection) Island. The Soviet military in the early 1950s selected this, at the time tiny, isolated island in the middle of the Aral Sea, as the primary testing ground for its super-secret biological weapons program (Bozheva et. al., 1999; Wijinsema, 2000;). From then until 1990, they tested various genetically modified and "weaponized" pathogens, including anthrax, plague, typhus, smallpox and other disease causing organisms. These programs stopped with the collapse of the USSR in 1991. Allegedly the departing Soviet military took measures to decontaminate the island. As the sea shrunk and shallowed since the 1960s, Vozrozhdenia grew in size and in 2001 united with the mainland to the south as a huge peninsula extending into the Aral Sea. The fear is that some weaponized organisms survived and could escape to the mainland via infected rodents or that terrorists might gain access to them. The U.S. has committed $6,000,000 to help the Government of Uzbekistan kill any surviving organisms (Science Scope, 2002).

4. IMPROVEMENT EFFORTS FOR THE ARAL SEA AND REGION

The Soviet Union launched programs to improve the Aral situation in the late 1980s when that government finally admitted the existence of a serious problem (Micklin, 1991, pp. 68-81). The fundamental aims, but not the major players, have remained remarkably consistent: improve medical and health services for the people living near the sea, ensure safe drinking water supplies, improve ecological conditions in the delta of the Amu Dar'ya, diversify the economy and improve local food supplies, and rebuild irrigation system to raise their efficiency in order to deliver more water to the Aral Sea.

After the collapse of the USSR in 1991, the new states of the region (Kyrgyzstan, Uzbekistan, Turkmenistan, Kazakhstan and Tajikistan) assumed responsibility for dealing with the Aral situation. In March 1993, the presidents of the five republics signed an agreement to promote cooperation in solving the key problems (Micklin, 2002). It established the Interstate Council on the Problems of the Aral Sea Basin (ICAS) [*Mezhgosudarstvennyy Sovet po problemam basseyna Aral'skogo mory*]. A major purpose of the new organization was to facilitate assistance from the World Bank and other international donors as well as assume responsibility for various Aral Sea Basin assistance programs. The presidents also created an International Fund for the Aral Sea (IFAS) [*Mezhdunarodnyy fond spaseniya Arala*] with the responsibility to collect revenue from each basin state for financing of rehabilitation efforts. ICAS was abolished in 1997 and merged its functions into a restructured IFAS The leadership of IFAS rotates in a two-year cycle among the Central Asian Heads of State; currently, Emomali Rakhmanov, the President of Tajikistan, heads the fund.

Following independence, international aid donors began to play a major role in promoting cooperation in the management of the transnational water resources in the Aral Sea Basin (Micklin, 1998, 2002). The World Bank (International Bank for Reconstruction and Development - IBRD) was the first major agency to become involved. In the early 1990s, the Bank worked with Aral Sea Basin governments to formulate an Aral Sea Basin Assistance Programme (ASBP) to be carried out over 15 to 20 years. Initially, planning and implementation of the program was estimated at a around 250 million USD, later upped to 470 million USD (World Bank, 1996, Annex 3). The main goals of the program were (1) rehabilitation and development of the Aral Sea disaster zone, (2) strategic planning and comprehensive management of the water resources of the Amu Dar'ya and Syr Dar'ya, and (3) building institutions for planning and implementing the above programs. The Bank encouraged the basin states to create ICAS and IFAS and has worked with and through these organizations to realize the ASBP. Afghanistan was invited to join the ASBP but did not respond to the overture (World Bank, 1998, p. 9).

The latest aspect of the Bank funded effort, supported through the Global Environmental Facility (GEF), is the Water and Environmental Management Project (World Bank, 1998, pp. 19-34; see also program website at grida.no/aral/wb/wemp). Work on this project began in 1998 and finished in 2003, at a cost of 21.5 USD million. In line with a new emphasis on regional responsibility for the ASBP, the Executive Committee of IFAS managed the program, with the Bank playing a cooperative/advisory role. Key tasks were (1) improvement of the management of water and soil salinity related to irrigation practices, (2) development of low-cost, local, on-farm water conservation measures, (3) reduction of the amount of irrigation drainage water flowing back into rivers, (4) strengthening the existing interstate water sharing agreements, (5) improving public awareness of critical water problems, (6) enhancing dam and reservoir management and safety, (7) monitoring of water quality and quantity at transboundary river crossings, and (8) implementing a program to restore wetlands in the lower Amu Dar'ya delta, particularly Lake Sudoch'ye, which has been recommended to be a RAMSAR (wetland of international importance) site.

A number of other international donors, directly or indirectly, have been contributing to Aral Sea region improvement. The United States Agency for International Development (USAID) funded the Environmental Policy and Technology (EPT) project, running from 1993 to 1998, which financed measures to improve drinking water supplies in the Amu Dar'ya delta, aided in the formulation and implementation of regional water management policies and agreements, and provided advice on water management issues to specific governments (Micklin 1998). A smaller-scale follow-up project in 1999 and 2000 provided further assistance.

USAID initiated a new, major effort in 2001 known as the Natural Resource Management Project (NRMP) (for information, see the project website at www.nrmp.uz). This is a 5-year effort focusing on providing assistance to Kazakhstan, Kyrgyzstan, Turkmenistan, Uzbekistan and, to a lesser extent Tajikstan, to improve management of water, energy, and land.

The European Union initiated a major aid program for the Aral Sea Basin states in 1995 known as the Water Resources Management and Agricultural Production in the Central Asian Republics Project (WARMAP) (Micklin, 1998). Key objectives of the program were to assist the five former Soviet republics to develop policies, strategies and development programs for utilization, allocation and management of the water resources of the basin;

and to assist at the regional level with the establishment of the institutional structure for allocation and management of interstate waters. Phase 1 and 2 were completed by mid-1997. Major accomplishments of this program were development of a GIS (Geographic Information System) based land and water database for the basin (WARMIS), providing help to the World Bank and ICAS (now IFAS) in their efforts to improve and legally codify the 1992 interstate water sharing agreement among the new states of the basin and funding of training seminars and workshops, and an attempt to gather detailed data on irrigated water use at the farm level through a Water Use and Farm Management Monitoring Survey (WUFMAS) (World Bank, 1998, pp. 8-9). A continuation/follow-on project to WARMAP began in 1998.

The United Nations has been providing assistance on the Aral Sea Crisis since 1990 when a joint UNEP/Soviet working group on the Aral was formed (Micklin, 1998). This aid has continued and expanded in scope in the Post-Soviet era. UNESCO (United Nations Educational, Scientific and Cultural Organization) funded a research and monitoring program for the near Aral region from 1992-1996 focusing on ecological research and monitoring in the Syr Dar'ya and Amu Dar'ya deltas (UNESCO, 1998). The overall intent was to model the terrestrial and aquatic ecosystems of the study area in order to provide a scientific basis for implementation of ecologically sustainable development policies. The project relied mainly on the expertise of scientists and technicians from the Central Asian Republics and Russia with limited involvement of foreign experts. UNICEF (United Nations Childrens' Fund) launched the Aral Sea Project for Environmental and Regional Assistance (ASPERA) in 1995. It provides assistance to the disaster zone around the sea and focuses on health, nutrition, health education, water and environmental sanitation, and support to NGOs.

UNDP (United Nations Development Program) has also been very active in Aral Sea region activities. This organization has had two primary foci: strengthening regional organizations that have been established to deal with the Aral Crisis (earlier ICAS and IFAS, now the reconstituted IFAS) and promoting sustainable development to improve conditions for the several million people in the parts of Kazakhstan, Uzbekistan, and Turkmenistan that are closest to the Aral Sea. UNDP also convened the International Conference on the Sustainable Development of the Aral Sea Basin, held in Nukus, Karakalpakstan in September 1995 which led to the signing by the five Central Asian presidents of a Declaration of Central Asian States and International Organizations on Sustainable Development of the Aral Sea Basin. The document commits the five states to pursue sustainable development in the management of land, water, biological resources and human capital.

The North Atlantic Treaty Organization (NATO) has become involved in Aral Sea region activities through its Scientific and

Environmental Affairs Division. The first NATO sponsored event was an Advanced Research Workshop (ARW) titled "Critical Scientific Issues of the Aral Sea basin: State of Knowledge and Future Research Needs" held in Tashkent, Uzbekistan during May 1994 (Micklin and Williams 1996). A second NATO ARW with an Aral theme took place in Wageningen, the Netherlands in January 1995. The focus was on irrigation, drainage and the environment in the Aral Sea Basin.

Since 1995, the NATO Science Division has been a main sponsor of work (along with the German Remote Sensing Center in Munich, Department of Geography, Western Michigan University, Kalamazoo, Michigan, Institutes of Geography and Water Problems, Russian Academy of Sciences, Moscow, and Department of Geography, Karakalpakstan State University, Nukus, Uzbekistan) to develop a comprehensive land and water GIS for the Amu Dar'ya delta and Aral Sea. This system is intended to serve as a key tool for decision-making on land, water, and environmental management in the delta. The project has been cooperating closely with the government of Karakalpakstan to establish indigenous GIS capabilities through continuing development of a GIS center at Karakalpakstan State University in Nukus. The Center serves as a training site for local specialists and scientists in GIS techniques and also operates a program for monitoring environmental conditions in the Amu Dar'ya delta and in the Aral Sea. The Science for Peace (SfP) program of the Science Division has been in recent years the main supporter of this work

SfP has also supported another project to develop an environmentally appropriate water management regime, implemented through a decision support system based on GIS and a set of hydrologic models, for the larger lakes/wetlands that have been created or restored in the Amu Dar'ya delta. This project involves cooperation between the Scientific Information Center of the ICWC in Tashkent and the private consulting firm Resource Analysis in the Netherlands. (For more information on NATO funded efforts see NATO/OTAN, 2003, pp. 153-154 and 189-190, SfP 974101 website at http://sfpp.nm.ru and SfP 974357 website at http://www.icwc-aral.uz).

5. THE FUTURE OF THE ARAL SEA

What could the future hold for the Aral and its adjacent area? Can the sea be returned to its pre-1960s size and it as well as the severely impacted adjacent zone, particularly the delta of the Amu Dar'ya, restored to their former ecological condition? If not, what improvement measures are rational and feasible to undertake.

5.1 Aral Sea Restoration

From 1911-1960, discharge to the Aral from the Amu Dar'ya and Syr Dar'ya averaged approximately 55 km^3/yr and supported the sea at around 53 meters above msl (mean sea level) with an area around 67,000 km^2 (Table 1). This is generally accepted as the modern (i.e., pre-desiccation) state of the lake, although, a longer historical perspective shows the Aral in one of its high phases during this period. Beginning in the mid-1960s, river inflow to the sea started a persistent decline owing to both natural and human influence, with the latter far and away the dominant factor (Micklin, 1991, pp. 44-45). The Aral shrank in parallel with the decrease in inflow. During the 1980s, discharge to the lake averaged only 6 km^3- 11% of the 1911-1960 figure.[2] For most of the 1990s, the Aral Sea basin experienced a return to a high-flow cycle. Average combined discharge of the two rivers exiting the mountains averaged 104 km^3 for 1990-98, compared to a 40 year (1959-98) figure of 94 km^3/yr.[3] Estimated inflow to the sea for this period averaged 14-15 km^3. Severe drought affected the mountainous, flow generating regions of the Aral Sea basin from 1999 through 2001. (Agrawala, 2001). Average annual discharge to the Aral Sea during this period was probably no more than 2 km^3, with most of the water entering the small Aral from the Syr Dar'ya and practically nothing flowing to the large Aral from the Amu Dar'ya. Assuming continuation of basin withdrawals typical of recent years, and average annual outflows from the zones of flow formation typical of recent decades, a conservative but reasonable estimate of long-term average annual future inflow to the sea is around 10 km^3 (the "business as usual" scenario).

Under the business as usual scenario, restoration of the Aral to its size during the first six decades of the 20th century would require raising average annual discharge to the sea by approximately 45 km^3, or 450 percent, bringing total inflow to 55 km^3. Supporting the Aral close to 50 meters above msl with an area near 59,000 km^2, according to historical, literary, and cartographical evidence, the lowest and smallest the sea was in the 200 years prior to the early 1970s, would entail raising inflow to 43 km^3, an increase of 33 km^3 or 330% (Bortnik, 1996). To stabilize the sea at its January 2004 size of 17,400 km^2 would require an increase in inflow to near 16 km^3, 60% above the business as usual figure. Even during the higher flow period of 1990-1998, after upstream withdrawals, there was insufficient inflow to the Aral to stop the sea's recession. In dry cycles such as most years in the 1980s and 1999 through 2001, practically no water reached the sea and desiccation proceeded alarmingly rapidly.

In a regional context, the only realistic means for substantially increasing inflow to the Aral is reducing the consumptive use of water for irrigation in the sea's drainage basin. The reason is simple: this water intensive activity, conducted on around 7.9 million hectares and the basis of agriculture here, accounts for 92% of withdrawals and an even larger share of consumptive use (Ruziev and Prikhod'ko, 2002;).[4] The largest irrigated hectarage in the basin is found in Uzbekistan, and Turkmenistan; these two nations, respectively, account for 54% and 22% of all irrigation withdrawals (Micklin, 2000, p. 37). It is irrigation that has depleted the flow of the Amu Dar'ya and Syr Dar'ya and led to the great reduction in discharge of these rivers to the Aral as well as the consequent desiccation of the water body with all its attendant negative ramifications.

Irrigation in the Aral Sea basin is inefficient. Substantial improvements to it, technical, economic, and institutional, could save considerable water. Attempts are underway to implement improvement measures, but the substantial and comprehensive program needed would be extremely costly and faces concerted opposition from forces within governments and from segments of the public. Taking costs as an example: complete renovation of irrigation systems on 6 million hectares could likely save 12 km^3/year but would cost at least 16 billion USD (Micklin, 2002). To reach the maximum potential savings of 28 km^3 (based on technically, economically, and institutionally reforming irrigation on the "Israeli" model) would cost multiples more. These figures are far beyond the willingness and ability of the basin states, in combination with international donors, to pay. Furthermore, the technical condition of irrigation systems in the basin, far from improving, is steadily deteriorating owing to inadequate funding for, and lack of management responsibility over, operation and maintenance activities.

Converting more of the irrigated area to less water intensive crops (e.g., substituting grains, soybeans, fruits, and vegetables for cotton and rice) and reduction of the irrigated area are other means of significantly reducing water usage in irrigation (Micklin, 2002). The former strategy is being employed. Between 1990 and 1998, the area of cotton as a percent of the total irrigated area dropped from 45% to 25% percent while the area of winter wheat rose to 28%. (Dukhovnyy and Sokolov, no date). This probably was a major factor in the drop in irrigation withdrawals from 109 to 92 km^3 (16%) at the same time the irrigated area increased 10%. However, there are limits to such a program as the two primary irrigating nations (Uzbekistan and Turkmenistan) are intent on keeping cotton as a major crop since it plays a key role in earning foreign currency. Reductions in the irrigated area are unlikely in the near to mid-term future. All the former Soviet republics, except Kazakhstan, intend to expand irrigation, mainly to meet food needs for a growing population.

Considering the above, it is extremely doubtful that the Aral could be restored to 53 meters (the 1910-1960 average) or even 50 meters (the pre-1970s 200 year minimum) in the foreseeable future. The former would require increasing average annual discharge to the sea by 45 km^3 over the estimated future inflow figure of 10 km^3. This would mean decreasing basin wide consumptive usage by a somewhat larger amount to compensate for natural losses of the net additions to flow before they reached the sea. Accepting an estimate of these at 14% (ICAS, 1996, chapter 7, Tables 7.1 and 7.2) means that the reduction in basin wide consumptive use would need to be 52 km^3. Similar figures to support a 50-meter sea level are 33 and 38 km^3. These amounts are far above even the most optimistic and costly scenario of water use efficiency improvements and could only be met by a major cut-back in irrigation that would wreak economic and social havoc on the countries of the basin that are the major users of water for irrigation..

Of course it is possible to bring water to the Aral Sea from outside Central Asia. During the latter part of the Soviet period, water managers in Moscow and in Central Asia proposed diversion of massive flow, up to 60 km^3, from Siberian rivers to the region as the panacea for perceived water shortage problems (Micklin, 1991, pp. 60-68). The initial stage of this project would have taken 27 km^3 from the Irtysh-Ob river system in the Western Siberian region of Russia. It was on the verge of implementation when stopped by the Gorbachev regime in 1986. Although real and serious potential ecological threats (of regional, not global magnitude as claimed by some opponents) were given as the chief reason for canceling the project, economic considerations were the fundamental factors in this decision (Micklin, 1987).

This grandiose scheme continues to be discussed and promoted in Central Asian water management and governmental circles and in more recent years has, again, found a sympathetic ear among some water management professionals and bureaucrats in Russia, including Yuri Luzhkov, mayor of Moscow and N.N. Mikheyev, the First Deputy Minister of Natural Resources (Mikheyev, 2002; Polad-Zade, 2002; Temirov, 2003). However, implementation of this project in any but the far term, if ever, seems a pipe dream. Costs would be enormous, at least 30 billion USD by latest estimates, and even if Russia were willing to help finance the project, it is doubtful sufficient funds could be accumulated for construction (Temirov, 2003). International donors, such as the World Bank, given their newfound sensitivity to environmental concerns, have stated opposition to such a project (Interfax Information Agency, 19 August, 2002). Finally, there is tremendous opposition among Russians to sending water from their precious Siberian rivers to Central Asia where, in their view, it would be wasted. Even if implemented, much less than the 27 km^3 diverted, probably less than 15, would reach the Aral owing to substantial evaporation and exfiltration losses in the transfer system, withdrawals along the route for irrigation and other purposes, and usage in Central Asia for irrigation. Certainly, it would be more rational to spend precious capital and effort on improving regional water management rather than importing water from Siberia (Kamalov, 2003).

5.2 Mitigation Scenarios

Although restoration of the Aral to, or near, its pre-1960s level and ecological state is not viable in the foreseeable future, partial restoration and rehabilitation of sections of the sea are feasible as is improvement of environmental conditions in the Amu Dar'ya delta. The primary candidate for restoration is the small (northern) Aral (Aladin and Plotnikov, 1995). Its level has been relatively stable (several meters variation) since its separation from the large, southern Aral in the late 1980s. Because its area is much smaller than the large sea, an inflow of 4.5 km^3 on an average annual basis (for 1987-98 the figure was around 4 km^3) could within 7 years (by 2011, assuming implementation of the project in 2004) raise its level to near 45 meters and area to near 4,000 km^2, while lowering salinity to less than 15 g/l (Table 1). By 2025 (Fig. 3), the sea could reach 47 meters, have an area of 4310 km^2, and reach salinity below 10 g/l (similar to 1960). This would provide optimal habitat conditions and allow for the flourishing of indigenous fishes that now live most of the time in the Syr Dar'ya with the concomitant restoration of a commercial fishery; restoration of limited commercial navigation might also be possible. With more inflow the sea could be restored in a shorter time and sustained at a higher level. If Syr Dar'ya water that has been periodically dumped into the artificial desert Lake Arnasay, averaging 3.6 km^3/yr for 1993-1997[5], were to be added to the flows reaching the small Aral, it should be no problem to maintain discharge to this waterbody at an average of 4.5 km^3/yr and probably considerably higher.[6]

Realization of this project would expensive. First of all, the flow from the small to large sea must be controlled. Local Kazakh authorities in the early 1990s constructed a crude dike to block the channel (river) that had formed between the two seas. However, the dike in the ensuing years periodically washed away. In 1997, they replaced this makeshift facility with a 20 km long, 26-meter wide dam that raised water levels by several meters by early 1999 (Optimism rises..., 1999). The existing dam was "overtopped" by wave and wind-running action in April 1999 and breached with a significant loss of water (Dukhovnyy, 1999). Secondly, the channel downstream of the Chardara reservoir on the Syr Dar'ya is incapable of handling heavier flows, particularly in winter when it is ice-choked, and must be cleared and improved to allow flows now sent to Arnasay to be diverted to the small Aral. The World Bank has approved funding (84 million

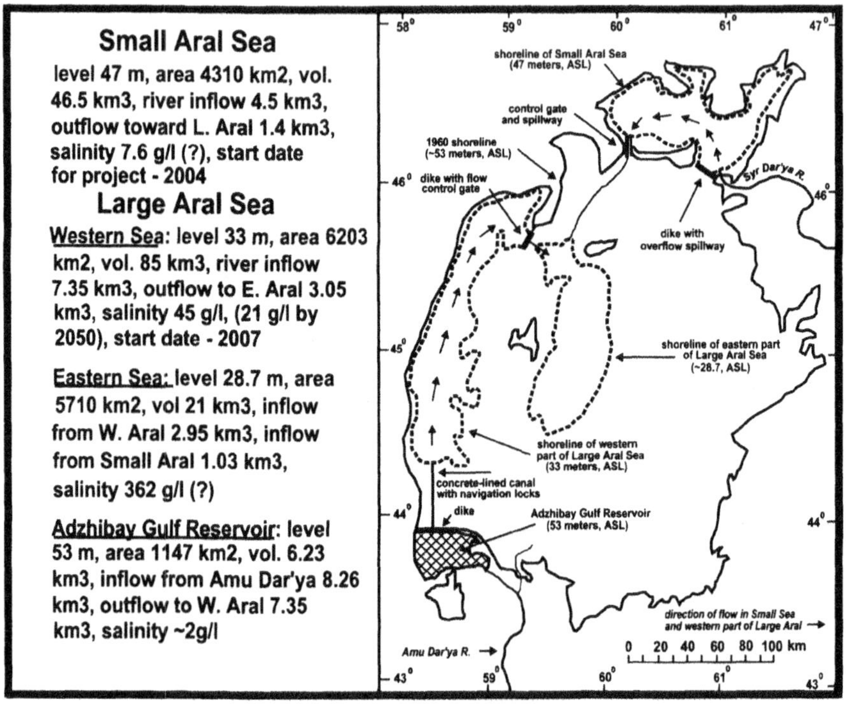

Figure 5. Projects Scenario for Aral Sea in 2025

USD) for construction of an engineeringly sound dike, a control gate and channel from the small sea to the large sea, and for improvements to the Chardara dam and lower Syr Dar'ya channel below the dam (World Bank, 1998, p. 10; Eurasianet, 2002; Pala, 2003).. Work is reportedly underway and will be completed in 2004. The project will raise the level of the small Aral by around three meters.

The southern, large Aral Sea is a different story. To even stabilize the Large Aral Sea at its January 2004 level of 30+ meters and area of around 17,400 km^2 would require increasing inflow to around 15 km^3. Perhaps around a kilometer of this could still arrive from the Small Aral after implementation of the diking project for that portion of the Aral, but the remaining 14 km^3 would need to be supplied by the Amu Dar'ya– an increase of 7-8 km^3 above the average annual discharge of this river to the sea for 1987-2002. Attainment of such an increase of Amu Dar'ya inflow is possible, but realistically improbable in the near or even mid-term future. However, there is a potential "silver lining" to this generally gloomy picture of the future Large Aral Sea: *Artemia parthogenetica.* This tiny brine shrimp proliferates in salinities from 100 to 500 g/l (Ptichnikov, 2002). The salinity of the Large Aral will soon enter this range. Already the shrimp are numerous in some parts of it. The shrimp eggs are gathered, dried, and sold as fish food (the dried eggs when rehydrated hatch into shrimp) at $11 to $33 per kilogram (Pala, 2003). The shrimp could be artificially raised in the Aral and provide a major new industry based on an otherwise economically useless waterbody. A consortium of companies based in Belgium is investigating the commercial viability of Artemia in the Large Aral.

A strong case may be made that it is wiser to use the residual flow of the Amu Dar'ya for rehabilitating and preserving deltaic ecosystems owing to their natural, economic and human benefits rather than being dumped into the large Aral Sea to evaporate. Rehabilitation and partial preservation of the Amu Delta and its wetlands has been a priority since the late 1980s, first by the Soviet government and subsequently by the new states of Central Asia, and international donors. The prime objective of the most recent program known as the Aral Sea Wetland Restoration Project (ASWRP), which was implemented by the International Fund for the Aral Sea and funded by the Global Environmental Facility, has been partial ecosystem rehabilitation through creation of artificial ponds and wetlands in the delta and on the dry bed of the Aral Sea (IFAS, 2000, pp. 19-23). Specific benefits of lake/wetland restoration are enhanced biodiversity, improved fisheries, greater forage production, treatment of wastewater by aquatic vegetation, and some reduction in salt and dust transfer from the dried sea bottom to arable lands (The Aral Sea Basin Sustainable Development Commission, 1998, pp.59-81). A companion measure is the revegetation/reforestation of parts of the dried bottom to stabilize them and lower their deflation potential. With the completion of parts 1 and 2 of the project, some 73,000 ha will enjoy improved conditions for both flora and fauna. The aggregate cost of parts 1 and 2 was $6 million USD.

Experts have estimated that 4-5 km^3 of water (mainly relatively clean river flow supplemented by irrigation drainage) are needed to support minimally acceptable "hydroecological conditions" in the lower delta of the Amu Dar'ya, including the natural and artificially created lakes and wetlands there (MKVK, 2002, p. 39). Thus, some water would be left, most years, to send to the large Aral Sea. If modest increases in flow to the lower delta could be implemented, allowing flow to the large Aral to average 7-10 km^3/yr., there would be the possibility to separate the deep western part from the shallow eastern by a dike of modest proportions, direct the flow of the Amu Dar'ya into the western part (a channel already exists for this, but would need considerable enlargement) and, over time, freshen and partially ecologically restore it by allowing a controlled flow of saline water from it to the eastern portion. The eastern part would then rapidly shrink into a residual brine lake. One scenario for this is shown on Figure 3. This alternative has been little studied so cost estimates are highly speculative, although it would likely be more expensive than the project to rehabilitate the small Aral Sea. Also, the range of potential negative environmental consequences is unknown (e.g., would the shrinking eastern sea leave a much larger "salt desert" that would significantly aggravate the problem of salt/dust storms? Also, this option would eliminate the possibility of raising brine shrimp in the western portion of the Large Aral, although it might enhance their production in the eastern portion whose salinity would increase substantially.

---

[1] I.M. Rubanov and N.M. Bogdanava, "Kolichestvennays otsenka solevoy deflyatsii na osushayushchemsya dne Aral'skogo moray," [A quantitative estimate of salt deflation from the dried bottom of the Aral Sea] *Problemy osvoyeniya pustyn'*, No. 3, 1987, pp. 9-16. The reason for the predicted decrease in exported dust and salt, in spite of a larger area of dried bottom, is an expected significant decrease in losses from "old" areas of the bottom where the loose material subject to transport would become largely depleted, carried by precipitation below the surface or formed into a hard, deflation resistant crust.

[2] Hydrologic data obtained by the author between 1984 - 2002 from a variety of sources, including the Gidroproyekt [Hydro Planning] Institute in Moscow, Glavgidromet [Main Administration of Hydrometeorology] in Tashkent, and the GIS Research Center in Nukus, Karakalpakstan and used to derive long-term time series inflows into the Aral Sea from the Amu Dar'ya and Syr Dar'ya and water balances for the Aral Sea.

[3] (based on calculations from data provided by Glavgidromet, Tashkent, Uzbekistan and from Iliya Zholdaova, "Fish population as an ecosystem component and economic object in the Aral Sea basin," M. Glantz (ed.), *Creeping Environmental Problems and Sustainable Development in the Aral Sea Basin* (Cambridge: Cambridge University Press, 1999), p. 205 (Table 10.1)).

[4] Withdrawals are a measure of the total water taken from sources (rivers and groundwater) for irrigation. Consumptive use is a measure of the water that is withdrawn that is lost to evaporation (from conveyance canals and fields) and transpired from or incorporated into crops. The difference between the two is termed return flow. Return flow includes filtration from canals, filtration from fields, and surface runoff from fields. Part of return flow ultimately reaches the river from which taken or adds to groundwater while another portion runs off into desert hollows to form lakes (the water from these is lost to evaporation).

[5] data from Glavgidromet, Uzbekistan

[6] Calculations are based on a simple, iterated annual water balance model for the small Aral which assumes sea surface evaporation at 1000 mm, sea surface precipitation at 198 mm, and net subsurface inflow of groundwater at 0.2 km^3.

REFERENCES

Abdirov, Ch. A. et. al (1993), "Stress reaktsiya zdorovykh detey Nukusa na vozdeystviya otritsatel'nykh ekologicheskikh faktorov v Priaral'ye," [Stress reaction of healthy children of Nukus from the effects of ecological factors in Priaral'ye] Vestnik Karakalpakskogo otdeliniya Akademii Nauk Respubliki Uzbekistana, No. 2, pp. 15-20

Agrawala, Shardul et. al,(2001) "The Drought and Humanitarian Crisis in Central and Southwest Asia: A Climate Perspective," The international Research Institute for Climate Prediction, IRI Special Report No. 01-11, November, 20 pp. (available at http://iri.columbia.edu)

Aladin, N.V. and I.S. Plotnikov (1995), "K voprosy a vozmozhnoy konservatsii i reabilitatsii malogo Aral'skogo morya," [Concerning the possibility of conservation and rehabilitationin of the small Aral Ses] Biologicheskiye i prirodovedcheskiye problemy Aralskogo morya i pryaral'ya, [Biological and natural science problems of the Aral Sea and near Aral region] Trudy Zoologicheskogo Instituta, RAN tom 262, 1995, pp. 3-16. For an English version of this study see Aladin, N.V., I.S. Plotnikov and W.T.W. Potts, "The Aral Sea Dessication and Possible way of Rehabilitating and Conserving its Northern Part, Environmetrics, Vol. 6, pp. 17-29.

Bel'gibayev, M. Ye. (1984), "Pylesolemep – pribor dlya ulavlivaniya pyli I soley v vozdushnom potoke," [A dust/salt meter – an apparatus for entraining dust and salt from air currents], Problemy osvoyeniya pustyn', No. 1, , pp.72-74.

Bortnik, V.N.(1996), "Changes in the water-level and hydrological balance of the Aral Sea," pp. 25-32 in Micklin, Philip P. and William D. Williams (eds.). The Aral Sea Basin (Proceedings of an Advanced Research Workshop, May 2-5, 1994, Tashkent, Uzbekistan), NATO ASI series, Vol. 12 (Springer-Verlag: Heidelberg,).

Bozheyeva, Gulbarshyn et al (1999), Former Soviet Biological Weapons Facilities in Kazakhstan: Past, Present, and Future, Monterrey Institute, Center for Nonproliferation Studies, Occasional Paper No. 1, June.

Bortnik, V.N and S.P. Chistyaevaya (eds.) (1990), Gidrometeorologiya i gidrokhimiya morey SSSR [Hydrometeorology and hydrochemistry of the seas of the USSR], vol VII, Aral'skoye more [Aral Sea] (Leningrad: Gidrometeoizdat,).

Chub, V. Ye. (2002), Izmeneniye klimata i ego vliyaniye na prirodno-resursnyy potentsial respublika Uzbekistana [Climate change and its influence on the natural resouce potential of the Republic of Uzbekistan] (Tashkent:Glavgidromet,).

Chub, V. Ye (1998), "Regional'nyye problemi otsenki vliyaniya aerosoley (anthropogenennogo i prirodnogo proiskhozhdeniya) na klimaticheskiya kharakteristiki v tsentral'noaziatskom regione, [Regional problems of estimating the influence of aerosols (of anthropogenic and natural occurrence) on the climatic characteristics of the Central Asiatic region pp. 3-4 in , Materialy k nauchnomy rerional'nomu seminaru, Eolovyy perenoc soley v basseyne Aral'skogo Morya [Materials for the regional scientific seminar, Aeolian tranport of salt in the Aral Sea basin], Tashkent 4-5 March,.

Dukhovnyy, V. (2003), email from Victor Dukhovnyy, 23 June.

Dukhovnyy, V. (1999), conversation with Victor Dukhovnyy, director of the SIC, ICWC, Delft, Netherlands, July 19.

Dukhovnyy, V. A. and V. I. Sokolov, "Integrated Water Resources Management in the Aral Sea Basin," no date.

Eurasianet (2002), "Kazakhstan: World Bank Help for Aral Sea," Feb. 24 (www.eurasianet.org/departments/environment).

Friedich, Jani and Heidi Oberhansli (2003), "The Aral Sea: New Data from a Changing Ecosystem," poster given at the Symposium on Dead and Dying Seas, Liege, Belgium, May 5-10.

Glazovskiy,.N.F., Aral'skiy krizis: prichiny vozniknoveniya i put' vykhoda [The Aral crisis: causative factors and means of solution] (Moskva, "Nauka," 1990), pp. 20-23.

IFAS (2000) (International Fund for Saving the Aral Sea and Priaralye), Projects of the Aral Sea Basin Program Recommended for Implementation, Tashkent.

ICAS (1996), Fundamental Provisions of Water Management in the Aral Sea Basin: A Common Strategy of Water Allocation, Rational Water Use and Protection of Water Resources, prepared with the assistance of the World Bank, October (no pagination).

Kamalov, Yusup (2003), "The Aral Sea: The Legends, the Problems, the Solutions, " Ecostan News, 10/1 (January-February), pp. 9-10.

Kec', A.S (1978), "Prichiny uzmeneniyy urobnya Arala v Goletsine," [Reasons for the Aral's level changes in the Holocene], Izvestiya Akamdemii Nauk SSSR, seriya Geograficheskaya, No. 1, , pp. 16.

Kes', A.S. and I.A. Klyukanova (1990), "O prichinakh kolebaniya urovnya Aral'skogo moray v proshlom," [The reasons for fluctuations of the Aral Sea in the Past], seriyaa geograficheskaya, No. 1, pp.78-86.

Khvorog (ed.) (1992), Ekologicheskaya karta priaral'ya [Ecological map of the near Aral region], Alma-Ata.

Kes', A.S., B.P. Andrianov and M.A. Itina (1980), "Dinamika gidrograficheskoy seti i izmeneniya urovnya Aral'skogo morya,"[Dynamics of the hydrographic network and changes of Aral Sea Levels] pp. 185-197 in Kolebaniya uvlazhnennosti Aralo-Kaspiyskogo regiona v golotsene [Fluctuation of the moistness of the Aral-Caspian region in the Holocene] (Moskva: "Nauka).

Kosarev, A.N., A.G. Kostyanoy and V.N. M. Mikhaylov (2003), "Aral, kotoryy my teryayem" [The Aral which we are losing], Zemlya i Vselennaya, No. 2, pp. 27-35.

Medecins sans Frontieres (2000), MSF in Uzbekistan and Turkmenistan, Activity Report 1999-2000.

Micklin, Philip (2002), "Water in the Aral Sea Basin of Central Asia: Cause of Conflict or Cooperation?" Eurasian Geography and Economics, 43, No. 7, pp. 505-528.

Micklin, Philip P. (2000), Managing Water in Central Asia (London: The Royal Institute of International Affairs, Central Asian and Caucasian Prospects).

Micklin, Philip P. (1998), "Regional and International Responses to the Aral Crisis," Post-Soviet Geography and Economics, Vol. 39, No. 7 (September), pp. 399-417.

Micklin, Philip P. (1994), "The Aral Sea Problem," Civil Engineering, Proceedings of the Institution of Civil Engineers, August, pp. 114-121.

Micklin, Philip P. (1992), "The Aral Crisis: Introduction to the Special Issue," Post Soviet Geography, Vol. XXXIII, No. 5 (May), pp. 269-283.

Micklin, Philip P. (1991), The Water Management Crisis in Soviet Central Asia, The Carl Beck Papers in Russian and East European Studies, No. 905 (Pittsburgh: The Center for Russian and East European Studies, August).

Micklin, P. (1987), "The Fate of 'Sibaral': Soviet Water Politics in the Gorbachev Era," Central Asian Survey, No. 2, pp. 67-88.

Mikheyev, N.N. (2002), "Voda bez granits,"[Water without limits] Melioratsiya i vodnoye khozyaystvo, No. 1, pp. 32-34.

Mirabdullayev et. al, (2003) "Succession of the ecosystems of the Aral Sea during its transition from oligohaline to polyhaline waterbody" presentation at the Symposium on Dead and Dying Seas, Liege, Belgium, May 5-10.

MKVK (Mezhgosudarstvennaya koordinatsionnaya vodokhozyaystvennaya komissiya) (2002) Byulleten, [Bulletin] No. 3 (31), August.

NATO/OTAN (2002), Science for Peace: SfP Progress Report No. 4 (through October), Scientific and Environmental Affairs, 2003.

Novikova, N. (1996), "The Tugai of the Aral Sea Region is dying: Can it be restored?" Russian Conservation News, February, No. 6, pp. 22-23.

Novikova, N. (1997), Printsipy sokhraneniya botanicheskogo raznoobraziya deltovykh ravnin Turana [Principles of preserving the botanical diversity of the deltaic plains of the Turan], .dissertatsiya v forme nauchnogo doklada na soiskaniye uchenoy stepeni doktora geograficheskikh nauk (Moskva).

O'Hara, Sarah L., Giles F S Wiggs, Batyr Mamedov, George Davidson, Richard B Hubbard (2000), "Exposure to airborne dust contaminated with pesticide in the Aral Sea region," Lancet Research letters Volume 355, Number 9204 19 February.

"Optimism rises, with water, in bid to revive Aral Sea" (1999), The Christian Science Monitor, February 5, p. 7.

Pala, Christopher, (2003), "$85 Million Project Begins for Revival of the Aral Sea," The New York Times, Tuesday, August 4, p. F3.

Palvaniyazov, M. (1989), "Vlyaniye pyl'nykh bur' na mestoobitaniya nekotorykh mlekopitayushchikh primorskoy zony Aral'skogo morya," [The influence of dust storms on the habitats of certain mammals of the coastal zone of the Aral Sea] Problemy osvoyeniya pustyn', No. 1, pp. 55-59.

Polad-Zade, P.A. (2002), "Soobrazeniye po upravleniyu vodnymi resurami v Tsentral'noy Azii: opyt proshlogo i realii sovremennosti," [Thoughts about the management of water resources in Central Asia: past experience and modern realities] Melioratsiya i vodnoye khozyaystvo, No. 1, pp. 34-37.

Ptichnikov, A. (ed.) (2002), Bulletin No. 3, , NATO Science for Peace Project 974101: Sustainable Development of Ecology, Land and Water Use through Implementation of a GIS *and Remote Sensing Centre in Karakalpakstan*, Uzbekistan (in Russian and English) (also available on line at http://www/sfpp.nm.ru/)

Puziev, M.T. and V.G. Prikhod'ko (2002), "Otsenka perspektiv ustoychivogo razbitiya gosudarstv basseyna Aral'skogo moray s pomoshch' model'nykh raschyotov," [Evalustion of the future of sustainable development of the governments of the Aral Sea basin with the help of model calculations] Melioratsiya i vodnoye khozyaystvo, No. 1, pp. 54-57.

Shivareva, S., Ye Ponenkova,. and B. Smerdov (1998), "Modelirovaniye urovnya Aral'skogo morya," [Modeling the level of the Aral Sea], pp. 5-10 in Problemy basseyna Aral'skogo Morya, issledovaniya, proyekty, predlozheniya [Problems of the Aral Sea Basin, research, projects, proposals] (Tashkent, "Chinor" ENK).

Temirov, Rustam (2003) "Lobbying Grows in Moscow for Siberia-Uzbekistan Water Scheme," Eurasianet, February 19 (www.eurasianet.org/departments/environment)

The Aral Sea Basin Sustainable Development Commisssion (1998), Reference Book, Training Course, Tashkent, 6-10 July.

Tursunov, A.A. (1989), "Aral'skoye more i ekologicheskaya obstanovka v Sredney Azii i Kazakhstane," [The Aral Sea and the ecological situation in Central Asia and Kazakhstan] Gidrotekhnicheskoya stroitel'stvo, No. 6, pp. 15-22.

UNESCO (1998), Ecological Research and Monitoring of the Aral Sea Deltas, Aral Sea Project 1992-1996, Final Scientific Reports (Paris: UNESCO).

Wijinsima, Petra (2000), "Deadly Soviet legacy threatens fishing village," De Volkskrant, 31 March (English translation).

Williams, W.D. and N.V. Aladin (1991), "The Aral Sea: recent limnological changes and their conservation significance," Aquatic Conservation: Marine and Freshwater Ecosystems, vol. 1 pp. 3-23.

World Bank, Aral Sea Basin Program (Kazahstan, Kyrgyz Republic, Tajikistan, Turkmenistan, and Uzbekistan) (Water and Environmental Management Project, Washington, D.C., May 1998).

World Bank (1996), Aral Sea Basin Program, Phase 1: Progress Report No. 3, Washington, D.C., February.

Zholdasova, I.M., L.P. Pavlovskaya, A.N. Urashbayev and S.K. Lubimova (1998), "Biological Bases of Fishery Development in the Waterbodies of the Southern Aral Region," in Ecological Research and Monitoring of the Aral Sea Deltas, UNESCO Aral Sea Project 1992-1996, Final Scientific Reports (Paris: UNESCO), pp.213-215.

Chapter 6

HYDROBIOLOGY OF THE ARAL SEA

N.V. Aladin, I.S. Plotnikov

Laboratory of Brackish Water Hydrobiology, Zoological Institute of RAS, Universiteskaya naberezhnaya, 1, 199034 St Petersburg, Russia

R. Letolle

RL, Department of Applied Geology, Box 123, University P.M. Curie, 75252, Paris Cedex 05 France

The ecological crisis of the Aral Sea has been widely discussed during recent years in both the scientific and popular literature. However, only the consequences of anthropogenic desiccation and increased salinitywere usually discussed with little note of the role played by introduced species in this ecosystem (Micklin, 1991; Williams, Aladin, 1991; Keyser, Aladin, 1991). Here, we review the role of introduced species during periods of varying salinity.

The Aral Sea is a giant (66000 km^2 in 1960) continental closed lake. This relatively young water body appeared in the early Holocene as a terminal reservoir of Syrdarya river (Fig. 1A). During the Pleistocene glacial epoch the Aral basin contained only small hypersaline ponds and marshes. After the Aral depression filled with water in the early Holocene, flora and fauna of Aral Lake derived from freshwater inputs of the Syrdarya river and then later of Amudarya river. The Amudarya had previously flowed into the

J.C.J. Nihoul et al. (eds.), Dying and Dead Seas, 125–157.

Caspian Sea (Fig. 1A). Inhabitants of water bodies in the Aral depression during transgressions of the ancient Caspian Sea, when Aral was a gulf of Akchagyl or Apsheron seas (Fig. 2), completely died out when regression followed transgressions of the Caspian waters. At the beginning of Holocene, the Aral Sea aquatic fauna was of mainly freshwater origin with new invaders from the Caspian Sea and other saline water bodies of Central Asia appearing only later.

Most ancient invaders from the Caspian Sea arrived at the Aral Sea via Uzboy about 5000 B.P., when run-off from the ancient Amudarya and Syrdarya rivers had filled the Great Aral. The Great Aral was a giant lake joining the depressions of Aral and Sarykamysh lakes, and its level was over +58-60 m asl (Fig. 1A). At this stage, water from the Great Aral drains to the Caspian Sea in the southwest from Sarykamysh depression (Fig. 1A). Ancestors of recent Aral thorn sturgeon, other relatives of Caspian fishes, and possibly some other Caspian hydrobionts were able to overcome the current in Uzboy and colonize the Aral Sea.

At the beginning of the 1960's anthropogenic desiccation of the Aral Sea begun. At that time the lake was inhabited by dozens of fish species and more free-living invertebrates (Table 1). Note that some of these species were recently introduced by humans into the lake ecosystem.

The fist introductions of exotic species into the Aral Sea occurred at the end of the 1920's, when *Alosa caspia* (Caspian shad) и *Acipenser stellatus* (starred sturgeon) were introduced from the Caspian Sea. This introduction cannot be considered as successful because these fishes did not naturalize in Aral Karpevich, 1975). Furthermore parasites of starred sturgeon roe (*Polypodium hydriforme*) and gills (*Nitzschia sturionis*) passed onto aboriginal thorn sturgeon and caused strong epizooties. Thorn sturgeon before introduction of starred sturgeon did not suffer from these parasites because they were absent from the Aral Sea (Dogel, Byhowsky, 1934; Dogel, Lutta, 1937). Thus, the first attempt of exotic species introduction to the Aral Sea can be considered extremely unsuccessful.

After the Second World War attempts to settle exotic species in the Aral Sea continued. The main basis of these actions was the idea that because there were few plankton-eating fishes and sturgeons in the Aral Sea, introduction of new consumers of plankton and benthos would increase fish productivity (Karpevich, 1947, 1948, 1953, 1960, 1975). On the basis of these considerations, from the Caspian Sea again starred sturgeon

(*Acipenser stellatus*) was again introduced in 1948-1963, and in 1958 a subspecies of thorn sturgeon (*A. nudiventris derjavini)* from Ural river was introduced. These sturgeon introductions were again unsuccessful. Both species failed to persist and only in 1958 were some individuals of starred sturgeon caught (Karpevich, 1975).

In the same years (1954-1956) mullets (*Mugil auratus*, *M. salensis*) were introduced from the Caspian. This attempt was also unsuccessful (Karpevich, 1975) probably because these planktophages new for the ecosystem could not find sufficient amount of convenient food to survive.

More successful was the introduction (1954-1959) and acclimatization of Baltic herring (*Clupea harengus membras*). This exotic planktophage from the Baltic Sea became naturalized in the Aral Sea, and caused significant changes in the zooplankton community. Beginning in 1957 Baltic herring appeared in large number in catches. The pressure on zooplankton increased sharply and the average summer biomass of zooplankton decreased more than 10 times – from 160 mg/m^3 to 10-15 mg/m^3 (Karpevich, 1975; Yablonskaya, Lukonina, 1962; Kortunova, 1975). Introduced plankton-eating fishes led to the extermination of such large organisms of zooplankton as *Arctodiaptomus salinus* and *Moina mongolica*. Decreased zooplankton abundance and biomass instantly affected the number of herring and their number decreased also (Kortunova, Lukonina, 1970).

During the aforesaid planned introductions, many fish and invertebrate species were introduced accidentally. Among them were many non-commercial fishes. For example, six species of gobies, three of them – bubyr (*Pomatoschistus caucasicus*), monkey goby (*Neogobius fluviatilis pallasi*) and round goby (*N. melanostomus officinus*) naturalized successfully. Also, the successful introduction of silverside (*Atherina mochon pontica*) and pipefish (*Syngnathus nigrolineatus*) quickly invaded the whole Aral Sea. During planned acclimatization of plankton-eating fishes in the Aral Sea, two shrimp species (*Palaemon elegans* and *P. adspersus*) were accidentally introduced and (*P. elegans*) acclimatized successfully (Karpevich, 1975). This established shrimp became a concurrent of aboriginal amphipod *Dikerogammarus aralensis* and gradually forced out it from the benthic associations (Andreeva, 1989; Aladin, Potts, 1992). So, even before anthropogenic desiccation and increase salinity the Aral Sea ecosystem had undergone significant changes due to planned and accidental introductions of exotic species.

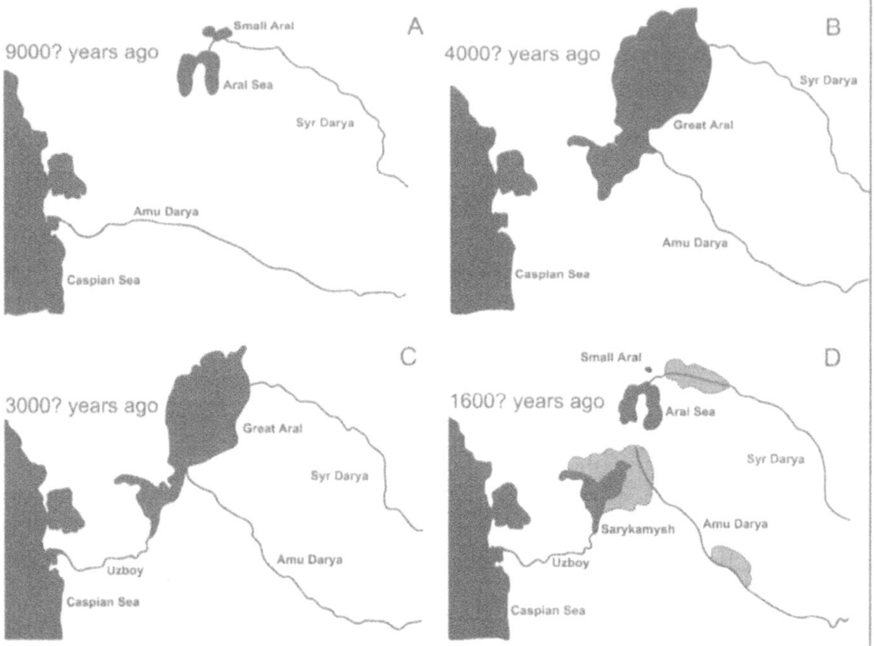

Fig. 1A. Early Holocene and late Holocene palaeohydrographical systems of Central Asia and Kazakhstan:
A — Paskevich stage; *B* —the Great Aral stage; *C* — stage of the first natural discharge throw Uzboy to the Caspian Sea; *D* — stage of the first anthropogenic desiccation.
Areas of irrigation are *shaded*.

At the same time (1958-1960), besides introductions into the Aral Sea proper, a complex of fishes and invertebrates introductions were carried out into deltaic areas of Syrdarya and Amudarya. From river Don, were introduced four mysid species (*Paramysis baeri, P. lacustris, P. intermedia, P. ullskyi*), two of them (*P. lacustris, P. intermedia*) became numerous in 1961, one species (*P. baeri*) did not acclimatize, and the third (*P. ullskyi*) has naturalized but remained uncommon (Karpevich, 1975). Also, introduced were three species of freshwater fishes from China: grass carp (*Ctenopharyngodon idella*), silver carp (*Hypophtalmichthys molitrix*) and *Aristichtys nobilis* along with incidental introductions into the deltaic areas of three other species of this complex: black carp (*Mylopharyngodon piceus*) and snakehead (*Ophiocephalus argus*). Except for *A. nobilis*, all these species naturalized successfully and became of commercial value (Karpevich, 1975). These naturalized Chinese fishes and mysids, which invaded estuaries of Amudarya and Syrdarya, migrated many kilometers from the deltas into the Aral Sea proper. When comparing consequences of introductions in deltaic areas with those in the Aral Sea proper one may note that the first were more successful and had fewer negative consequences than the second. However, even in the case of the deltas, there were no significant rises in catches of commercial fishes or increase food resources.

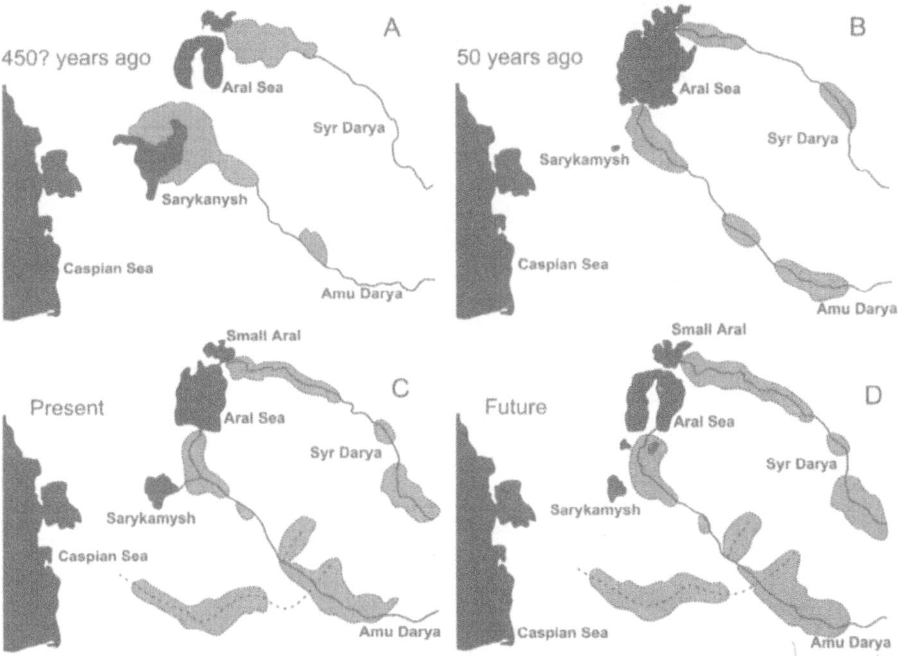

Fig. 1B. Medieval, in the middle of XX century, modern and possible future hydrographical systems in Central Asia and Kazakhstan:
A — stage of medieval anthropogenic desiccation; *B* — stage of the modern Aral before anthropogenic desiccation; *C* — stage of the present anthropogenic desiccation; *D* — stage of the Aral stabilisation in future. areas of irrigation are *shadowed*

Table 1.
List of free-living invertebrates in the Aral Sea.

No.	Species
	Coelenterata
1.	*Protohydra leuckarti* Greef
	Turbellaria
1.	*Mecynostomum agile* (Beklemischev)
2.	*Macrostomum hystricinum* Beklemischev
3.	*M. minimum* (Luther)
4.	*Promonotus orientalis* Beklemischev
5.	*Kirgisella forcipata* Beklemischev
6.	*Gieysztoria bergi* (Beklemischev)
7.	*Byrsophlebs geniculata* Beklemischev
8.	*Beklemischeviella contorta* (Beklemischev)
9.	*Phonorhynchoides flagellatus* Beklemischev
10.	*Gyratrix hermaphroditus*
11.	*Pontaralia relicta* (Beklemischev)
12.	*Placorhynchus octaculeatus dimorphis* Karling
	Nematodes
1.	*Adoncolaimus aralensis* Filipjev
	Rotatoria
1.	*Eosphora ehrenbergi* Weber
2.	*Trichocerca (Diurella) heterodactyla* Tschugunoff
3.	*T. (D.) similis* (Wierzejski)
4.	*T. (D.) porcellus*
5.	*T. s. str. elongata* (Gosse)
6.	*T. s. str. pusilla* (Lauterborn)
7.	*T. s. str. longiseta* (Schrank)
8.	*T. caspica* Tschugunoff
9.	*Synchaeta stylata* Wierzejski
10.	*S. vorax* Rousselet
11.	*S. tremula* (Müller)
12.	*S. pectinata* Ehrenberg
13.	*Polyarthra euryptera* Wierzejski
14.	*P. luminosa* Kutikova
15.	*P. vulgaris* Carlin
16.	*P. longiremis* Carlin
17.	*Lindia torulosa* Dujardin
18.	*Encentrum limicola* Otto

19.	*Asplanchna priodonta* Gosse
20.	*A. girodi* Guerne
21.	*Brachionus angularis* Gosse
22.	*B. calyciflorus* Pallas
23.	*B. quadridentatus* Hermann
24.	*B. plicatilis* Müller
25.	*B. rubens* Ehrenberg
26.	*B. urceus* (Linnaeus)
27.	*Platyias quadricornis* (Ehrenberg)
28.	*P. palustris* (Müller)
29.	*Keratella cochlearis* (Gosse)
30.	*K. tropica* (Apstein)
31.	*K. quadrata* (Müller)
32.	*K. valga* (Ehrenberg)
33.	*Notholca squamala* (Müller)
34.	*N. acuminata* (Ehrenberg)
36.	*Kellicottia longispina* (Kellicott)
37.	*Euchlanis dilatata* Ehrenberg
38.	*E. triquerta* Ehrenberg
39.	*Trichotria pocillum* (Müller)
40.	*T. tetractis* (Ehrenberg)
41.	*Mytilina ventralis* (Ehrenberg)
42.	*Lecane* (*Lecane*) *luna* (Müller)
43.	*L.* (*L.*) *ungulata* (Gosse)
44.	*L.* (*Monostyla*) *lamellata* (Daday)
45.	*L.* (*M.*) *stenroosi* (Meissner)
46.	*L.* (*M.*) *bulla* (Gosse)
47.	*L.* (*M.*) *lunaris* (Ehrenberg)
48.	*Colurella obtusa* (Gosse)
49.	*C. adriatica* (Ehrenberg)
50.	*C. uncinata* (Müller)
51.	*C. colurus* (Ehrenberg)
52.	*Hexarthra fennica* (Levander)
53.	*H. oxyuris* (Zernov)
54.	*H. mira* ()
55.	*Testudinella patina* (Hermann)
56.	*T. bidentata* (Ternetz)
57.	*Filinia longiseta* (Ehrenberg)
58.	*Collotheca mutabilis* (Hudson)

	Olligochaeta
1.	*Aeolosoma hemprichi* Ehrenberg
2.	*Nais elingius* Müller
3.	*N. communis* Piguet
4.	*Paranais simplex* Hrabe
5.	*Amphichaeta sannio* Kallstenius
6.	*Chaetogaster* sp.
7.	*Limnodrilus helveticus* Piguet
8.	*Potamothrix bavaricus* (Oeschmann)
9.	*Psammorhyctides albicola* (Michaelsen)
10.	*Lumbriculus lineatus* (Müller)
	Cladocera
1.	*Diaphanosoma brachyurum* Lievin
2.	*Chydorus sphaericus* (O. F. Müller)
3.	*Alona rectangula* G. Sars
4.	*Bosmina longirostris* (O. F. Müller)
5.	*Daphnia longispina* (O. F. Müller)
6.	*Ceriodaphnia reticulata* (Jurine)
7.	*C. cornuta* G. Sars
8.	*C. pulchella* G. Sars
9.	*Moina mongolica* Daday
10.	*M. micrura* Kurz
11.	*Podonevadne camptonyx* (G. Sars)
12.	*P. angusta* (G. Sars)
13.	*Evadne anonyx* G. Sars
14.	*Cercopagis pengoi aralensis* M.-Boltovskoi
	Copepoda
1.	*Phyllodiaptomus blanci* (Guerne et Richard)
2.	*Arctodiaptomus salinus* (Daday)
3.	*Halicyclops rotundipes aralensis* Borutzky
4.	*Cyclops vicinus* Uljanin
5.	*Acanthocyclops viridis* Kiefer
6.	*Mesocyclops leuckarti* (Claus)
7.	*Thermocyclops crassus* (Fischer)
	Harpacticoida
1.	*Halectinsoma abrau* (Kritchagin)
2.	*Schizopera aralensis* Borutzky
3.	*S. jugurtha* (Blanchard et Rich.)
4.	*S. reducta* Borutzky
5.	*Nitocra lacustris* (Schmankewitsch)

6.	*N. hibernica* (Brady)
7.	*Mesochra aestuarii* Gurney
8.	*Onychocamptus mohammed* (Blanchard et Rich.)
9.	*Cletocamptus retrogressus* Schmankewitsch
10.	*C. confluens* (Schmeil.)
11.	*Limnocletodes behningi* Borutzky
12.	*Nannopus palustris* Brady
13.	*Enchydrosoma birstein* Borutzky
14.	*Leptocaris brevicornis* (Van Douwe)
15.	*Paraleptastacus spinicauda* Noodt
	Ostracoda
1.	*Darwinula stevensoni* (Brady et Robertson)
2.	*Candona marchica* Hartwig
3.	*Cyclicypris laevis* (O. F. Müller)
4.	*Plesiocypris newtoni* (Brady et Robertson)
5.	*Cyprideis torosa* (Jones)
6.	*Amnicythere cymbula* (Livental)
7.	*Tyrrenicythere amnicola donetziensis* (Dubowsky)
8.	*Limnocythere* (*Limnocythere*) *dubiosa* Daday
9.	*L.* (*L.*) *inopinata* (Baird)
10.	*L.* (*Galolimnocythere*) *aralensis* Schornikov
11.	*L.* (*Loxocaspia*) *immodulata* Stepanitys
	Malacostraca
1.	*Dikerogammarus aralensis* (Uljanin)
	Hydracarina
1.	*Eylais rimosa* Piersig
2.	*Hydriphantes s. str. crassipalpis* Könike
3.	*H.* (*Polyhydriphantes*) *flexuosus* Koenike
4.	*Hydrodroma despiciens* (O. Müller)
5.	*Limnesia undulata* (O. F. Müller)
6.	*Arrenurus s. str. tricuspidator* (O. F. Müller)
7.	*Copidognathus s. str. oxianus* Viets
	Bivalvia
1.	*Dreissena polymorpha* (Pall.)
2.	*D. p. aralensis* (Andr.)
3.	*D. p. obtusicarinata* (Andr.)
4.	*D. caspia* (Eichwald)
5.	*D. c. pallasi* (Andr.)
6.	*Cerastoderma rhomboides rhomboides* (Lam.)
7.	*C. isthmicum* Issel

8.	*Hypanis vitrea* (Eichw.)
9.	*H. v. bergi* Starobogatov
10.	*H. minima* (Ostr.)
11.	*H. m. sidorovi* Starobogatov
12.	*H. m. minima* (Ostr.)
	Gastropoda
1.	*Theodoxus pallasi* Ldh.
2.	*Caspiohydrobia conica* (Logv. et Star.)
3.	*C. husainovae* Starobogatov

As seen above, by 1961 the Aral Sea, its deltaic areas and estuaries had been transformed due to planned and accidental introductions. On the eve of oncoming ecological catastrophe connected with diversions of riverine waters for extensive irrigation development, the Aral Sea had been subject to many exotic species introductions. While biodiversity had increased byfourteen species of fishes and four species of invertebrates, only a few of these species had commercial value or could serve as a food for fishes. Many species of fishes were introduced accidentally and only increased the pressure on the nutritional chain without giving benefits to the fisheries. Promised increase of commercial catches and raised nutritional value of invertebrate associations did not occur. At the same time, due to predation by introduced fishes or competition with introduced invertebrates, two aboriginal species of zooplankton (*Moina mongolica* и *Arctodiaptomus salinus*) and one species of zoobenthos (*Dikerogammarus aralensis*) disappeared completely. Thus the whole series of planned introductions into the Aral Sea and deltas between 1927 and 1961 must be considered unjustified and in some cases even harmful. The scientific-journalistic expedition "Aral-88" in 1988 conducted after "perestroika" also noted the negative consequences of planned and accidental introductions at the Aral Sea during the aforesaid period (Seliunin, 1989).

Since 1961 anthropogenic regression of the Aral Sea has continued unabated except for a single year in 1968, when water input was enormous (Fig. 3). Only in that year total amount of river run-off and precipitation exceeded evaporation from the surface (The Aral Sea, 1990). This rapid anthropogenic desiccation and salinization of the Aral Sea prompted efforts to introduce euryhaline species able to endure constantly increasing salinity.

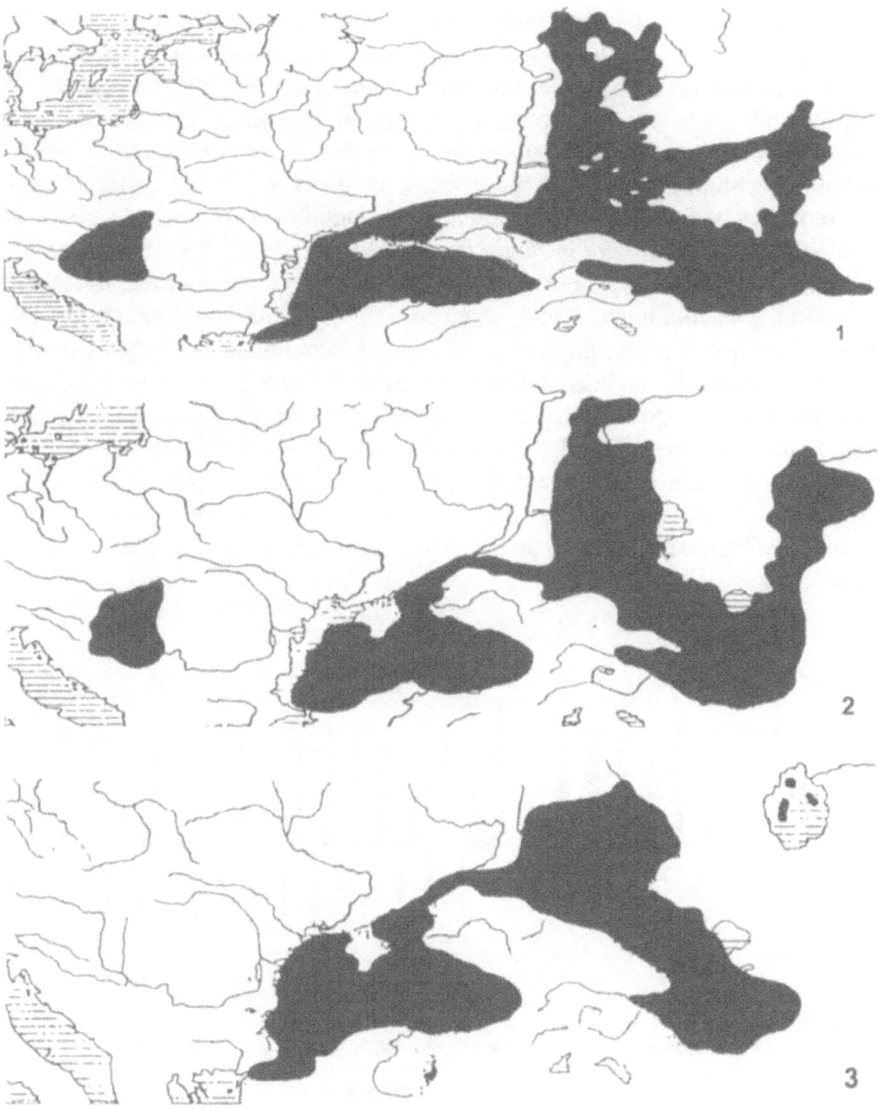

Fig. 2. Palaeohydrography of southeastern Europe and southwestern Central Asia in late Miocene-Pleistocene.:

1 — Akchagyl and Kuyalnik lake-seas (3 mln. years B.P.); *2* — Apsheronian and Gurian lake-seas; *3* — Anciant Euxinian and Hazarian lake-seas (0.4 mln. years B.P.)

At the beginning of 1960's a polychaete, *Nereis diversicolor,* and a bivalve, *Abra ovata,* were introduced from the Sea of Azov. The first species became numerous since 1963 and the second since 1967. In the middle of 1960's there was an unsuccessful attempt to introduce bivalve mollusk *Monodacna colorata* (Karpevich, 1975). In the middle and end of 1960's and in the beginning 1970's there were attempts to introduce planktonic invertebrates. Candidate for introduction included two euryhaline copepods – *Calanipeda aquaedulcis* and *Heterocope caspia.* The first species naturalized successfully and since 1970 is a dominant zooplankter of the Aral Sea (Kazakhbaev, 1974; Andreev, 1978) and substituted for the former dominant *Arctodiaptomus salinus*, exterminated by Baltic herring. The second exotic species did not naturalize. During these introductions ofplanktonic copepods, larvae (zoëa) of crab *Rhithropanopeus harrisii tridentata* were accidentally introduced into the Aral. Since 1976, this benthic crustacean is widespread in the southern water area of Aral Sea (Andreev, Andreeva, 1988), but due to certain causes (they will be considered below) is absent in the northern water area – in the so-called Small Aral.

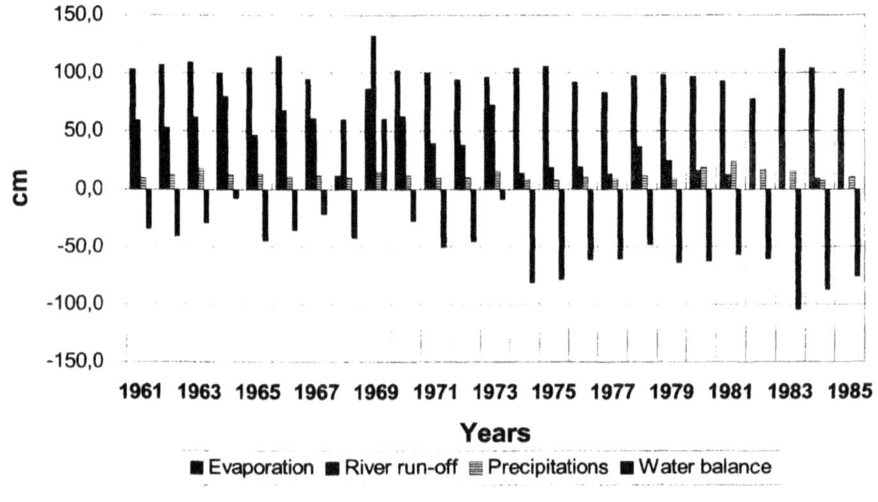

Fig. 3. Water balance of the Aral Sea in 1961-1985.

Acclimatization of euryhaline planktonic and benthic invertebrates could be regarded as an example of well thought-out and successful introduction.

These euryhaline species succeeded to save the feeding value of zooplankton and benthos under conditions of the Aral Sea anthropogenic salinization. A positive role of euryhaline acclimatizants is especially clear since the beginning and middle of 1970's when salinity of the Aral water exceeded 12-14 g/l and fresh and brackish water organisms of plankton and benthos began to die out. However, rising salinity negatively affected the ichthyofauna. The early stages of ontogenesis in freshwater fishes originally dominating in the Aral Sea were particularly vulnerable. Survival of larvae and fries of these fishes sharply began to decrease even at salinity increases of 1-2 g/l from 8-10 g/l. Nearly all freshwater fishes and invertebrates existed in the Aral Sea at the upper limit of their salinity tolerance range, and this explains why they have disappeared so quickly (Karpevich, 1975; Aladin, Kotov, 1989; Plotnikov et al., 1991). During only one decade, since the anthropogenic desiccation began, more than 50-70% of fishes and free-living invertebrates became extinct in the Aral Sea. Under these extreme conditions, when acclimatization of euryhaline invertebrates was successful, the idea of introducing euryhaline commercial fishes was suggested. At the end of 1970's plaice (*Platichthys flesus*) from the Sea of Azov were introduced. Since 1981 this exotic for the Aral Sea commercial fish is ubiquitous in catches (Lim, Markova, 1981). Of the 20 aboriginal fish species in the Aral Sea, only the euryhaline stickleback (*Pungitius platygaster aralensis*) remains. All other aboriginal fishes disappeared due to salinization and only some of them remain in deltas and deltaic water bodies of Amudarya and Syrdarya. Successful acclimatization of plaice allowed fisheries to continue on the Aral Sea. At the beginning of the1980's, besides stickleback and plaice, the accidentally introduced atherine and 2-3 species of gobies were also present. While the Baltic herring was also present, it was present in small numbers and did not formed large schools.

At the end of 1970's and in the beginning 1980's the last attempt to introduce sturgeons in the Aral Sea was undertaken. In this case Russian sturgeon (*Acipenser guldenstadti*) was introduced (Lim, Markova, 1981). But this attempt could not be successful because the salinity of the Aral Sea reached 18-20 g/l, which is very high for this species. Besides, natural anadromous migration for spawning was prevented because the deltas of Amudarya and Syrdarya had become very shallow.

In the middle of 1980's attempts to introduce euryhaline invertebrates into Aral Sea continued. One tried to introduce two species of bivalves (*Mytilus galloprovincialis* and *Mya arenaria*) from the Sea of Azov. Both introductions were unsuccessful. In the first case, it was because of the absence of solid bottom substrates essential for mussel attachment. The second species was released in shallows, which dried up within months due to continuous lake level lowering. If this quick desiccation of shallows had been taken into consideration, successful introduction of *Mya arenaria* may have been possible. In the same years was introduced planktonic copepod *Acartia clausi*, but it did not naturalized in the Aral Sea. Possibly due to insufficient number of released individuals but perhaps also because the ecological niche was already occupied by acclimatized *Calanipeda aquaedulcis*.

At the end of 1980's the history of planned and accidental introductions in the Aral Sea was ended (Table 2). Since then, only natural colonizations, unconnected with human activity, have occurred.

In 1989 continued desiccation of the Aral Sea led to its division into two lakes (Fig. 4), which have evolved in different ways. The Small Aral Sea, located in the North, receives run-off of the Syrdarya River and began to overfill due to positive water balance. The surface area of this lake is small, and evaporation from its surface is less than inflows from the Syrdarya, atmospheric precipitation and ground waters. As for the Large Aral Sea in the south, its water balance is negative, and evaporation from its huge surface is still higher than the small inputs of the Amudarya River, atmospheric precipitation and ground waters (Aladin, Plotnikov, Potts, 1995). These difference in the hydrological regimes of the two new lakes has led to stabilization of the Small Aral Sea level and continued desiccation and salilnization of the Large Aral Sea.

The salinity of the Aral Sea was about 28-30 g/l. at the moment it divided into two lakes at about +40 m asl (Aladin, Plotnikov, Potts, 1995; Aladin, 1995) and their fauna and flora were similar. But biological differences between this two water bodies appeared very quickly due to different hydrological regimes. In 1961 before anthropogenic desiccation and salinization the Aral Sea was a brackish lake with average salinity 8-10 g/l, and its level was about +53 m asl (Zenkevich, 1963). Its ecosystem was characterized by low biodiversity and biological productivity. With salinization and level fall biodiversity and productivity decreased and the ecosystem was transformed from brackish water into mesohaline where surviving aboriginal and introduced euryhaline and marine species of fishes and invertebrates predominated (Plotnikov et al., 1991). At the time dividing into two lakes, only 7 species of fish, 10 common zooplankton species, and 11 common benthos species were present.

Table 2. Exotic species in the Aral Sea

N	Taxonomic Group	Species	Source	Year(s) of introduction	Year of first finding
1	Pisces	*Alosa caspia* (Eichw.)	Caspian Sea	1929—1932	-
2		*Acipenser stellatus* Pallas	Caspian Sea	1927—1934 /1948-1963	1958
3		*Acipenser nudiventris derjavini* Borz.	Ural River Delta	1958	-
4		*Acipenser guldenstadti* Brandt	?	1978-1980	1981
5		*Clupea harengus membras* (L.)	Balthic Sea	1954-1959	1957
6		*Mugil auratus* (Risso)	Caspian Sea	1954-1956	-
7		*Mugil saliens* (Risso)	Caspian Sea	1954-1956	-
8		*Ctenopharyngodon idella* (Val.)	China	1960-1961	1963
9		*Hypophtalmichthys molifrix* (Val.)	China	1960-1961	1963
10		*Aristichtys nobilis* (Rich.)	China	1960-1961	?
11		*Platichthys flesus* Pallas	Azov Sea	1979-1987	1981
12		*Mylopharyngodon piceus* (Rich.)	China	1960—1961	1963
13		*Syngnathus abaster caspius* Eichw.	Caspian Sea	1954—1956	?
14		*Atherina boyeri caspia* (Eichw.)	Caspian Sea	1954—1956	1959
15		*Pomatoschistus caucasicus* Berg	Caspian Sea	1954—1956	1958
16		*Neogobius fluviatilis* (Pallas)	Caspian Sea	1954—1956	1958
17		*Neogobius melanostomus* (Pallas)	Caspian Sea	1954—1956	1959
18		*Neogobius syrman* (Nordm.)	Caspian Sea	1954—1956	1959
19		*Proterorchinus marmoratus* (Pallas)	Caspian Sea	1954—1966	1959
20		*Neogobius kessleri* (Gunter)	Caspian Sea	1954—1956	1959

N	Taxonomic Group	Species	Source	Year(s) of introduction	Year of first finding
21		*Ophicephalus(Channa) argus* Cantor	Karakum canal	1960s	1965
22	Monogenea	*Nitzschia sturionis* (Abil.)	Caspian Sea	1927—1934	?
23	Coelenterata	*Polipodium hydriforme* Ussov	Caspian Sea	1927—1934	?
24	Mysidacea	*Paramysis baeri* (Czern.)	River Don	1958—1960	-
25		*Paramysis lacustris* (Czern.)	River Don	1958—1960	1961
26		*Paramysis intermedia* (Czern.)	River Don	1958—1960	1961
27		*Paramysis ullskyi* (Czern.)	River Don	1958—1960	1963
28		*Limnomysis benedeni* (Czern.)	?	?	1975
29	Decapoda	*Palaemon elegans* Rathke	Caspian Sea	1954—1966	1957
30		*Palaemon adspersus* Rathke	Caspian Sea	1954—1966	-
31		*Rhithropanopeus harrisii tridentata* (Maitland)	Azov Sea	1965, 1966,	1976
32	Copepoda	*Calanipeda aquaedulcis* Kritsch.	Azov Sea	1965, 1966/1970	1970
33		*Heterocope caspia* Sars	?	1971	-
34		*Acartia clausi* Giesb.	?	1985, 1986	-
35	Polychaeta	*Nereis diversicolor* Muller	Azov Sea	1960—1961	1963
36	Bivalvia	*Abra ovata* Phil.	Azov Sea	1960, 1961,1963	1967
37		*Monodacna colorata* (Eichw.)	?	1964,1965	-
38		*Mytilus galloprovincialis* Lam.	Azov Sea	1984-1986	-
39		*Mya arenaria* Linne	Azov Sea	1984-1986	-

N	Taxonomic Group	Species	Status after acclimatization	Status in 1990s
1	Pisces	*Alosa caspia* (Eichw.)	-	-
2		*Acipenser stellatus* Pallas	-	-
3		*Acipenser nudiventris derjavini* Borz.	-	-
4		*Acipenser guldenstadti* Brandt	Rare	-
5		*Clupea harengus membras* (L.)	Rare	?
6		*Mugil auratus* (Risso)	-	-
7		*Mugil saliens* (Risso)	-	-
8		*Ctenopharyngodon idella* (Val.)	Commercial fish	-
9		*Hypophtalmichthys molifrix* (Val.)	Commercial fish	-
10		*Aristichtys nobilis* (Rich.)	Rare	-
11		*Platichthys flesus* Pallas	Commercial fish	Commercial fish
12		*Mylopharyngodon piceus* (Rich.)	Commercial fish	-
13		*Syngnathus abaster caspius* Eichw.	Rare	-
14		*Atherina boyeri caspia* (Eichw.)	Numerous	Limited number
15		*Pomatoschistus caucasicus* Berg	Numerous	?
16		*Neogobius fluviatilis* (Pallas)	Numerous	?
17		*Neogobius melanostomus* (Pallas)	Numerous	-
18		*Neogobius syrman* (Nordm.)	Limited number	-
19		*Proterorchinus marmoratus* (Pallas)	Limited number	?
20		*Neogobius kessleri* (Gunter)	Limited number	-

N	Taxonomic Group	Species	Status after acclimatization	Status in 1990s
21		*Ophicephalus(Channa) argus* Cantor	Commercial fish	Commercial fish in river delta
22	Monogenea	*Nitzschia sturionis* (Abil.)	Common	-
23	Coelenterata	*Polipodium hydriforme* Ussov	Common	-
24	Mysidacea	*Paramysis baeri* (Czern.)	?	-
25		*Paramysis lacustris* (Czern.)	Numerous	In river deltas
26		*Paramysis intermedia* (Czern.)	Numerous	-
27		*Paramysis ullskyi* (Czern.)	Limited number	-
28		*Limnomysis benedeni* (Czern.)	Limited number	-
29	Decapoda	*Palaemon elegans* Rathke	Numerous	Numerous
30		*Palaemon adspersus* Rathke	?	-
31		*Rhithropanopeus harrisii tridentata* (Maitland)	Numerous	Numerous
32	Copepoda	*Calanipeda aquaedulcis* Kritsch.	Numerous	Numerous
33		*Heterocope caspia* Sars	-	-
34		*Acartia clausi* Giesb.	-	-
35	Polychaeta	*Nereis diversicolor* Muller	Numerous	Numerous
36	Bivalvia	*Abra ovata* Phil.	Numerous	Numerous
37		*Monodacna colorata* (Eichw.)	-	-
38		*Mytilus galloprovincialis* Lam.	-	-
39		*Mya arenaria* Linne	-	-

N	Taxonomic Group	Species	Ecologycal status	Way of introduction	Effect
1	Pisces	*Alosa caspia* (Eichw.)	N	del	0
2		*Acipenser stellatus* Pallas	N	del	0
3		*Acipenser nudiventris derjavini* Borz.	N	del	-
4		*Acipenser guldenstadti* Brandt	N	del	0
5		*Clupea harengus membras* (L.)	N	del	+
6		*Mugil auratus* (Risso)	N	del	0
7		*Mugil saliens* (Risso)	N	del	0
8		*Ctenopharyngodon idella* (Val.)	N	del	+
9		*Hypophtalmichthys molifrix* (Val.)	N	del	+
10		*Aristichtys nobilis* (Rich.)	N	del	+
11		*Platichthys flesus* Pallas	N	del	+
12		*Mylopharyngodon piceus* (Rich.)	N	assoc	0
13		*Syngnathus abaster caspius* Eichw.	N	assoc	-
14		*Atherina boyeri caspia* (Eichw.)	N	assoc	-
15		*Pomatoschistus caucasicus* Berg	N	assoc	-
16		*Neogobius fluviatilis* (Pallas)	N	assoc	-
17		*Neogobius melanostomus* (Pallas)	N	assoc	-
18		*Neogobius syrman* (Nordm.)	N	assoc	-
19		*Proterorchinus marmoratus* (Pallas)	N	assoc	-
20		*Neogobius kessleri* (Gunter)	N	assoc	-

N	Taxonomic Group	Species	Ecologycal status	Way of introduction	Effect
21		*Ophicephalus(Channa) argus* Cantor	N	assoc	+
22	Monogenea	*Nitzschia sturionis* (Abil.)	Par	assoc	-
23	Coelenterata	*Polipodium hydriforme* Ussov	Par	assoc	-
24	Mysidacea	*Paramysis baeri* (Czern.)	N/B	del	0
25		*Paramysis lacustris* (Czern.)	N/B	del	+
26		*Paramysis intermedia* (Czern.)	N/B	del	+
27		*Paramysis ullskyi* (Czern.)	N/B	inc	+
28		*Limnomysis benedeni* (Czern.)	N/B	inc	+
29	Decapoda	*Palaemon elegans* Rathke	N/B	assoc	?
30		*Palaemon adspersus* Rathke	N/B	assoc	?
31		*Rhithropanopeus harrisii tridentata* (Maitland)	B	assoc	+
32	Copepoda	*Calanipeda aquaedulcis* Kritsch.	P	del	+
33		*Heterocope caspia* Sars	P	del	0
34		*Acartia clausi* Giesb.	P	del	0
35	Polychaeta	*Nereis diversicolor* Muller	B	del	+
36	Bivalvia	*Abra ovata* Phil.	B	del	+
37		*Monodacna colorata* (Eichw.)	B	del	0
38		*Mytilus galloprovincialis* Lam.	B	del	0
39		*Mya arenaria* Linne	B	del	0

Way of introduction: del – deliberately, inc – incidentally, assoc – in association with deliberate acclimatizants Ecologycal status: N – necton, B – benthos, N/B – nectobenthos, P – plankton Effect: - negative; + positive; 0 none; ? unknown

The recent history of the Aral Sea can be viewed as including three critical periods (Plotnikov et al., 1991) followed by the current period in which two distinct lakes are evolving differently. As seen above, the first crisis in 1957-1960 was associated with planned and accidental introductions into the Aral Sea ecosystem. The second crisis period took place in 1971-1976 when salinity of Aral increased to above 12-14 g/l and brackish water species of fresh water origin disappeared. The third crisis was initiated in 1986 when salinity exceeded 23-25 g/l and lasted until the Aral Sea division in 1989. During this time brackish water species of Caspian origin became extinct (Plotnikov et al., 1991). The current period began with partition of the Aral Sea in which both parts inherited a common fauna. The zooplankton community consisted eleven species (Rotifera – 5, Cladocera – 1, Copepoda – 5). Rotifers included 5 euryhaline widespread species: *Synchaeta vorax*, *S. cecilia*, *Notholca acuminata*, *N. squamula* and *Brachionus plicatilis*. The single Cladoceran representative was the euryhaline species of caspian Onychopoda – *Podonevadne camptonyx* surviving at the limit of its salinity tolerance. The Copepoda were dominated by two euryhaline species: *Calanipeda aquaedulcis* and *Halicyclops rotundipes aralensis*, but also included three species from Harpacticoida. Besides these species, zooplankton also included larvae of benthic invertebrates and protozoans, mainly Tintinida. Thus, at partition the Aral Sea zooplankton included one recent invader (*Calanipeda aquaedulcis*), one ancient invader (*Podonevadne camptonyx*) and nine euryhaline species, some of which were widespread in the region while others could be considered as aboriginal. As one can see the portion of recent invaders in zooplankton was 9%.

Eight species (Bivalva – 2, Gastropoda – 2, Polychaeta – 1, Ostracoda – 1, Decapoda – 2) remained in the zoobenthos. Bivalves were represented by 2 species: the recently introduced *Abra ovata* and the ancient invader, *Cerastoderma isthmicum*. Gastropods were represented by two widely euryhaline species of genus *Caspiohydrobia*. Polychaetes were represented by the introduced euryhaline *Nereis diversicolor*. Ostracods included only one euryhaline species *Cyprideis torosa*. Decapoda were represented by two accidental invaders – shrimp *Palaemon elegans* and crab *Rhithropanopeus harrisii tridentata*. Besides these species, the zoobenthos included some Protozoan species. Thus at partition the Aral Sea zoobenthos included four recent invaders (*Abra ovata*, *Nereis diversicolor*, *Palaemon elegans* и *Rhithropanopeus harrisii tridentata*), one ancient invader (*Cerastoderma isthmicum*) and three euryhaline species (*Cyprideis torosa* and two species of *Caspiohydrobia*). Part of the last group is widespread in the region and could be considered as aboriginal. Thus, 50% of the zoobenthic species were recent invaders.

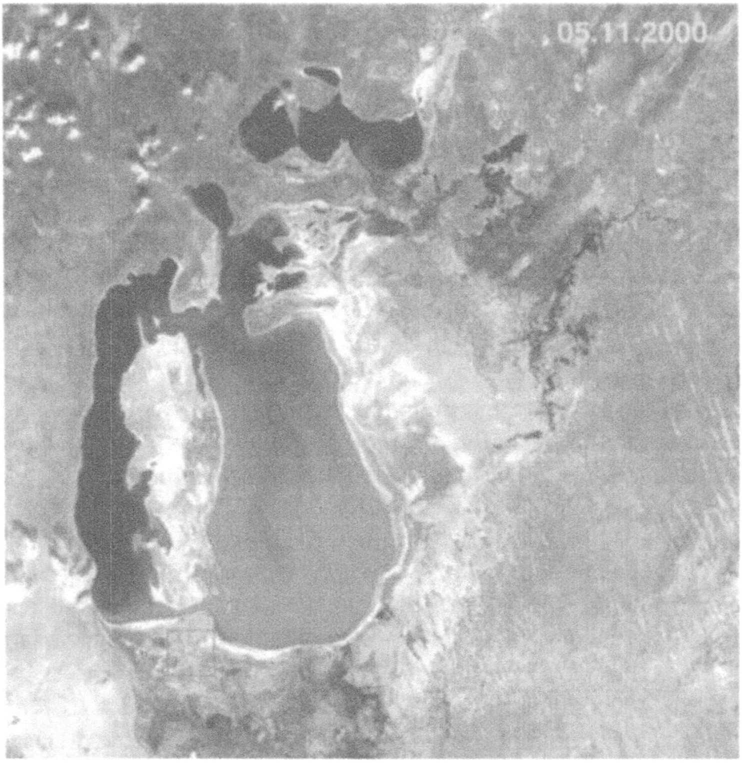

Fig. 4. The Aral Sea in 2000.

An exception to sharing of a common fauna between the two lakes is that since 1976, crabs became abundant in the Large Aral Sea following their accidental introduction. Because currents spread their larvae, settling of this crab was slow and followed the current patterns. As the Small Aral generally did not receive water from the Large Aral, the crab's larvae did not colonize the northern water area of the Aral Sea prior to partition in 1989.

Note also that some authors (Starobogatov, Andreeva, 1981) describe more than twenty species of gastropods from the genus *Caspiohydrobia*, instead of the two listed here. This difference arises from their taxanomic splitting which we consider insufficiently proved and unwarranted. Future application of new molecular methods could partly solve this issue, and thus provide a more exact enumeration of *Caspiohydrobia* species existing in the Aral Sea.

Seven ichthyofauna species were present at the division. From aboriginal species only stickleback (*Pungitius platygaster aralensis*) remained. Among intentionally introduced fishes only two species survived: Baltic herring (*Clupea harengus membras*) and plaice (*Platichthys flesus*). In accidentally introduced fishes only silverside (*Atherina mochon pontica*) and three species of gobies – bubyr (*Pomatoschistus caucasicus*), monkey goby (*Neogobius fluviatilis pallasi*) and round goby (*N. melanostomus officinus*) remained. Thus, the portion of recent invaders in ichthyofauna was 86%.

After division in 1989 the Small Aral Sea stabilized at +40 m asl and began to rise due to positive water balance (Aladin, 1995; Aladin, Potts, Plotnikov, 1995). As a result waters of the Small Sea began to flow southward into the Large Aral. This outflow did not occur over all the surface of the dried bottom of former Berg's strait but only in its central part, which was earlier dredged. This dredging had begun in 1980's when water level in Berg's strait has fallen so much as to cause troubles for shipping. At that time a navigation canal was cut between the northern and southern basins. In spring 1989, this canal was visible and a slow southward current was present in autumn. This flow was due to declining lake levels in the Large Aral. The flow sharply increased with continuing desiccation of the Large Aral and reached 100 m^3/sec as the Large Aral level fell to +37.1 m, a difference between the two lakes reaching 3 m. This strong stream eroded the bottom and threatened to almost completely drain the Small Aral Sea (Aladin et al., 1995). To prevent this, the canal between the Large and Small Aral was dammed in July-August 1992 and the flow stopped. In the next years this dam in Berg's strait was partly destroyed by floods and restored several times. The dam existence allowed to raise the Small Aral Sea level up to +42.8 m at April 1999 and to decrease salinity from 29.2 g/l (at division) to 18.2 g/l. Unfortunately, in late April 1999 the dam was completely destroyed by waves due to the rise of Small Aral level. After 7 years the level returned to the mark +40 m. Dam restoration has not been undertaken and waters of Small Aral are again flowing out to the south. They do not reach the Large Aral and are lost in sands and salt marshes north of past Barsakelmes Island. Now the Large Aral has dried so much that its modern shoreline here is far (many km) from the modern Small Aral Sea (Fig. 4).
After the dam was built in 1992 rising lake levels and declining salinity partially restored the Small Aral. Biodiversity increased, desiccated Bolshoy Sary-Cheganak gulf was filled with water again, and rehabilitation processes began in Syrdarya delta. Reeds (*Phragmites australis*) began to grow again, forming an environment for hydrobionts and amphibiotic

animals (Aladin, 1995; Aladin, Plotnikov, Potts, 1995). Increasing depth of Syrdarya delta resulted from the Small Aral level raise and allowed for aboriginal and introduced freshwater fishes to forage in the estuary as before. The peak of such foraging was at the end of 1990's when the Small Aral level reached more than + 42 m. The foraging of fresh water fishes also was favored by the average salinity decrease to about 18 g/l. Before the dam in Berg's strait was built the Syrdarya estuary was poorly developed, and the zone of fresh and saline water mixing practically absent because most of the fresh water moved directly to the canal between Small and Large Seas. After construction of the dam, fresh water was retained in the Small Aral and its average salinity decreased down to 11 g/l.

In spite of the significant decrease in salinity which allowed fishes from Syrdarya to forage in the Small Aral, the number of fish species resident in the Small Aral remained at seven, the same ones present at the prior partitioning of the Aral Sea. Of these only plaice is of commercial value. Some fishes from Syrdarya, pike-perch (*Lucioperca lucioperca*) for example, that can now forage in a large part of the Small Aral cannot be considered as the salinity of the Small Aral is still high for reproduction of these fishes.

Level raise and salinity decrease were favorable not only for ichthyofauna of Small Aral, but for its zooplankton and zoobenthos as well. Thus, two species of Cladocera: *Moina mongolica* and *Evadne anonyx* reappeared, and the number of *Podonevadne camptonyx* increased. Appearance of *E. anonyx* could be explained by peculiarities if its life cycle. Cladocera from family Podonidae, to which *E. anonyx* belongs, have latent (dormant) eggs, sinking in water and capable of surviving under unfavorable conditions for some years in a stage of diapausing embryos. Before the Aral Sea division *E. anonyx* was observed for the last time in the northern water area in summer 1988 when salinity exceeded 25 g/l. Later this species was not found in zooplankton during some years. But, when in 1993 salinity of Small Aral decreased below 25 g/l, *E. anonyx* probably hatched from dormant eggs surviving on the sea bottom.

The re-appearance of *Moina mongolica* in the Small Aral also appears to be from dormant eggs. Dormant eggs of these crustaceans do not sink as in thecase of *E. anonyx,* but can survive on the shoreline. Dormant eggs of *M. mongolica* are capable of enduring extended periods of drying and freezing of up to tens of years (Makrushin, 1985). Dormant eggs of *M. mongolica* would be brought into the Small Aral by wind from nearby saline ponds during dust-salt storms. *M. mongolica* probably did not become a permanent member of the zooplankton community of the recently isolated Small Aral as it was before extermination by Baltic herring in the beginning of 1960's. Prior to the 1960s *M. mongolica* was found in summer plankton of the Aral Sea (from spring till autumn), and latent eggs were always present in recently exposed sediments areas. This planktonic crustacean was widespread in all Aral Sea from estuaries to saline gulfs because it is widely euryhaline and endures salinity from fresh water up to 97 g/l. Presently, *M. mongolica* in the Small Aral Sea is only sporadically observed, and its dormant eggs are not found on the shore. Obviously, spring dust/salt storms are bringing latent eggs, from which crustaceans later hatch. After parthenogenetic reproduction, herring and silverside exterminate them, without leaving dormant eggs. In subsequent springs, this situation is repeated, as in nearby fishless salt lakes and ponds, *M. mongolica* is common and produces large quantities of dormant eggs. Thus, zooplankton biodiversity in the Small Aral has increased by only one species to the present twelve which consists of the one recent invader (*Calanipeda aquaedulcis*), two ancient invaders (*Podonevadne camptonyx*, *Evadne anonyx*), and nine euryhaline species, part of which are widespread in Aral region and others could be considered aboriginal. Therefore, the portion of recent invaders in zooplankton has decreased to 8%.

New species also appeared in the Small Aral zoobenthos. Two species of Ostracoda – *Eucypris inflata* и *Heterocypris salina* were added to remaining *Cyprideis torosa*. They were never recorded in the Aral Sea before and were first noted in 1995 in Bolshoy Sary-Chaganak bay after refilling due to construction of the dam. Invasion of these species of Ostracoda, as in the case of *M. mongolica*, was evidently the result of their spreading by dust-salt storms. Both these euryhaline species have latent stages, enduring freezing and desiccation and are easily transported by wind. The frequency of dust-salt storms in the Aral region has increased following anthropogenic desiccation. Meteorological data from the Small Aral region indicate the annual frequency of dust storms has increased from about 60 in the mid-1960's to almost double that in the mid-1980's (The Aral Sea, 1990). Given their frequency and strength, we conclude that aeolian transfer is becoming a significant factor in maintaining and introducing new species into water bodies of Aral Sea region.

At the end of 1990's, when the average salinity of Small Aral decreased to 18 g/l, larvae of Chironomidae appeared in the benthos again. Before anthropogenic desiccation and increased salinity some species of

Chironomidae were main components of the zoobenthos. Now, following more than 30 years absence, larvae of *Chironomus halophilus* have re-appeared (Aladin et al., 2002). In the near future, larvae of other Chironomidae may appear in the Small Aral, as the deltaic water bodies of Syrdarya and others saline water bodies of the Aral Sea region contain many species of Chironomidae, imagoes of which are able to actively (flight) or passively (aeolian transfer) reach the Small Aral area and lay eggs. Return of Chironomidae larvae into Small Aral (natural reintroduction) is a sign of increased benthic productivity. The >10 g/l salinity decrease that occurred during the period of dam in the Berg's strait, positively affected other components of the zoobenthos as well. Other species not only survived, but also increased in abundance and biomass despite active predation by plaice, providing further evidence of increased benthic productivity.

Thus, the zoobenthic community of the modern Small Aral rose to 10 species; three recent invaders (*Abra ovata*, *Nereis diversicolor* and *Palaemon elegans*), one ancient invader (*Cerastoderma isthmicum*), three euryhaline species, two new species of Ostracoda (*Eucypris inflata* and *Heterocypris salina*), and *Chironomus halophilus*. Thus, the portion of recent invaders in the zoobenthos has increased slightly to 60%.

The dam collapse in late April 1999 reestablished outflows from the Small Aral Sea and the Bolshoy Sarycheganak bay practically dried up and the straits connecting Shevchenko and Butakov bays with Small Aral became shallow. Nevertheless, there is no threat of the Syrdarya changing course to flow into the Large Aral, as in the early 1990's, because in the late 1990s the Syrdarya had its flow artificially channeled and it enters the Small Aral north of its former natural mouth. Meanwhile, quick restoration of the dam in Berg's strait is required to maintain and enhance biodiversity and productivity of the Small Aral.

After division in 1989, the Large Aral Sea level continued to decline due to a negative water balance and salinity rapidly increased. After the dam in Berg's strait was built in 1992, the Large Aral level declined slightly faster, because inflows from Small Aral ceased. Nevertheless, the increased rate of desiccation due to dam construction was small as indicated by comparative measures of Large and Small Aral levels by satellites. The increasing salinity is negatively impacting the biota and biodiversity is decreasing.

The recent salinity increase in the Large Aral has caused extinction of almost all marine and euryhaline fish and invertebrate species except a few remaining halophiles. Of seven fish species present at partition of the Aral Sea none were present in autumn 2002 when salinity exceeded 70 g/l. Along the shoreline there were a lot of dead decaying bodies of plaices and silversides. But there is a possibility that in Chernyshov and Tsche-Bas bays and near Aktumsyk cape, where there is increased outcome of freshened subterranean waters, adult plaices still may survive during some years. Unfortunately, the output of ground waters is so little that it has influence on the salinity only near the bottom; so, plaices will die sooner or later. However, it is possible to say with certainty there is no natural reproduction of fishes in the Large Aral.

Of eleven invertebrate zooplankton species only the widely euryhaline rotifer, *Brachionus plicatilis* has survived. However, three new halophylic species appeared apparently due to aeolian transfer; the cladoceran *Moina mongolica,* the brine shrimp *Artemia salina* and of the infusorium *Fabrea salina.* Thus only four species remain in the Large Aral Sea zooplankton. In contrast to the Small Aral, *Moina mongolica* has became a permanent component of summer plankton in the Large Aral, but has not settled over the whole water area. *Artemia salina* has invaded the Large Aral Sea and in some areas reaches high abundance. There is no doubt that the Large Aral may become an important center of harvesting brine shrimp cysts for use in aquaculture and thus provide economic value.

The zooplankton of the modern Large Aral Sea includes four euryhaline species, widespread in the region. *Brachionus plicatilis* and reintroduced *Moina mongolica* cannot be considered invaders, as they were present before anthropogenic desiccation. However, *Artemia salina* and *Fabrea salina* are invaders and constitute 50% of the zooplankton species.

Of eight zoobenthic species only two species of widely euryhaline gastropods from the genus *Caspiohydrobia* and one euryhaline species of ostracods *Cyprideis torosa* remain. All other bottom inhabitants, present at partition of the Aral Sea, such as Gastropoda, Polychaeta and Decapoda, have disappeared due to increased salinity or are near extinction. As in the case of zooplankton, the Large Aral Sea zoobenthos was enriched by aeolian transfer of new halophylic invaders. Euryhaline ostracod *Eucypris inflata* and halophylic protozoans appeared in zoobenthos and along with larvae of halophylic Chironomidae. Thus, due to new invaders the number of dominating species has reached six.

Higher zoobenthic biodiversity in Tsche-Bas and Chernyshov bays deserves special note. Here and probably near Aktumsyk cape biodiversity is higher than in the rest of the Large Aral Sea. As mentioned above, near the bottom of these bays and at Aktumsyk cape, inflowing underground freshwaters from under cliffs of Ustjurt plateau occur and reduced salinity provides more favorable benthic conditions than in other areas of Large Aral. Field samples collected from these bays in August-September 2002 contained not only species of *Caspiohydrobia*, Chironomidae and euryhaline ostracod *Cyprideis torosa*, but also some recent (*Abra ovata*) and ancient (*Cerastoderma isthmicum*) invaders. Also in Tsche-Bas bay, where salinity was somewhat lower than in Chernyshov bay, adult *Cerastoderma isthmicum* and *Abra ovata* were present. Also, the presence of *A. ovata* juveniles suggests continuing reproduction of this species. However, *Nereis diversicolor* was not found on any stations in Tsche-Bas bay. As for more saline Chernyshov bay, no bivalves were present, but *Nereis diversicolor* was found.

We also have to note that not only high salinity is killing bottom animals and fishes. In many places near the bottom oxygen concentration is very low. In some other places H_2S makes bottom environment completely lifeless. So, not only salinity, but also anoxic conditions are controlling the bottom communities (Aladin et al., 2002).

These data indicate that after partition of the Aral Sea, the southern part was quickly transformed from a mesohaline to a hyperhaline water body. Biodiversity of Large Aral changed with typical hyperhaline species becoming dominant and most of its former inhabitants, including fishes, extinct. The phytoplankton of modern Large Aral is the halophylic alga, Dunaliella, which has become the dominate autotrophic organism of this hyperhaline water body. This alga came into Large Aral from neighboring hyperhaline water bodies. As in the case of Small Aral, the Large Aral fauna is enriched mainly by aeolian transfer of resting stages of hydrobionts from other water bodies of Aral region.

The rapid decline of the Large Aral level actually destroyed the delta of Amudarya. Unlike the delta of Syrdarya, where, natural rehabilitation processes began after the dam was built, rapid degradation of Amudarya delta continues. Moreover, deltaic water bodies of the Syrdarya are near the Small Aral and are regularly fed with fluvial waters, while those of Amudarya are far from the Large Aral and receive no regular flows. Thus the ecological situation in the south is more complicated than on the northern Aral Sea.

Restoration of the Small Aral is possible and depends on construction of a new dam with water locks. Increased biodiversity and productivity would accompany rising lake level and decreasing salinity. Apparently, natural migration of euryhaline species with fluvial waters from artificial and natural water bodies located in the delta and lower reaches of Syrdarya will also occur. It could be expedient to speed up this natural process by the introduction of food species of some valuable of invertebrates from lakes Kamyslybas, Zhalanash, Tuschibas etc., directly into the Small Aral. Many aboriginal and introduced species that perished in Aral survived in deltaic water bodies and, after the dam restoration, could be re-introduced into the Small Sea. However, one must stress that these actions could succeed only after the average salinity of Small Aral decreases to below 14 g/l. Reintroduction at higher salinity is of no avail.

Continued desiccation of the Large Aral is almost assured. In a few years its water area will inevitably be divided into at least 3 parts separate lakes (Fig. 1B). Tsche-Bas bay will soon be separated in the north, with a deep basin in the west and a shallow water body in the east basin. The latter could dry up completely by 2010 or even earlier. Separated Tsche-Bas bay will become saline slowly more, if underground fresh waters income noted by some authors (Radjabov, Tahirov, personal communication) are significant. Nevertheless, sooner (2020) or later (2025), Tsche-Bas bay will salinize anyway, because low mineralized underground waters in arid climate lakes couldn't compensate evaporation for the long time.

The deepwater basin of the north will obviously exist the longest, because it has the largest water volume and the lower area/volume ratio, and as with Tsche-Bas bay, has some subterranean inputs from Ustjurt plateau. Such inflows were found at Aktumsyk cape (Radjabov, Tahirov, personal communication). It is also probable that analogous underground inflows occur at other points along the steep shore of Large Aral, but as usual in arid climate lakes ground waters couldn't compensate evaporation for the long time. So, year after year the last part of the Large Aral will become smaller and more saline until the stability will be reached.

Before salinity increases to 200-300 g/l in all these water bodies, there will be only euryhaline halophylic species, and their number will decrease as salinity increases. As salinity reaches 300-350 g/l, only bacteria will survive. No introductions into the Large Aral are necessary or warranted. All hydrobionts able to survive in it are already present or could easily come into it naturally, as resting stages or by aeolian transfer or with migrating birds. It is well known that flamingos, eating zooplankton of hyperhaline lakes, often transfer cysts of euryhaline hydrobionts on its feathers and muddy feet.

Restoration and rehabilitation of Large Aral is practically impossible as it would require large amounts of both the Syrdarya and Amudarya waters which are diverted for irrigation. Syrdarya inflows to the Aral Sea have been greatly reduced and almost nothing remains of Amudarya inflows because all countries in the upper basin continue to divert almost all of its flows for irrigation. The withdrawal of river water during the next years will increase, as peace and economic development return to Afghanistan bringing further development of irrigated agriculture in this country.

Interest in restoration of Large Aral Sea is decreasing not only because of deficit of Amudarya's water, but also due to the discovery of large oil and gas deposits in this region. Extracting mineral resources is easier and cheaper from playas than from marine platforms. Economic benefits of oil development in Kazakhstan and Uzbekistan have reduced their interest in restoration and rehabilitation of the Large Aral.

Another evidence that interest to restore the Large Aral is low is coming from recently published documents on the project of river Ob' redirection to the Central Asia. In this project, strongly supported by Moscow major Yu. Luzhkov, all water is giving only for irrigation and social needs. Absolutely nothing left to the Large Aral Sea itself.

Finally, Large Aral reconstruction is complicated by political division stemming from belonging to two countries and intergovernmental accords require much time.

Fortunately, the situation at Small Aral lying entirely within Kazakhstan is not so despairing and there is a real possibility of rehabilitation. We hope that the dam in Berg's strait will be restored and that Small Aral will rise again. Rising water and decreasing salinity result in increased biodiversity due to natural and possibly intentional reintroduction of fishes and invertebrates from deltaic water bodies of Syrdarya. If these plans are realized, in some distant future the Small Aral could be a donor to any restoration of the Large Aral. Such a possibility is testified to by medieval desiccation. In 15-16th centuries the Large Aral was desiccated as now due to irrigation development, but by the 19th century had returned from +30 m to the +53 m. Let us hope that future generations could admire not only Small Aral but also Large Aral Sea.

The present study was supported by INTAS grants: Aral 2000-1030, 1039 and 1053.

References

1. Aladin N.V. (1995) The conservation ecology of the Podonidae from the Caspian and Aral Seas, *Hydrobiologia*, **307**, 85-97.
2. Aladin N.V., Filippov A.A., Plotnikov I.S., Egorov A.N. (2001) Modern ecological state of the Small Aral Sea. Ecological Research and monitoring of the Aral Sea Deltas. A basis for restoration. Book 2. UNESCO Aral Sea Project. 1997-1999 Final Scientific Reports, 73-82.
3. Aladin N.V., Filippov A.A., Plotnikov I.S. (2002) Zoobenthos and Zooplankot of the Northern Aral Sea and Possible Ways of the Sea Ecosystem Rehabilitation, *Lake Issyk-Kul: Its Natural Environment*, Kluwer, NATO Science Series. Series IV: Earth and Environmental Sciences – vol. 13, 169-190.
4. Aladin N.V., Kotov S.V. (1989) Natural condition of the Aral Sea ecosystem and the effect of manmade pollution, *Proc. of Zool. Inst. Of Acad of Sci USSR* **199**, 110-114 (in Russian).
5. Aladin N.V., Plotnikov I.S., Potts W.T.W. (1995). The Aral Sea desiccation and possible ways of rehabilitating and conserving its Northern part, *Int. J. Environmetrics* **6**, 17-29.
6. Aladin N.V., Potts W.T.W. (1992) Changes in the Aral Sea ecosystem during the period 1960-1990, *Hydrobiologia*, **237**, 67-79.
7. Aladin N.V., Potts W.T.W., 1992. Changes in the Aral Sea ecosystem during the period 1960-1990, *Hydrobiologia*, **237**, 67-79.
8. Andreev N.I. (1978) Results of *Calanipeda aquea-dulcis* Kritschagin acclimatization in the Aral Sea, *Biol. osnovy rybn. khoz-va vodoemov Credney Azii I Kazakhstana*, Ilim, Frunze, 6-9 (in Russian).
9. Andreev N.I., Andreeva S.I. (1988) Crab Rhithropanopeus harrisii tridentatus (Decapoda, Xanthidae) in the Aral Sea, *Zool. zh.* **(67) 1**, 131-132 (in Russian).
10. Andreeva S.I. (1989) Zoobenthos of the Aral Sea in the initial period of its salinization, *Proc. of Zool. Inst. Of Acad of Sci USSR* **199**, 53-82 (in Russian).
11. Bortnik V.N., Chistyaeva S.P. (eds.) (1990) *The Aral Sea. Hydrometeorology and Hydrochemistry of the USSR Seas. VII.*, Gidrometeoizdat Leningrad, 11-196 (in Russian).
12. Dogel V.A., Byhowsky B.E. (1934) Fauna of fish parasites of the Aral Sea, *Parazitol, sb. Zool. In-ta AN SSSR*, **4**, 241-346 (in Russian).
13. Dogel V.A., Lutta A.S. (1937) On the mortality of the thorn sturgeon in the Aral Sea in 1936, *Rybnoe Khoz-vo*, **12**, 26-27 (in Russian).
14. Fedorov, P.V. (1983) Problems of the Aral and Caspian Seas in the Late Pliocene and Pleistocene, *Palaeogeography of the Caspian and Aral Seas in the Cainozoic. Part I.*, 19-21 (in Russian).

15. Karpevich A.F. (1947) Prerequisites for acclimatization of new forms in the Aral Sea, *Doklady VNIRO,* **6**, 13-17 (in Russian).
16. Karpevich A.F. (1948) Results and perspectives of acclimatization of fishes and invertebrates in the USSR, Zool. zh., **27 (6)**, 469-480 (in Russian).
17. Karpevich A.F. (1953) State of food base of the southern seas after riverine run-off regulation, Trudy sovesch. ichtiol. komissii AN SSSR, **1**, 124-150 (in Russian).
18. Karpevich A.F. (1960) Basing of aquatic organisms acclimatization in the Aral Sea, *Trudy VNIRO,* **43 (1)**, 76-115 (in Russian).
19. Karpevich A.F. (1975) *Theory and practice of aquatic organisms acclimatization*, Pischevaya pomyshlennost, Moscow (in Russian).
20. Keyser D., Aladin N. (1991) Vom Meer zur Salzwüste: Der Aralsee, *WüstenErde. Der Kampf gegen Durst, Dürre und Desertification. Ökozid 7*, Focus-Verl., Giessen, 213-228.
21. Kortunova T.A. (1975) O the changes in the Aral Sea zooplankton in 1959-1968, *Zool. Zh.*, **54 (5)**, 567-669 (in Russian).
22. Kortunova T.A., Lukonina N.K. (1970) Quantitative characteristic of the Aral Sea zooplankton, *Rybnye resursy vodoemov Kazakhstana i ikh ispolzovanie*, **6**, Nauka, Alma-Ata, 52-60 (in Russian).
23. Lim R.M., Markova E.A. (1981) Results of introducing sturgeons and plaice to the Aral Sea, *Rybnoe Khoz-vo*, **9**, 26-26 (in Russian).
24. Makrushin A.V. (1985) Abhydrobiosis in primary fresh water invertebrates, 1-100 (in Russian).
25. Micklin Ph. (1991) The water Management Crisis in Soviet Central Asia. *The Carl Beck paper in Russian and East European Studies, No. 905*. 1-120.
26. Plotnikov I.S, Aladin N.V., Filippov A.A. (1991) The past and present of the Aral Sea fauna. *Zool. zh.* **70 (4)**, 5–15 (in Russian).
27. Seliunin V. (1989) The burden to act, *Novy Mir*, **5**, 213-241 (in Russian).
28. Starobogatov Ya.I., Andreeva S.I. (1981) New Species of mollusks of the family Pyrgulidae (Gastropoda, Pectinibranchia) from the Aral Sea, *Zool. Zh.*, **60 (1)**, 29-35 (in Russian).
29. Williams W.D., Aladin N.V. (1991) The Aral Sea: recent limnological changes and their conservation significance, *Aquatic Conservation: Marine and Freshwater Ecosystems*, **1**, 3-23.
30. Williams W.D., Aladin N.V. (1991) The Aral Sea: Recent limnological changes and their conservation significance, *Aquatic conservation: Marine and Freshwater Ecosystems*, **1**, 3-23.

31. Yablonskaya E.A, Lukonina N.K. (1962) On the question of productivity of the Aral Sea, *Okeanologiya*, **2 (2)**, 298-304 (in Russian).
32. Zenkevich L.A. (1963) *Biology of the seas of the USSR*, Izd. AN SSSR, Moscow (in Russian).
33. Kazakhbaev (1974) Calanipeda in the southern part of the Aral Sea, *Gidrobiol. Zh.*, **10 (1)**, 89–91.
34. Aladin N.V., Filippov A.A., Plotnikov I.S. (2002) Zoobenthos and zooplankton of the Northern Aral Sea and possible ways of the ecosystem rehabilitation, *Lake Issyk-Kul: Its natural environment. NATO Science Series. IV. Earth and Environmental Sciences*, **13**, 181-190.

Chapter 7

MAIN FEATURES OF THE CASPIAN SEA HYDROLOGY

Aleksey N. Kosarev

Moscow State University, Vorobievy Gory, Moscow, 119899, Russia

Valentin S. Tuzhilkin

State Oceanographic Institute, 6 Kropotkinsky Per., Moscow, 119838, Russia

Andrey G. Kostianoy

P.P. Shirshov Institute of Oceanology, Russian Academy of Sciences, 36 Nakhimovsky Ave., Moscow, 117997, Russia

1. INTRODUCTION

The Caspian Sea constantly attracts considerable attention thanks to its natural uniqueness, resource abundance, great historical value and vital importance to human societies of the vast Caspian region. In these circumstances, improving theoretical and applied knowledge of the sea is indispensable for addressing many complex issues. Specifically, there has been increasing environmental concern over expanding extraction of hydrocarbons off and on the Caspian shores. To cope effectively with increasing release of oil and other contaminants into marine and coastal ecosystems, thorough research of the Caspian Sea in the context of water and wildlife conservation should receive the attention it deserves. Since the Caspian is also affected by events that occur outside the region, it could be

J.C.J. Nihoul et al. (eds.), Dying and Dead Seas, 159–184.

successfully used as an indicator of regional and even global climatic changes, whether natural or human-induced.

The Caspian is the world's largest isolated water reservoir. Currently its level is at –27 m measured against the World Sea Level. The sea occupies an area of 392,600 km^2, with mean and maximum depths being 208 m and 1025 m, respectively (Kosarev, 1975; Terziev et al., 1992; Kosarev and Yablonskaya, 1994). The Caspian's longitudinal extent is three times larger than its latitudinal one (1000 km vs. 200-400 km), resulting in great variability of climatic conditions over the sea. The biologically most productive Northern Caspian is some 1% of the total Caspian water volume (78,600 km^3), while the deep water of the Middle and Southern Caspian accounts for as much as 99%.

Being a relic of the ancient Tethys Ocean, the Caspian Sea is dated at about 10 million years in age and inherited all the characteristic features of marine basins. It differs from the 'true seas' only by its isolation from the world's ocean. The hydrological regime and water circulation of the inland Caspian Sea are strongly driven by atmospheric processes above the sea and its extensive catchment basin. The Caspian thermohaline structure and its time variation depend heavily on heat and water fluxes through the sea surface as well as on river run-off. The action of winds that could be expressed in terms of momentum fluxes and relative vorticity generates three-dimensional water circulation, which further complicates the impact of thermohydrodynamic processes on the sea.

2. WATER BUDGET AND SEA LEVEL

To analyze seasonal and interannual variability of the Caspian hydrological regime, the most complete data set of 55,000 shipboard temperature (T) and salinity (S) profile measurements for the 1950-2000 period has been used (Fig. 1) (Kosarev and Tuzhilkin, 1995). In addition, data on the distribution of hydrometeorological parameters, such as air temperature, atmospheric pressure and river run-off, for the same period at the main coastal hydrometeorological stations, have been employed.

The Caspian external freshwater budget is determined mostly by relationships between river run-off, precipitation and evaporation from the sea surface, whose annual mean values for the 1900-1990 period are 300, 76 and –376 km^3 yr^{-1}, respectively (Terziev et al., 1992; Kosarev and Yablonskaya, 1994). These three components differ in the spatial and temporal distribution of their contribution to the Caspian Sea water budget.

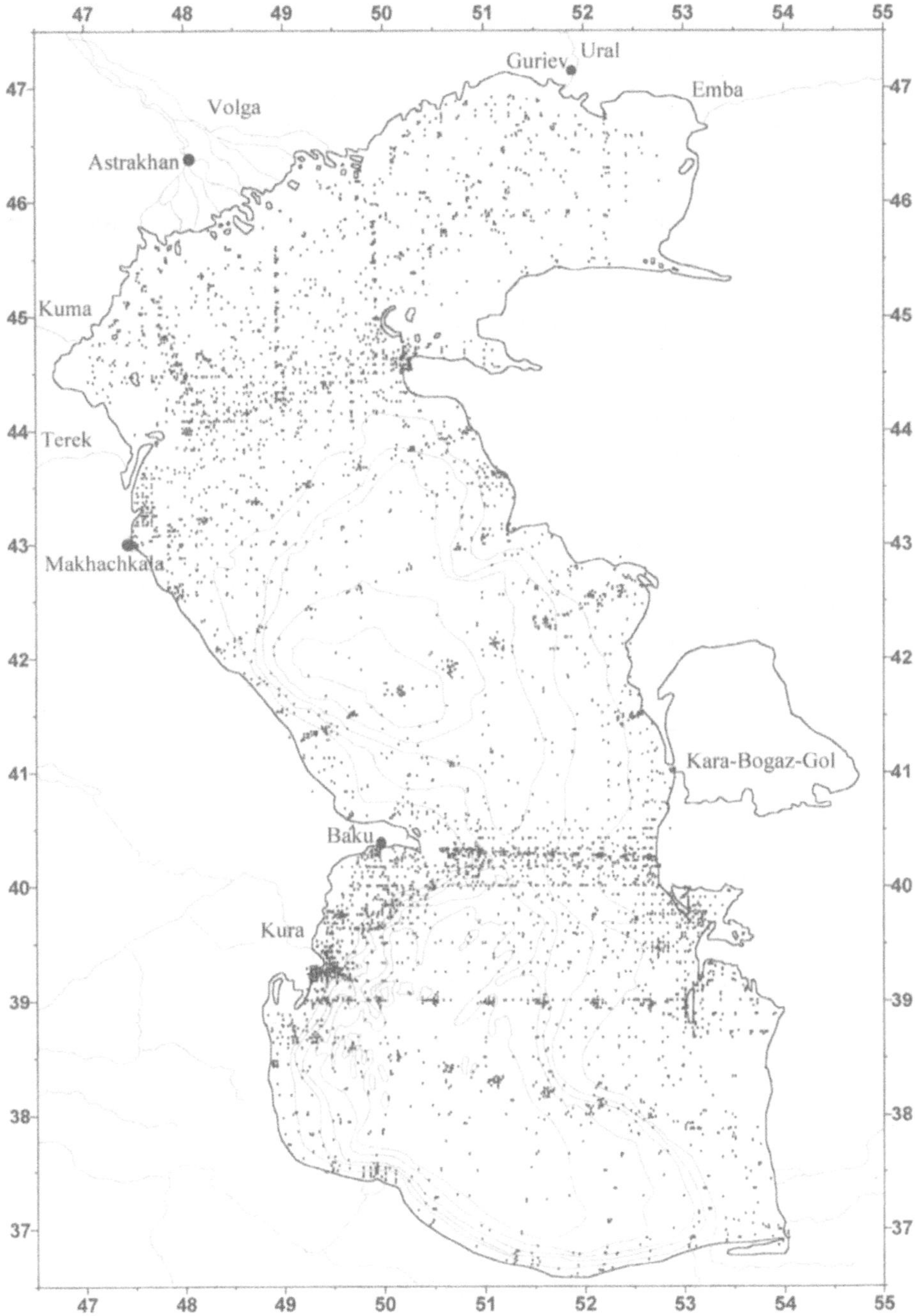

Figure 1. Location of the Caspian Sea shipboard stations making up the T,S-profile database.

Fresh waters flow mostly into the Northern Caspian, with the Volga and Ural contributing 80% and 3% of the total river run-off, respectively. Most other rivers drain to the Caspian western shore. Up to 30% of the freshwater riverine flux to the sea occurs within May and June, the two high water months. In contrast, evaporation and precipitation are spatially more uniform. Some 70% of the annual evaporation occurs during the May to October period with maximum values being recorded in June-August. The seasonal distribution of precipitation is opposite to that of evaporation. From November to April, the Caspian Sea gets about 65% of its annual precipitation. Coupled with lower evaporation, this results in a positive water budget for the sea during autumn and winter.

Components of the Caspian heat and freshwater budgets, such as river run-off and autumn-winter cooling, that could be expressed in terms of a winter severity index (calculated as a sum of daily mean temperatures below zero at Makhachkala), are subject to strong interannual variability that is comparable to their mean annual values (Fig. 2) (Kosarev and Tuzhilkin, 2000). For example, the Volga mean annual run-off varied in the range 150 to 350 $km^3\ yr^{-1}$ during the 1900-1990 period. Variability of the winter severity index for the same time interval was much greater – of the order of 10^1 to 10^2. Thus, significant year-to-year changes of the river run-off were responsible for large-scale fluctuations in the freshwater budget ($\pm 50\ km^3\ yr^{-1}$ for the 20th century), as well as in the level of the Caspian Sea.

Over the past half-century, there was a regression of the Caspian until 1977 when the sea level lowered to –29 m. This drop is considered to be the deepest for the last 400 years. In 1978 the water level started to rise rapidly, and now it has stabilized near the –27 m level (Fig. 3). There has been increasing concern over the Caspian Sea level fluctuations. Estimates provide support for the view of these fluctuations as climatically conditioned and show their intimate connection with components of the Caspian water budget, especially Volga River run-off (Terziev et al., 1992; Kosarev, 2002). As indicated by retrospective analysis, long-term changes in the Caspian Sea level with an amplitude of 2-3 meters are in full accordance with laws of nature (Kosarev, 2002). Such a manifestation of the 'sea breathing' should be given proper weight in planning further offshore and coastal activities. When assessing long-term fluctuations of the Caspian Sea level, adverse human interventions must also be taken into account. Among them are the irreversible loss of water from rivers within the Caspian catchment basin and temporal shutoff of the water flow from the sea to the Kara-Bogaz-Gol Bay. The latter is believed to be responsible for up to 20% of the Caspian transgression in the 1980s.

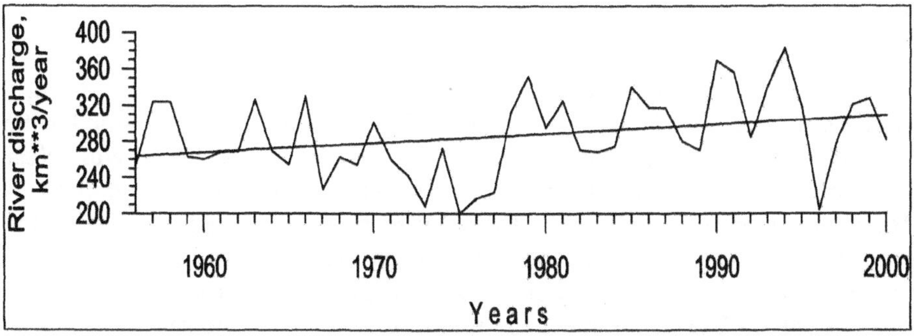

(a)

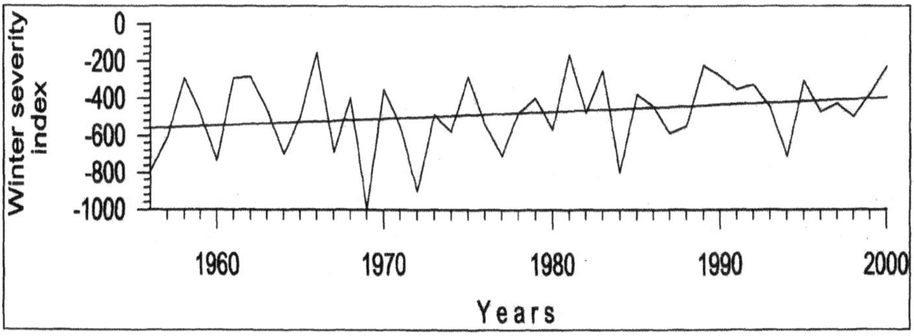

(b)

Figure 2. Interannual variability of (a) mean annual river run-off ($km^3\ yr^{-1}$) and (b) winter severity index (°C) during the 1956-2000 period (Kosarev and Tuzhilkin, 2000).

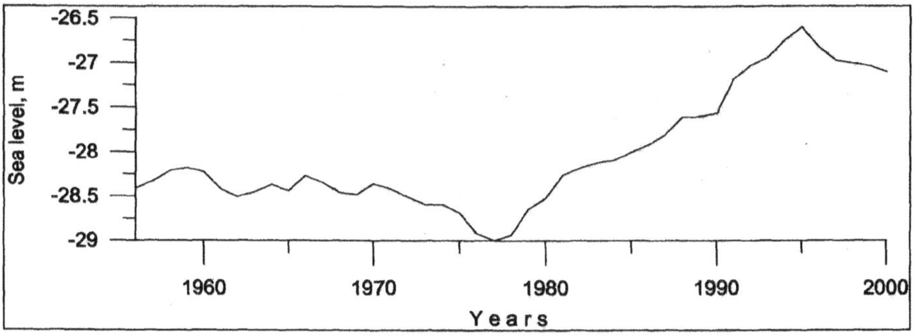

Figure 3. Interannual variability of the Caspian Sea mean level (in meters) during the 1956-2000 period.

3. CIRCULATION

The pattern of atmospheric circulation above the Caspian Sea is strongly dependent upon season (Terziev et al., 1992). In winter, the sea is under the influence of the south-western part of the Siberian anticyclone with clear dominance of south-eastern and eastern winds. The summer pattern is exactly the opposite, because it forms in response to a high pressure zone extending eastwards from the Azores anticyclone and causing north-western and northern winds to prevail above the Caspian Sea, particularly along its western shore. The structure of the pressure fields over the sea is favorable for year-round inflow of cyclonic wind vorticity. This is reflected in the water circulation, whose features, including spatial and seasonal variations, are largely responsible for the distribution of hydrological parameters and suspended matter.

In the mid-1990s, new data on the seasonal and mean annual fields of the Caspian water circulation were obtained on the basis of a non-linear hydrodynamic model (Trukhchev et al., 1995). It was found that the overall water circulation can be characterized as a set of meso-scale vortices (Fig. 4). For the Middle Caspian, there is a perennial dipole system consisting of cyclonic water circulation within its north-western part and anticyclonic within the south-eastern part. An oppositely directed vortical dipole structure develops in the Southern Caspian. Seasonal variability of these two structures manifests itself as interrelated variations of location, size and intensity of these gyres, which are quite evident up to 100 m depth. To gain an impression of water circulation within the deeper layers, additional observations are needed. Mean velocities of currents near the centers of the sub-basin gyres and along the contacts of oppositely directed gyres are, respectively, 5-10 cm s^{-1} and up to 20 cm s^{-1}.

According to field observations, strong northern and southern winds typical of the Caspian Sea are able to accelerate currents periodically up to 50-60 cm s^{-1} (Terziev et al., 1992; Kosarev and Yablonskaya, 1994). This should be taken into account when planning offshore development activities and assessing environmental impacts. Within the limits of the shallow-water Northern Caspian, having low gradients of the bottom topography and gently sloping shores, storm surges are among the most environmentally important manifestations of water dynamics (Terziev et al., 1992). Both the western and eastern shores of the Northern Caspian are affected by storm surges of 2.5-3 m or even higher, with inundated areas extending as far as 30 to 50 km from the shoreline. Such catastrophic events often result in pollution of the sea water by oil products, agricultural chemicals and other substances.

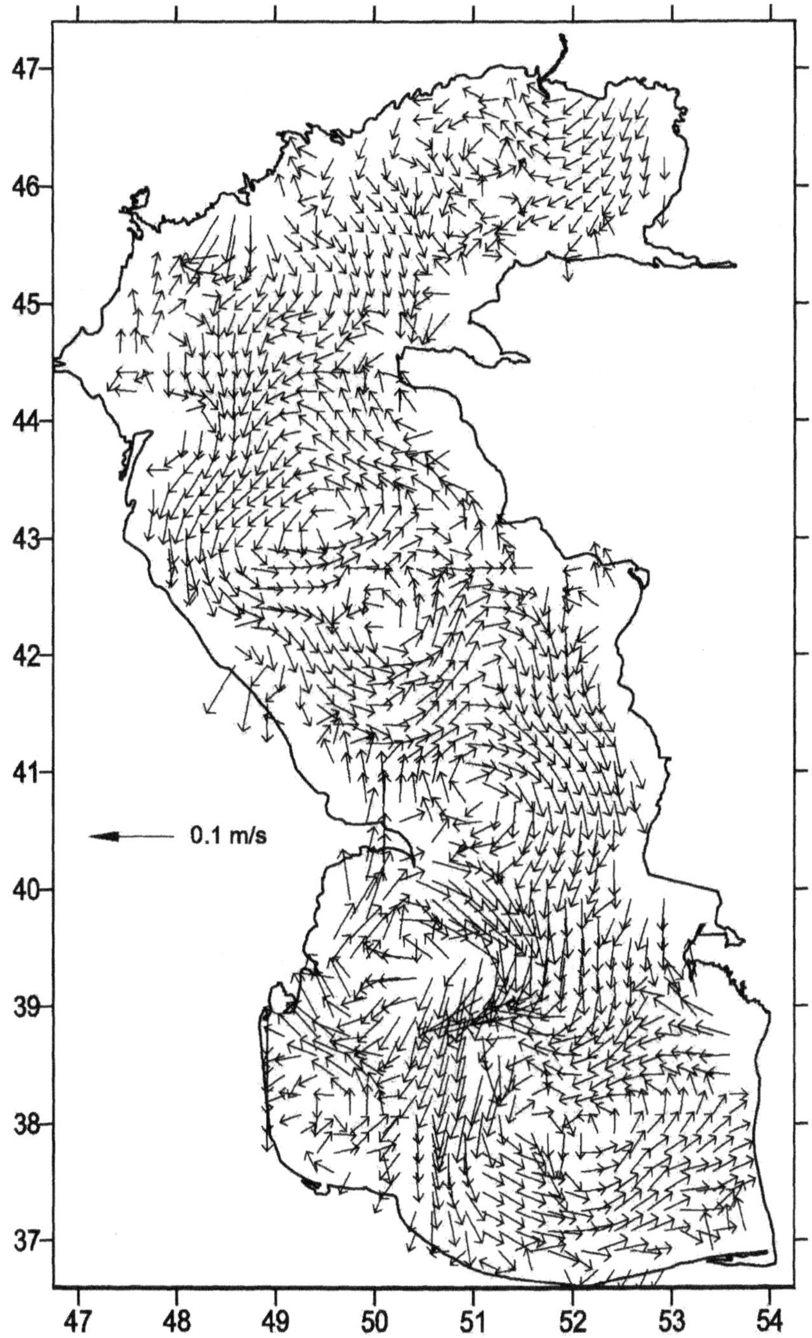

Figure 4. Mean annual climatic field of the Caspian Sea upper layer currents according to numerical model (Trukhchev et al., 1995).

4. SEASONAL VARIABILITY OF THERMOHALINE FIELDS

External factors control the formation of clear annual temperature and salinity cycles within the upper water layer (30 to 50 m) (Kosarev and Tuzhilkin, 1997). Winter and summer conditions of the Caspian Sea differ not only in values of surface water temperatures but also in the spatial pattern of the temperature field.

In winter (December-March), surface water temperature varies from 0°C within the northern part of the sea, which is partly covered by ice, to 10-11°C within the Southern Caspian (Fig. 5a). The Middle Caspian is fed by colder waters moving southward along the western shore. This water inflow is compensated by warmer intrusions from the Southern Caspian extending mostly along the eastern shore.

Summer temperatures of the Caspian upper water layer are 23-26°C. However, extensive coastal upwelling zone occurs in July and August over the eastern shelf of the Middle Caspian, where the water temperatures decrease to 18-20°C (see Fig. 5b) and even lower. The upwelling is most severe in July and August when it impedes heating of the surface waters and destroys the pattern of seasonal temperature trends. On the whole, upwelling affects the dynamic regime of the entire Caspian deepwater basin by promoting baroclinic processes.

Vertical distribution of water temperatures in winter is relatively uniform due to vigorous convective mixing. A sharp thermocline at a depth of 20-30 m is typical for summer conditions. It separates the upper heated layer from the rest of the water column and tends to disappear in autumn as the near-surface waters cool. The temperatures of the bottom layers within the deep-sea zones of the Middle and Southern Caspian are equal to 4.5-5.5°C and 5.8-6.5°C, respectively (Kosarev and Tuzhilkin, 2000).

Large volumes of river influx freshen the Caspian waters (Fig. 6), whose salinity averages 12.8-12.9 ‰, almost three times lower than mean ocean salinity (35 ‰). The Northern Caspian is characterized by considerable variations of water salinity. These range from 0.1-0.2 ‰ close to Volga and Ural mouths to 10-11 ‰ along the boundary between the Northern and Middle Caspian. The quasi-latitudinal frontal zone separating river and marine waters is noted by maximal horizontal salinity gradients reaching up to 0.5-1.0 ‰ km^{-1}. In the north-western part of the Middle Caspian, the seasonal change of water salinity is characterized by a decrease in June and

July due to the spread of freshened waters in the upper layer of the sea (see Fig. 6b); during August and September the surface salinity in this region increases and then drops again by 0.2-0.3 ‰ to its February minimum. By June, the freshening effect of river discharges can be found as far as south of Makhachkala (43°N). Further southward the advance of the low salinity waters becomes much slower as evaporation increases. A marked fresh tongue penetrating to the Southern Caspian occupies only a narrow zone along the western coast.

In the Middle and Southern Caspian, fluctuations of water salinity are largely insignificant, with most values lying in the 12.5-13.4 ‰ range. Water salinity generally tends to increase from the north-west to the south-east in response to the southward flow of freshened water along the western shore and high aridity of the eastern shore. Maximum salinity levels are endemic to the south-eastern shelf waters.

Salt distribution in most of the Caspian water column is rather uniform (Kosarev and Tuzhilkin, 1995, 2000). The difference in salinity between surface and saltier bottom layers is, at most, 0.2-0.3 ‰, due to effective vertical mixing and ventilation of the deep waters. Therein lies an outstanding dissimilarity of the Caspian Sea from the Black Sea.

Autumn and winter convection present an effective ventilation mechanism for deep Caspian waters. In winter, vertical circulation involves layers to a depth of 150-200 m in the Middle Caspian and of 80-100 m in the Southern Caspian (Kosarev and Yablonskaya, 1994). The colder the winter, the deeper the water circulation. Bottom layers are mostly ventilated due to the sinking of dense and cold waters formed over the northern and eastern shelf (such a process is known as slope convection). It results in sinking near surface waters to the deep basins of the Middle Caspian. Then the waters appear to flow over the Apsheron threshold into the deeper Southern Caspian. During the cooling period, the deep basin of the Southern Caspian also gains water with salinity near 13 ‰ sinking from the extensive eastern shelf. Intense convection promotes a compensating uplift of nutrients to the euphotic layer and provides sufficient oxygenation of the Caspian waters, with top layers containing 7 to 10 $ml \cdot l^{-1}$ of dissolved oxygen in winter and 5 to 6 $ml \cdot l^{-1}$ in summer. In the near-bottom layers the average content of oxygen is 2-5 $ml \cdot l^{-1}$. However, these concentrations are subject to interannual variability depending on the depth of mixing, which is controlled by ventilation of the bottom waters.

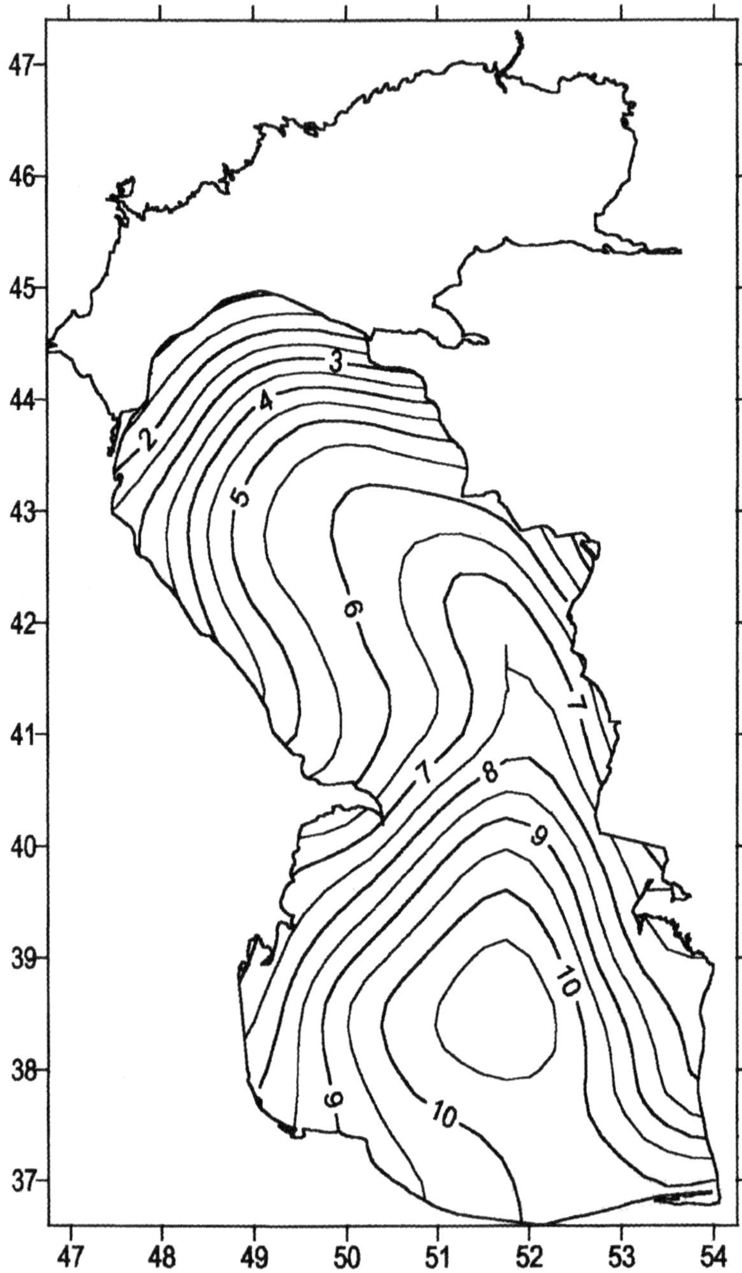

(a)

Figure 5. Caspian surface temperature in February (a) and August (b). The February field is confined in the north by the floating ice edge.

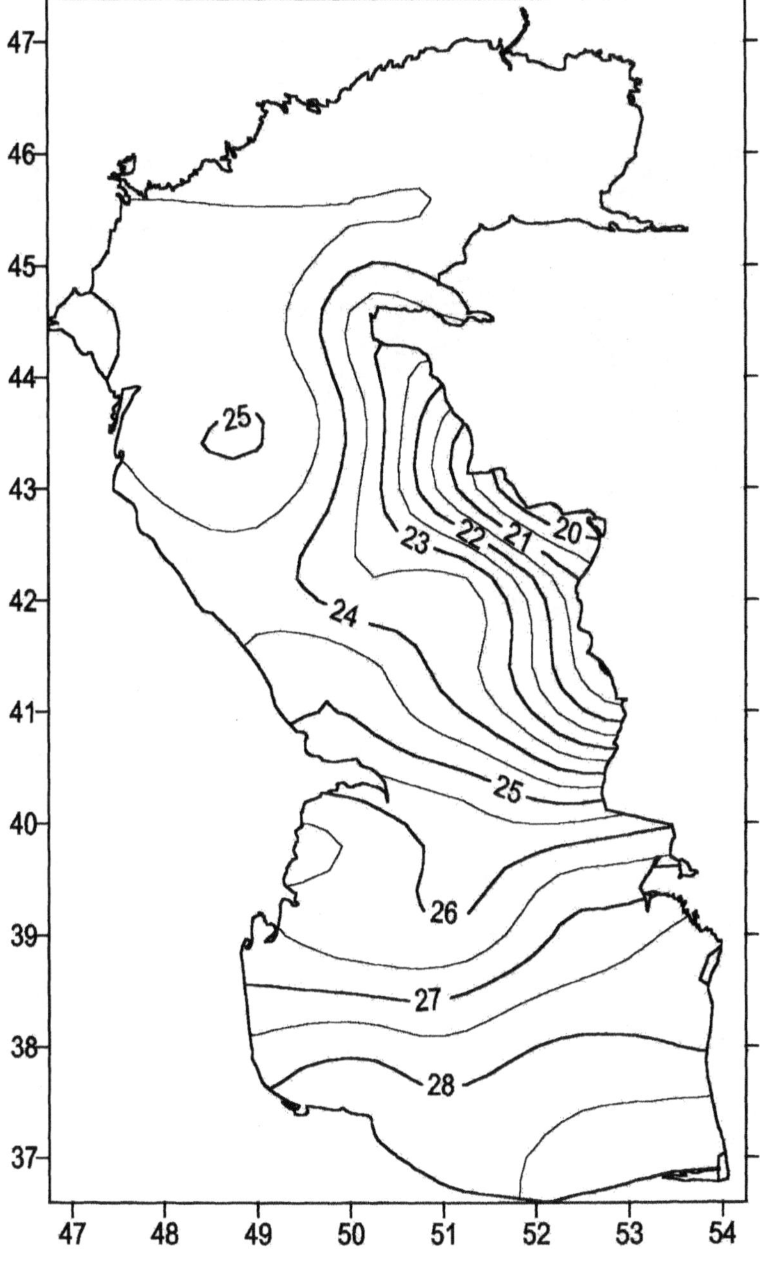

(b)

Figure 5 (Continued).

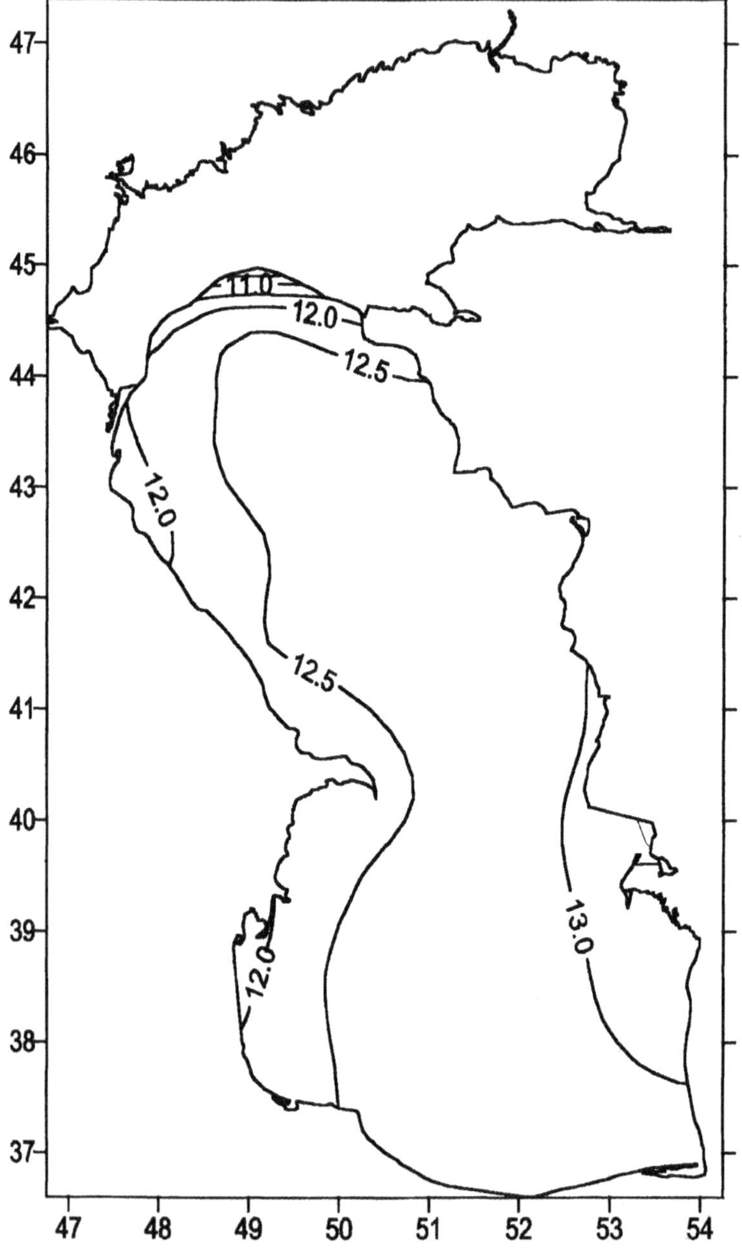

(a)

Figure 6. Caspian surface salinity in February (a) and August (b). The February field is confined in the north by the floating ice edge.

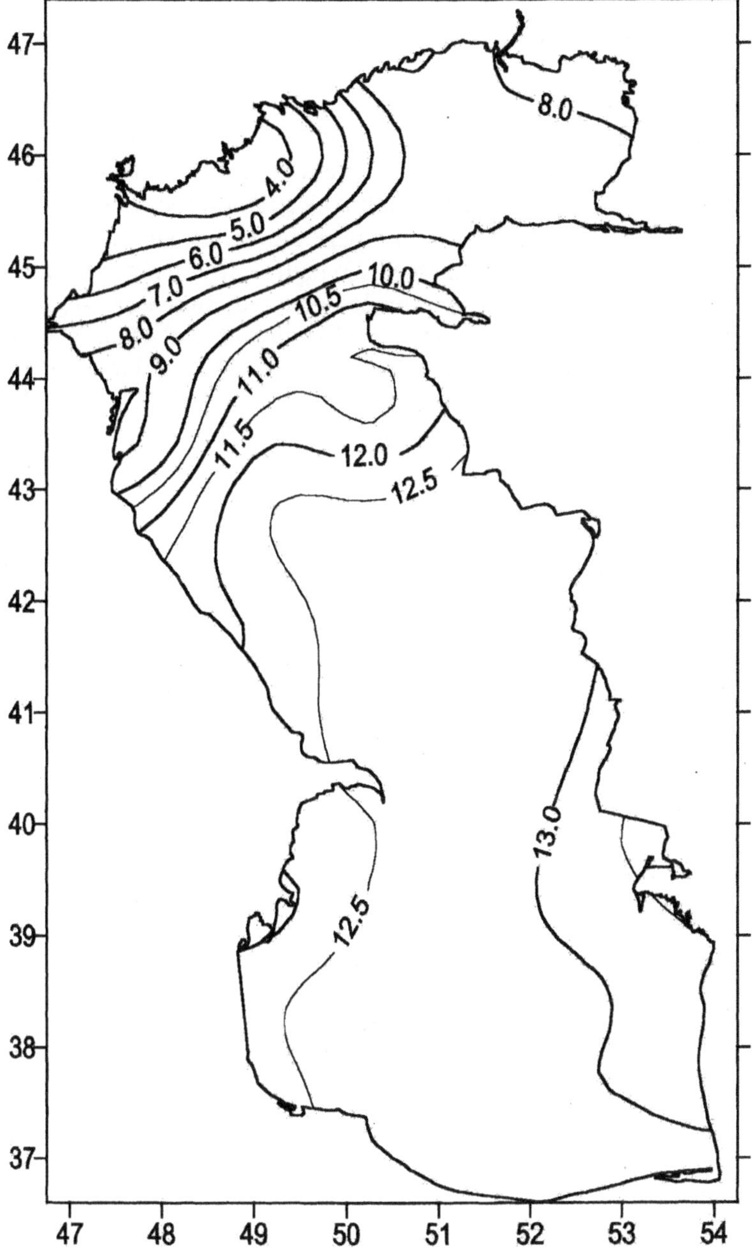

(b)

Figure 6 (Continued).

5. INTERANNUAL VARIABILITY OF THERMOHALINE FIELDS

Investigation of different aspects of the interannual variability of the state of the World's Ocean waters is one of the focal points of modern oceanography. Along with global, quasi-cyclic changes like the El-Nino – Southern Oscillation or the North Atlantic Oscillation (Harrisson et al., 1998, Kaplan et al., 1998), recently several aperiodic variations in the water state have been revealed in the World Ocean, which were interpreted as transition processes between stable regimes or as their temporal blocking. The best known of such oceanic processes is a blocking of the Atlantic meridional water circulation in the Labrador Sea in the surface (Houghton and Visbeck, 2002) and intermediate (Dobrolyubov et al., 2002) layers under high production of freshened water masses. This process is related to the passive phase of the North Atlantic Oscillation, but has its own development dynamics and is not fully periodic.

The analysis of long-term records of hydrometeorological characteristics in the enclosed seas shows that besides cyclic-type interannual variations there are aperiodic changes of the sea waters in the form of trends of different duration. For example, during the last century, the bottom water salinity in the Gottland hollow of the Baltic Sea experienced six temporal periods of 10-20 years duration with different tendencies (Bukhanovskiy et al., 2001).

In the beginning of the 1990s in the Eastern Mediterranean Sea, an important change in the deep water formation process occurred, which resulted in the appearance of a new region of deep water formation northward of Crete (the Crete Sea) instead of the Adriatic Sea (Klein et al., 1999, 2000; Theocharis et al., 1999). This led to the significant re-establishment of the thermohaline structure and circulation in the deep and intermediate layers at the most part of the Mediterranean Sea (Malanotte-Rizzoli et al., 1999, Wu et al., 2000, Blankart and Pinardi, 2001).

Recently, another important transition process was revealed in the Japan Sea (Gamo et al., 2001, Kim et al., 2001). In the mid-1990s, in the deep basins of the Japan Sea, hydrophysical and hydrochemical evidences of blocking of the deep-water autumn-winter ventilation were obtained. During ventilation, the surface waters reached intermediate layers only (200-1500 m), which resulted in the disappearance of the well-known intermediate oxygen minimum. The main reason for this was long-term warming of surface waters in winter. The process was recognized to be similar to the above mentioned blocking in the Labrador Sea, but the latter one has a haline nature.

In the Caspian Sea the interannual river run-off dynamics and values of winter severity index were found to be the key variables determining changes in hydrological parameters of the Caspian thermohaline pattern. The most significant shift in external thermohydrodynamic effects on the Caspian Sea was between 1968-1977 and 1978-1987 (Kosarev and Tuzhilkin, 2000). For these two decades, respectively, mean annual volumes of the river run-off were 240±13 km^3 and 306±11 km^3, decade-averaged indices of winter severity were −154±48°C and −90±16°C, and decade-averaged indices of summer longitudinal air transport (calculated as atmospheric pressure difference between western and eastern shores) were 0.01±0.11 Gpa and −0.39±0.15 Gpa. Thus, the 1978-1987 decade exhibited a considerable increase in river inflows, decrease in winter severity and enhancement of atmospheric processes stimulating the upwelling along the Caspian eastern shore.

The above described variations of the external factors were also reflected in distribution of the Caspian hydrology, as observed since 1950. According to Figure 7, mean February temperature of the surface water layer was 0.5-2.0°C higher in the 1978-1987 period compared with the previous decade.

Along the eastern shore, a decrease of water temperature by 0.8-1.2°C was observed in summer, with enhanced upwelling being considered as a main reason. At the same time, within the north-western part of the sea, mean temperatures for the later decade were higher (by 1°C near the Volga delta) as compared with the 1968-1977 period. This appeared to be a result of effective warming of large volume of riverine freshwaters.

In 1978-1987, both winter and summer values of surface salinity were much lower than those of the previous decade (Fig. 8). This interdecadal difference varied from 0.2-0.3 ‰ in the Southern Caspian to 1.0-1.5 ‰ in the Northern Caspian.

Vertical thermohaline structure of the Caspian waters underwent pronounced changes too, as clearly is demonstrated by the example of longitudinal vertical T and S sections based on shipboard observations during August-September 1976 and September 1995 (Kosarev and Tuzhilkin, 1995; Data Report IAEA research, 1995; Tuzhilkin and Kosarev, 2003) (Fig. 9, 10).

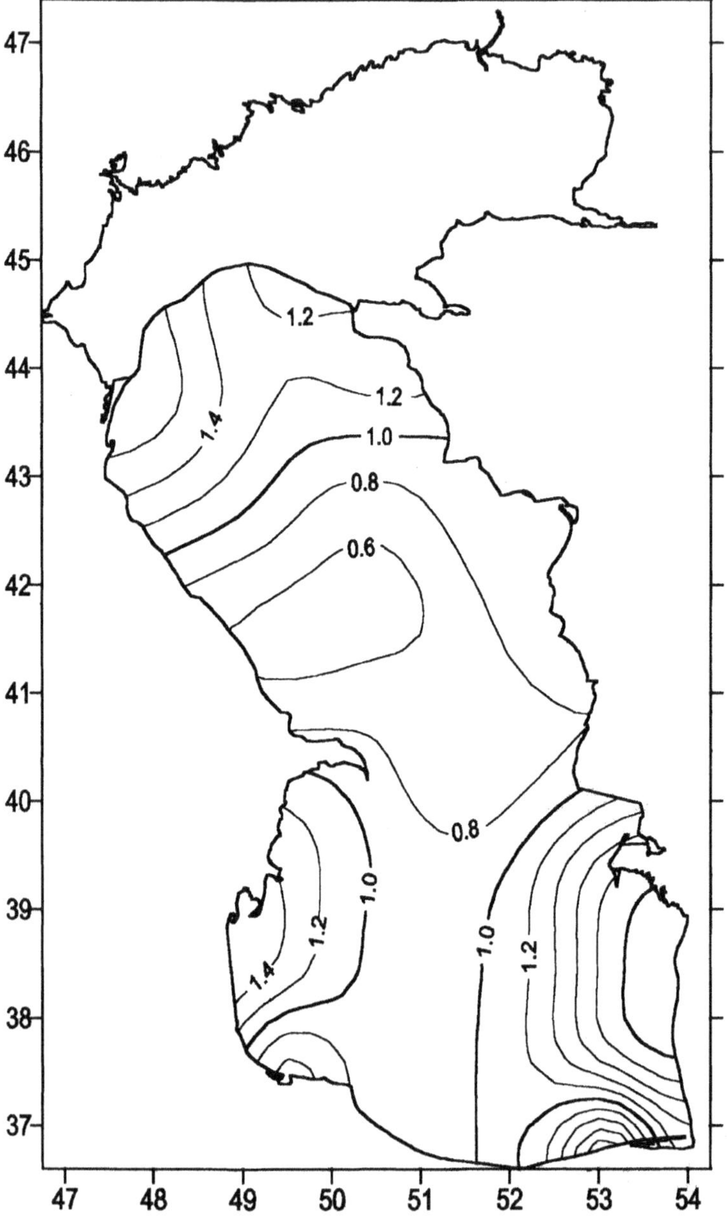

(a)

Figure 7. Caspian surface water temperature difference (°C) between transgression (1978-1987) and regression (1968-1977) periods: (a) February and (b) August (Kosarev and Tuzhilkin, 2000).

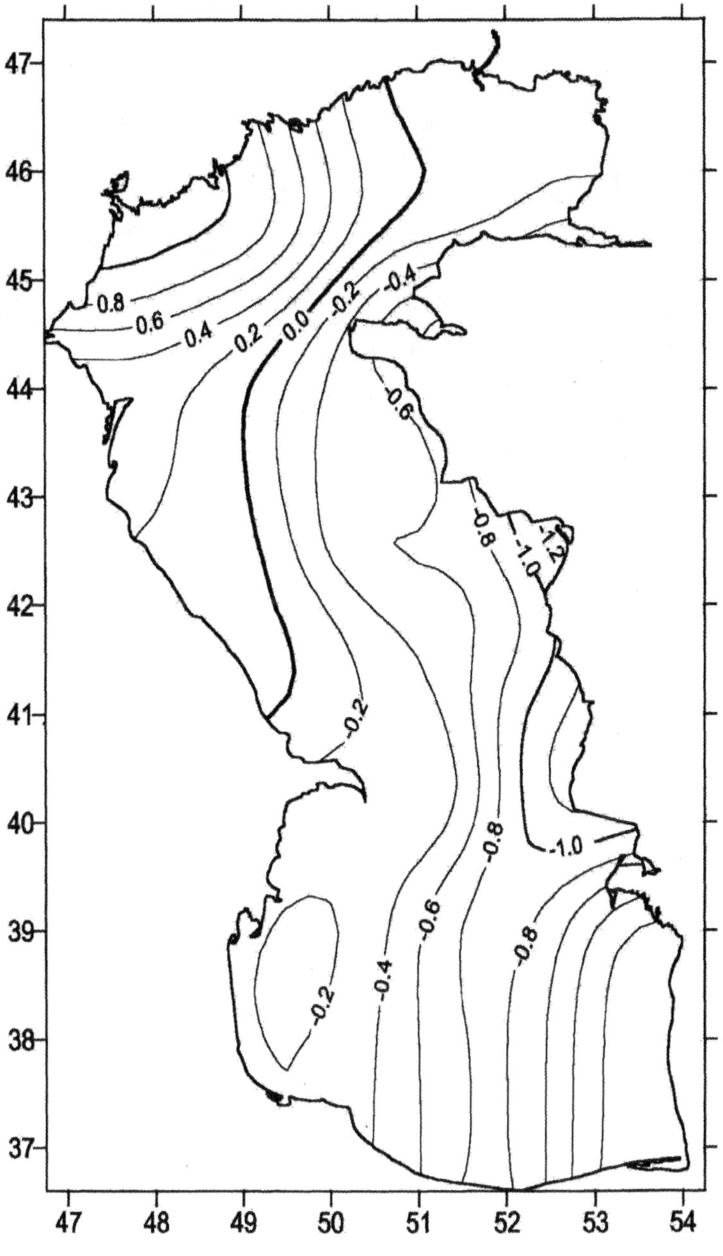

(b)

Figure 7 (Continued).

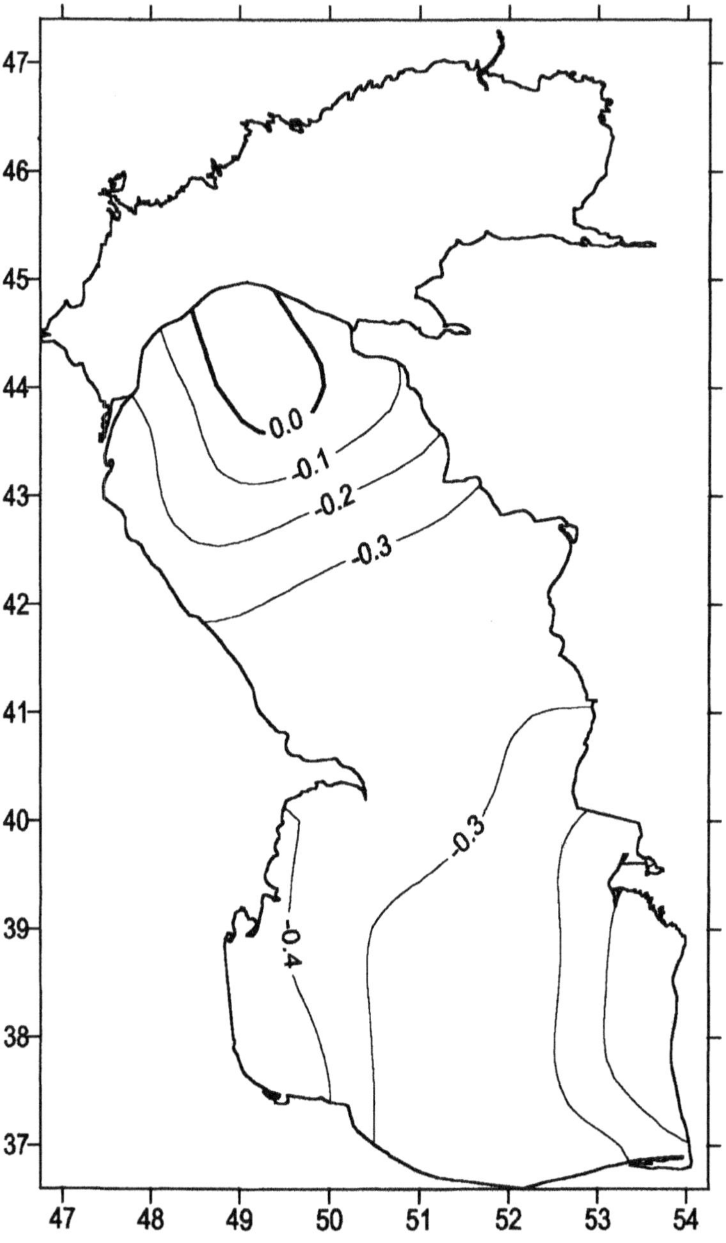

(a)

Figure 8. Caspian surface water salinity difference (‰) between transgression (1978-1987) and regression (1968-1977) periods: (a) February and (b) August (Kosarev and Tuzhilkin, 2000).

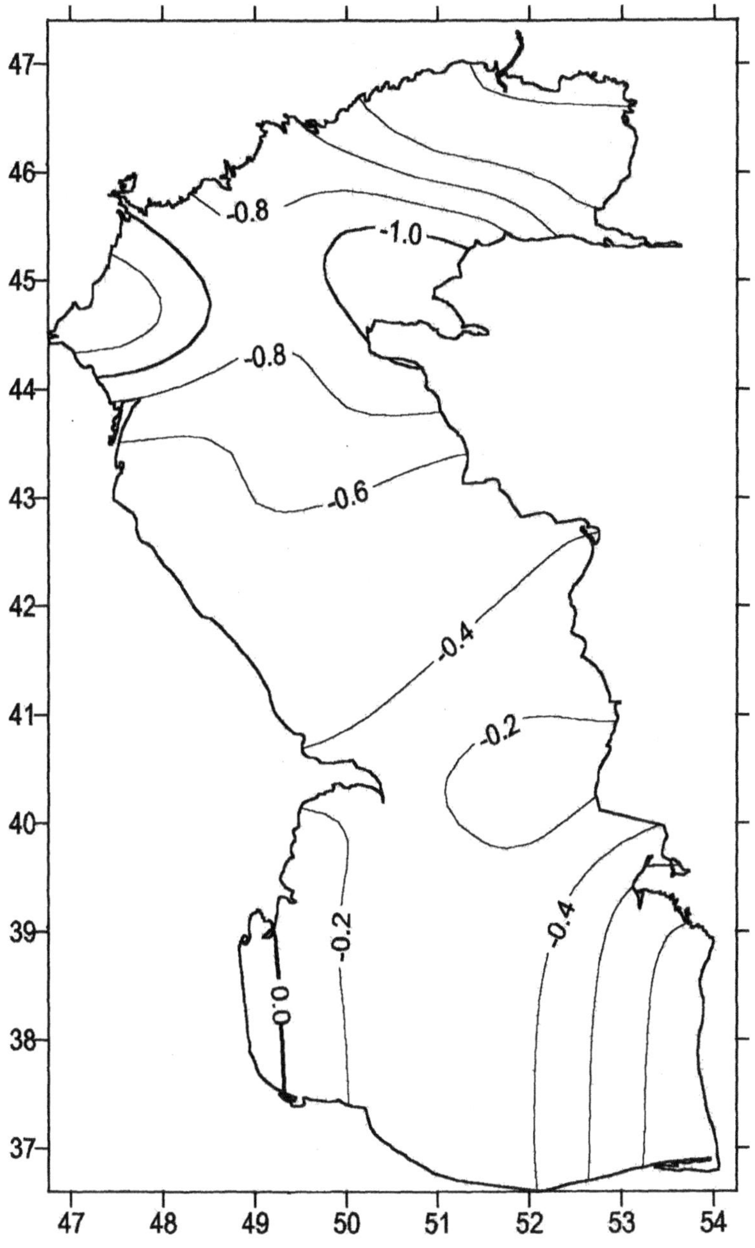

(b)

Figure 8 (Continued).

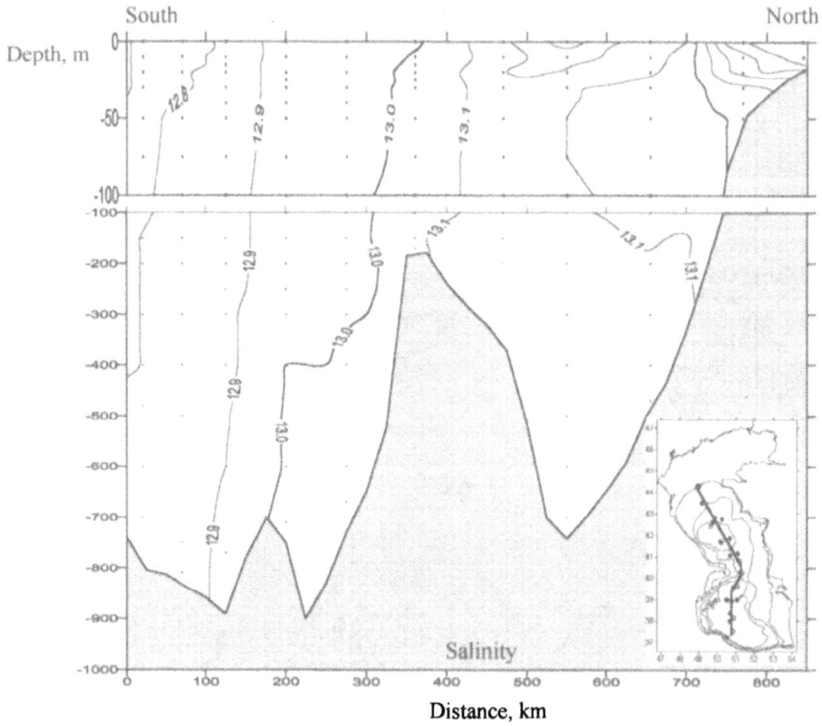

(a)

Figure 9. Quasi-longitudinal vertical cross-section in salinity (‰) of the Caspian Sea as measured in August 1976 (a) and September 1995 (b).

The 1976 sections reflect conditions of relatively long duration when water loss dominated over water inflow. Among the major consequences of this situation was a nearly uniform vertical distribution of salinity within the Caspian deepwater basin (see Fig. 9a). In winter, the absence of sustainable haline stratification in turn stimulated intense convective mixing, especially in the case of the so-called haline inversions, when high-salinity surface waters overlaid the summer thermocline. Subsequent cooling resulted in rapid destruction of the thermocline and sinking of the high-salinity surface waters to deeper layers. In response to this, all the Caspian water column below the thermocline was characterized by a uniform distribution of temperature and salinity. The 1995 sections characterize conditions at the end of a nearly 20-year period of the Caspian Sea when the delivery of freshwaters dominated over the water loss. It is not surprising that the salinity tended to decrease throughout the whole water column and became vertically stably stratified (see Fig. 9b).

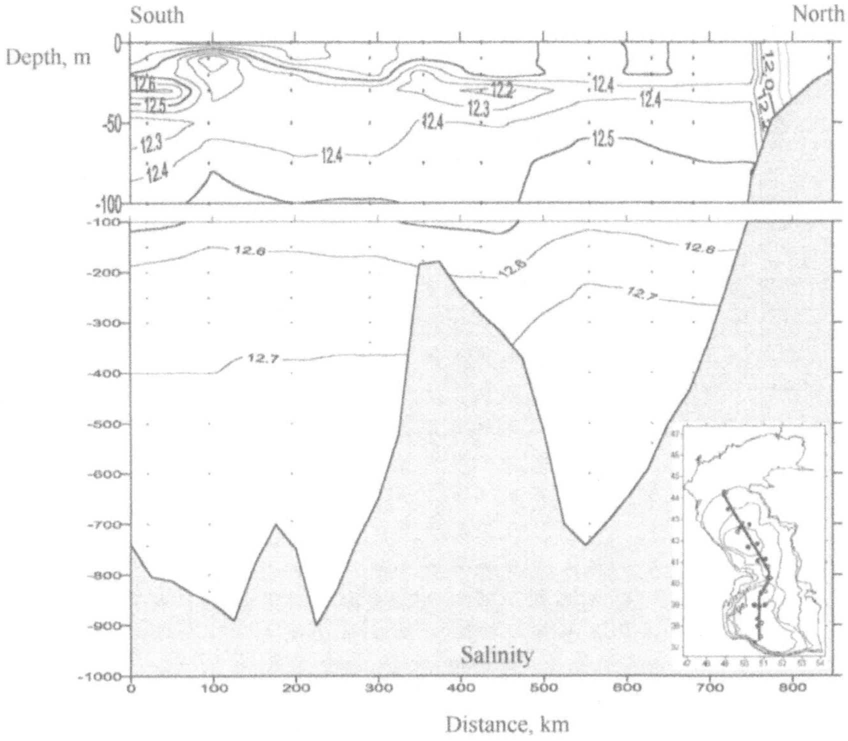

(b)

Figure 9 (Continued).

Profound changes were also recorded in vertical thermal structure of the Caspian waters. Owing to uplift of the lower boundary of the seasonal thermocline, its thickness was more than two times less in 1995 than it was in 1976 (see Fig. 10). At the same time, the vertical temperature and density gradients within the seasonal thermocline increased by factors of 2 and 1.5, respectively. The most significant drop of temperatures (more than by 1°C) and salinity (by 0.1-0.6 ‰) was registered within the intermediate layer (50-200 m). Potential density of both intermediate and deep waters markedly dropped (by 0.2-0.3 kg m^{-3}). All the described changes in mean values are statistically significant with a probability of 90 %.

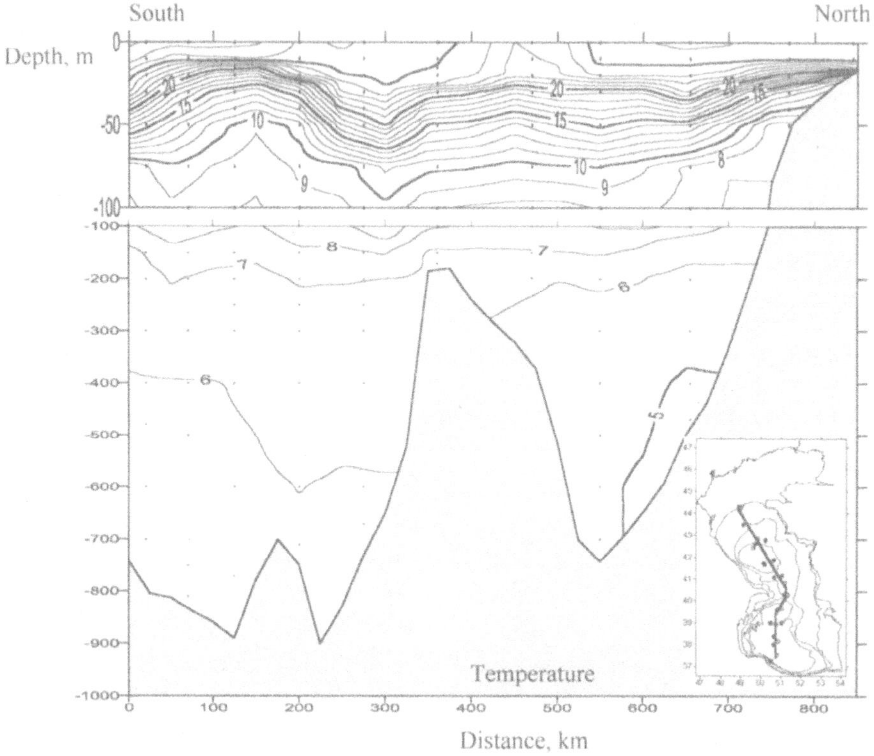

(a)

Figure 10. Quasi-longitudinal vertical cross-section in temperature (°C) of the Caspian Sea as measured in August 1976 (a) and September 1995 (b).

6. CONCLUSIONS

The research undertaken suggests that the Caspian thermohaline structure responds clearly to changes in large-scale external factors, such as heat and water fluxes through the sea surface and river run-off. Fundamental transformations of the Caspian hydrological parameters inevitably impact the functioning of marine ecosystems.

The analysis of long-term (for the 1956-2000 period) alterations in the Caspian's hydrology and interrelations within the system of external thermohydrodynamic factors and thermohaline structure enabled identification of two main states of this system. During the first one, reduced

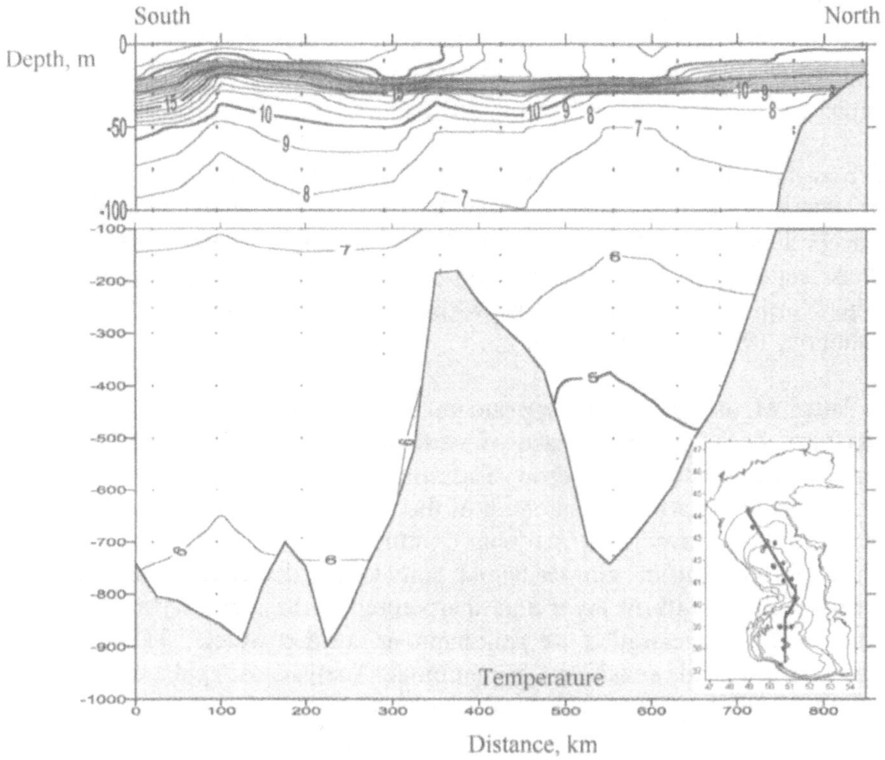

(b)

Figure 10 (Continued).

river run-off coupled with relatively high winter severity results in increased ventilation, which involves the whole water column (Kosarev and Tuzhilkin, 2000). This results in a water salinity increase and vertical salinity distribution homogenization. In parallel with vertical expansion of the seasonal thermocline, its temperature gradient becomes weaker, whereas deep waters become colder.

The other state corresponds to the opposite combination of external factors (i.e., increased river run-off and milder winters) and is associated with stable vertical salinity stratification, which impedes vertical diffusive exchanges of heat and salt and disrupts regular ventilation of the near-bottom layers. The lower boundary of the actively ventilated zone is located at 200 - 300 m level. In winter, intermediate layers are intensively fed by colder surface waters. Coupled with a decrease of vertical turbulent diffusion, this results in a drop of water temperatures within the lower part of the thermocline. The thermocline itself becomes thinner, while its vertical gradients become stronger.

The two states lasted for different durations of the Caspian history. While the first state was observed during 10 years, from the late 1960s to late 1970s, the second state persisted through 1995 (i.e., during 17 years). The duration of these states is likely to be controlled by macro-circulation processes (such as the North-Atlantic Oscillation) within the "ocean–atmosphere" system (Rodionov, 1994; Nesterov, 2001).

The latter of the two states appears to be less favorable for the Caspian ecosystem, because the increase of vertical water density stratification is normally accompanied by a drop of dissolved oxygen concentrations in the deep waters along with a weakening of the turbulence-induced supply of the surface euphotic layer with nutrients (Katunin et al., 2000). The processes such as diminishing of autumn-winter convection, decrease in thickness of the upper quasi-uniform layer and sharpening of the summer thermocline, contribute to concentration of pollutants in surface waters. This poses a serious hazard to the sensitive Caspian biota. Analysis of rapid and dramatic changes in the Caspian hydrological regime highlights the need for complex environmental monitoring and an adequate environmental protection system.

Acknowledgments. This work was supported by the Russian Federal Program “World Ocean”, the “Caspian Sea” Project of the Russian Ministry of Industry, Science, and Technologies, and by the Russian Foundation for Basic Research Grant № 03-05-96630. A.N.K. and A.G.K. are grateful to NATO for supporting their participation in the *“Dying and Dead Seas”* ARW as invited speakers.

7. REFERENCES

Blankart J.-M., Pinardi N. (2001) Abrupt cooling of the Mediterranean Levantine Intermediate water at the beginning of the 1980s: Observational evidence and model simulation. J. Phys. Oceanogr. 31(8): 2307-2320.

Bukhanovskiy A.V., Davidan I.N., Rozhkov V.A. (2001) A model of interannual variability of the Baltic Sea salinity. Izvestiya Russian Geogr. Society. 133(1): 54-62 (in Russian).

Data Report IAEA Research (1995) *Training cruise on the Caspian Sea 5-28 September 1995*. Vienna, IAEA, 96 pp.

Dobrolyubov S.A., Lappo S.S., Morozov E.G., Sokov A.V., Tereschenkov V.P., Shapovalov S.M. (2002) The North Atlantic water structure in 2001 based on the cross-atlantic section along 53°N. Doklady Russian Acad. Sci. 382(4): 543-546 (in Russian).

Gamo T., Momoshima N., Tolmachyov S. (2001) Recent upward shift of the deep convection system in the Japan Sea, as inferred from the geochemical tracers tritium, oxygen, and nutrients. Geophys. Res. Lett. 28(21): 4143-4146.

Harrisson D.E., Larkin N.K. (1998) El Nino - Southern Oscillation sea surface temperature and wind anomalies, 1946-1993. Rev. Geophys. 36(3): 353-399.

Houghton R.W., Visbeck M. H. (2002) Quasi-decadal salinity fluctuation in the Labrador Sea. J. Phys. Oceanogr. 32(2): 687-701.

Kaplan A., Cane M., Kushnir Y., Clement A.C., Blumenthal M.B., Rajagopalan B. (1998) Analyses of global sea surface temperature 1956-1991. J. Geophys. Res. 103(C9): 18567-18589.

Katunin D.N., Bespartochniy N.P., Sapozhnikov V.V., Torgunova N.I., Mikhaylovskiy Yu.A., Bulanov V.V. (2000) Complex oceanographic investigations of the Caspian Sea in the R/V "Issledovatel Kaspiya". Okeanologiya. 40(1): 156-158 (in Russian).

Kim K., Kim K.-R., Min D.-H., Volkov Yu., Yoon J.-H., Takematsu M. (2001) Warming and structural changes in the East (Japan) Sea: A clue to future changes in Global Oceans? Geophys. Res. Lett. 28(17): 3293-3296.

Klein B., Roether W., Manca B.B., Bregant D., Beitzel V., Kovacevic V., Luchetta A. (1999) The large deep water transient in the Eastern Mediterranean. Deep-Sea Res.-I. 46: 371-414.

Klein B., Roether W., Civitarese G., Gacic M., Manca B.B., d'Alcala M.R. (2000) Is the Adriatic returning to dominate the production of Eastern Mediterranean Deep Water? Geophys. Res. Lett. 27(20): 3377-3380.

Kosarev A.N. (1975) *Hydrology of the Caspian and Aral Seas*. Moscow, Moscow State University, 272 pp. (in Russian).

Kosarev A.N. (2002) The present-day state of the Caspian Sea environmental issues. The European Union and the Caspian Sea Region in Dialogue, 1st International Symposium, Abstracts, Berlin, 19-20.

Kosarev A.N., Tuzhilkin V.S. (1995) *The Caspian Sea Climatic Thermohaline Fields*. Moscow: State Oceanographic Institute, 96 pp. (in Russian).

Kosarev A.N., Tuzhilkin V.S. (1997) Climatic annual cycle of the Caspian Sea thermohaline structure variability. Water Resources. 2: 104-112 (in Russian).

Kosarev A.N., Tuzhilkin V.S. (2000) The main features of the Caspian Sea water thermohaline structure and their decade variability. Caspian Floating University Research Bulletin. 1: 26-40.

Kosarev A.N., Yablonskaya E.A. (1994) *The Caspian Sea*. Academic Publishing, The Hague, 259 pp.

Malanotte-Rizzoli P., Manca B.B., d'Alcala M.R., Theocharis A., Brenner S., Budillon G., Ozsoy E. (1999) The Eastern Mediterranean in the 80s and in the 90s: The big transition in the intermediate and deep circulation. Dyn. Atmos. & Oceans. 29: 365-395.

Nesterov E.S. (2001) Low-frequency variability of atmospheric circulation and the Caspian Sea level in the second half of the 20th century. Meteorology and Hydrology. 11: 27-36 (in Russian).

Rodionov S.N. (1994) *Global and Regional Climate Interaction: The Caspian Sea Experience*. Kluwer Academic Publishers, Dordrecht, the Netherlands, 243 pp.

Terziev F.S., Kosarev A.N., Kerimov A.A. (Eds.) (1992) *Hydrometeorology and Hydrochemistry of the Seas. Volume VI: The Caspian Sea: Issue 1: Hydrometeorological Conditions*, Hydrometeoizdat, St. Petersburg, 360 pp. (in Russian).

Theocharis A., Nittis K., Kontoyiannis H., Papageorgiou E., Balopoulos E. (1999) Climatic changes in the Aegean Sea influence the Eastern Mediterranean thermochaline circulation (1986-1997). Geophys. Res. Lett. 26(11): 1617-1620.

Trukhchev D., Kosarev A., Ivanova D., Tuzhilkin V. (1995) Numerical analysis of the general circulation in the Caspian Sea. Comptes rendus de l"Academie Bulgare des sciences. 11-12: 31-34.

Tuzhilkin V.S., Kosarev A.N. (2003) Long-term variability of vertical termohaline structure of the Caspian Sea deep basins waters. Water Resources (in press, in Russian).

Wu P., Haines K., Pinardi N. (2000) Toward an understanding of deep-water renewal in the Eastern Mediterranean. J. Phys. Oceanogr. 30(2): 443-458.

Chapter 8

Hydrobiology of the Caspian Sea

N.V. Aladin, I.S. Plotnikov

Laboratory of Brackish Water Hydrobiology, Zoological Institute of RAS, Universiteskaya naberezhnaya, 1, 199034 St Petersburg, Russia

The Caspian Sea is the world's largest lake in area. Hence it is possible to assume, that it should have the largest biodiversity as well. The larger area of lake, the large number of plants and animals species lives in it. In our opinion this correlation with the area is not direct but indirect. Certainly, the larger area of reservoir, the higher probability, that it will contain more species of hydrobionts. However, there are no doubts, that determining parameters in this case are also favourable conditions for water organisms, age of lake, diversity of biotic and abiotic conditions, presence of necessary and balanced amount of substance and energy (Aladin et al., 2002).

The Caspian Sea is one of the most ancient continental reservoirs on our planet. This lake is the remnant of the ancient, mainly epicontinental, Parathethys Sea. Just for this reason the age of this reservoir is estimated quite differently by the different researchers. Some authors give figure of 5 millions years and count the age from Balakhanian time, others consider it being 15 million years old and count from Low-Sarmatian time, and some indicate even older age of 25–35 millions years (Fig. 1, 2). Even accepting a minimal value of 5–6 millions years, as the Caspian Sea age, it will appear rather long time for formation of unique and diverse fauna and flora (Zenkevich, 1963; Karpevich, 1975; Aladin et al., 2002).

J.C.J. Nihoul et al. (eds.), Dying and Dead Seas, 185–226.

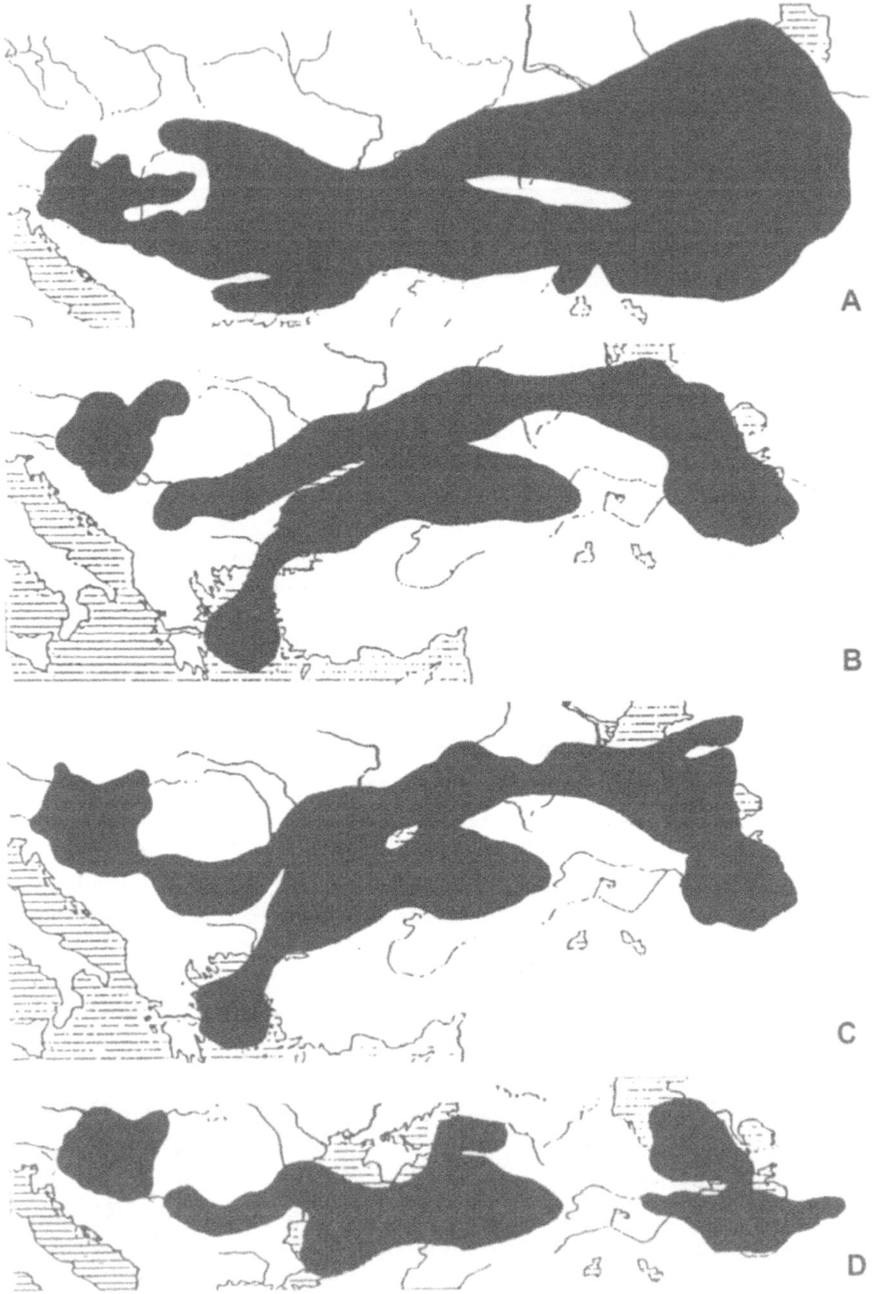
A
B
C
D

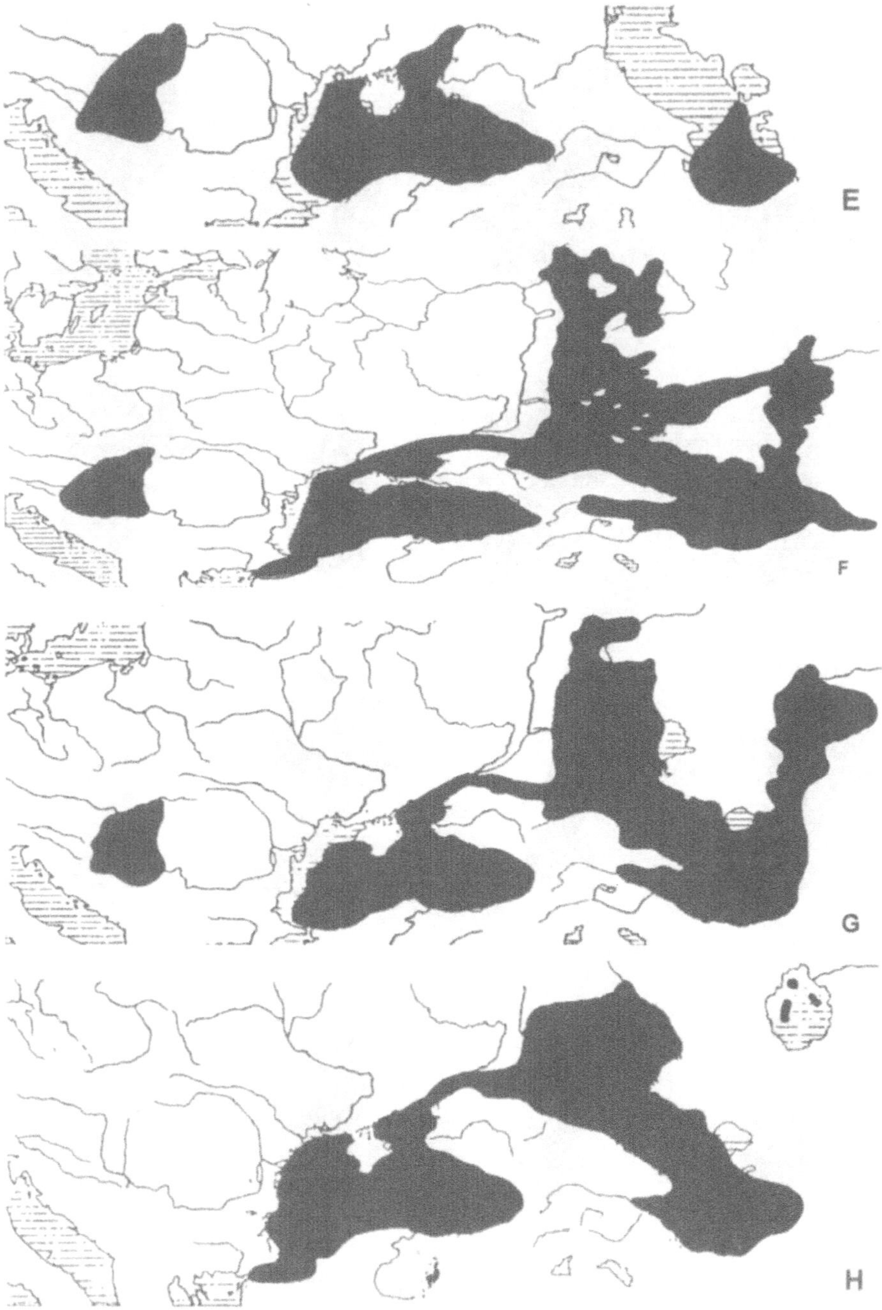

Fig 1. Palaeohydrography of south-eastern Europe ad south-western Central Asia in the late Miocene-Pleistocene (by Starobogatov, 1994 with changes and additions).

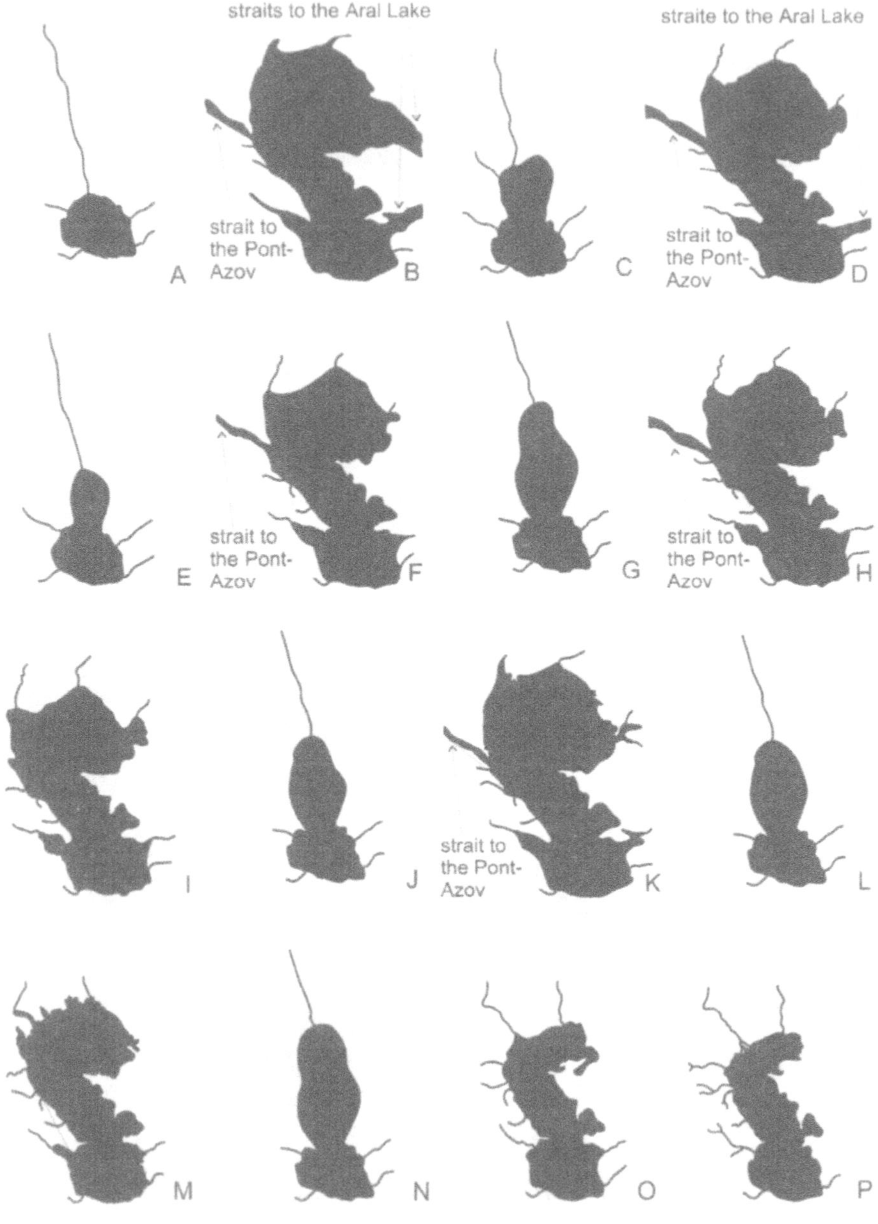

Fig. 2. Water bodies of the Palaecaspian (by Aladin, Plotnikov, 2000). A – Balakhanian; B – Akchagylian; C – Postakchagylian; D – Apsheronian; E – Turkianian; F – Bakuvian; G – Venedian or Ushtalian; H – the Early Khazarian; I – the Late Khazarian; J – Atelian; K – the Early Khvalynian; L – Enotaevian; M – the Late Khvalynian; N – Mangyshlakian; O – the New Caspian; P – the present.

Other important feature of the Caspian Sea is an extreme diversity of biotops, biotic and abiotic conditions (Zenkevich, 1963). First of all water strongly differs by salinity in different parts of this lake (Kosarev, Tuzhilkin, 1995). In some areas, especially those adjoining river deltas, water is practically fresh. In other places, first of all in the shallow gulfs of the eastern coast (striking example is gulf Kara Bogaz Gol), water contains large amount of salts and is even hyperhaline frequently, down to salinity of salt sedimentation. In open areas of the Northern Caspian, along north-south transsect, salinity is smoothly increasing from fresh water in the deltas of Volga and Ural rivers up to 10–12 g/l at the boundary between the Northern and Middle Caspian. Down the east-west transsect salinity is very sharply decreasing from 60–40 g/l in eastern shoals of Kazakhstan up to 6–7 g/l in the centre of the Northern Caspian and up to fresh water at the western coast near the Terek and Sulak river deltas. Average salinities of the Northern Caspian Sea is possible to consider as being 5–8 g/l, and this area, according to Venetian classification of waters by salinity, should be considered as mesohaline type. In open areas of the Middle and Southern Caspian salinity conditions are more stable in comparison with the Northern Caspian and salinity varies from 11 up to 14 g/l and only in areas, adjoining to the river deltas becomes much lower. Average salinity of the Middle Caspian is possible to consider 11–12 g/l, and in the Southern Caspian – 12–13 g/l. Thus these areas should be considered as mesohaline according to the Venetian classification. Shallow gulfs of eastern coast (Kara Bogaz Gol, Kaydak, Komsomoletz, Mertviy Kultuk, etc.) have salinity ranged from 13–14 g/l at entrance in gulfs up to values where salt sedimentation occurs (more than 200-300 g/l) in the tops of gulfs. Thus, these areas should be considered as reservoirs of hyperhaline type according to the Venetian classification (Fig 3).

Variable salinity conditions facilitate increase of the Caspian Sea biodiversity. Due to full salinity spectrum freshwater, brackishwater, euryhaline and hyperhaline hydrobionts can live here, and due to the similarity of marine and Caspian waters in relation of chemical structure many marine organisms feel well here. It is necessary especially to emphasise, that actually in the boundaries of the Caspian Sea there are three parallel coexisting ecosystems: freshwater, brackishwater (mesohaline) and hyperhaline, what promotes increase of biological diversity in this huge lake (Fig. 4-5).

Fig. 3. Caspian Sea water area (by Rodionov, 1994).

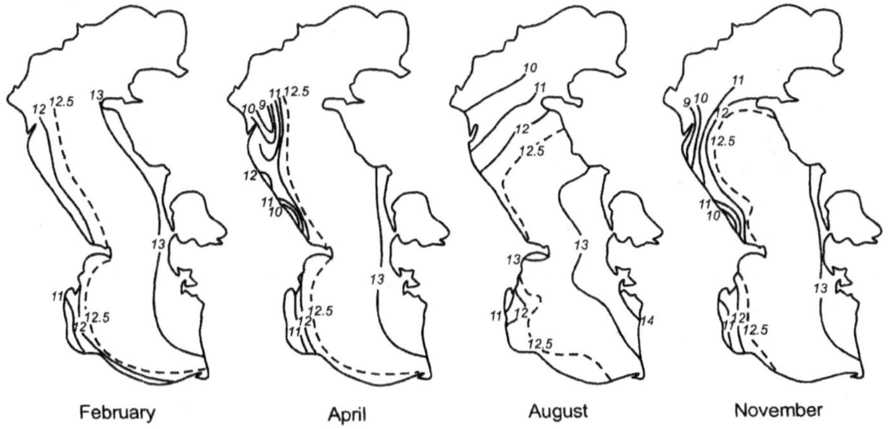

Fig. 4. Mean salinity (g/l) on the surface in the Caspian Sea: February, April, August, November.

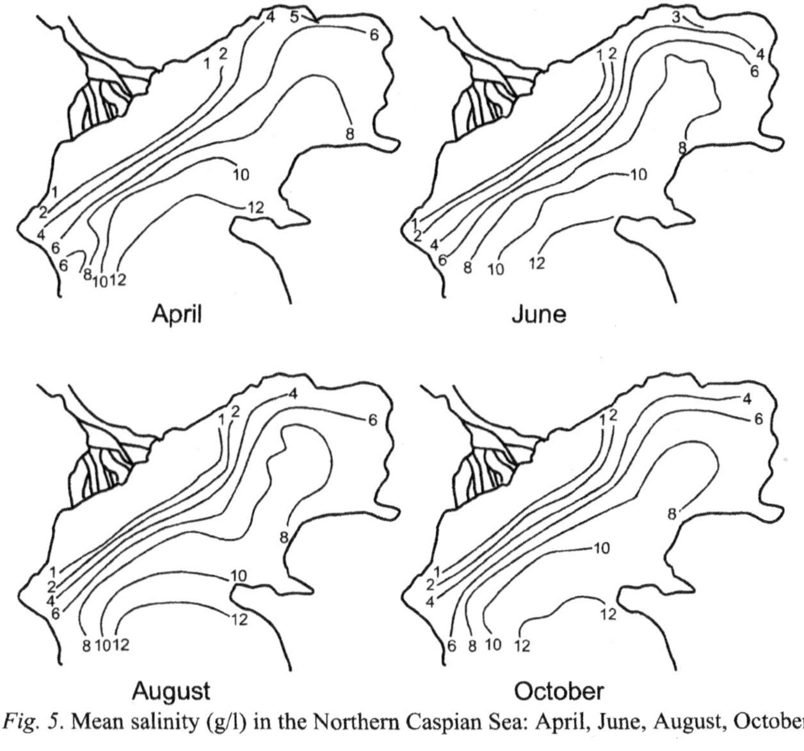

Fig. 5. Mean salinity (g/l) in the Northern Caspian Sea: April, June, August, October.

Vertical stratification of waters by salinity in the modern Caspian Sea is weakly expressed, though salinity values at bottom differ from those on the surface quite a little (Fig. 6). The lack of any vertical salinity stratification facilitate good water mixing, and as a result, relatively high oxygen content is observed now near the bottom (Fig. 7). However earlier, when the Caspian Sea level was much higher than nowadays, the strong vertical salinity stratification was observed, and there was practically no oxygen near bottom (Kosarev, Yablonskaya, 1994; Dumont, 1998). Probably because of this, life in a depth of more than 100-150 m is very poor now in the Caspian Sea. True abyssal fauna and flora are absent in this lake. Probably today's animals and plants «remember genetically» recent situations of deep anoxia. It is possible to assume, that this anoxic condition in the past has strongly reduced the modern diversity of the Caspian fauna and flora. If there were no mass «suffocation» of deep-water inhabitants in the periods of the ancient Caspian Sea level increases, deep-water biodiversity of this lake cold probably be comparable with that of the lake of Baikal.

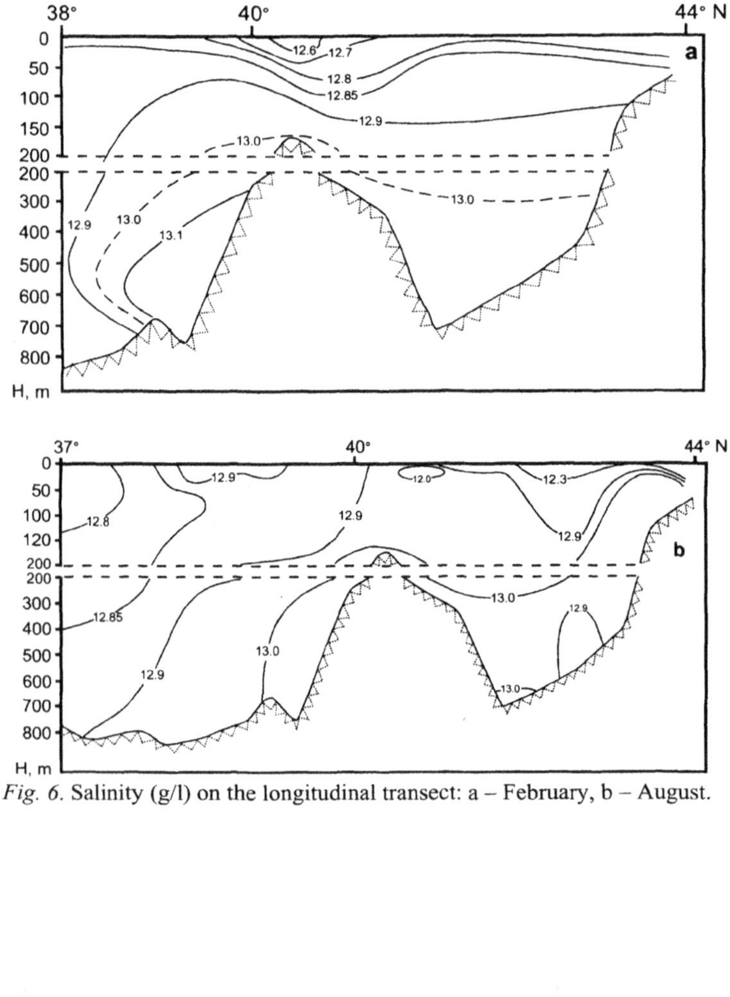

Fig. 6. Salinity (g/l) on the longitudinal transect: a – February, b – August.

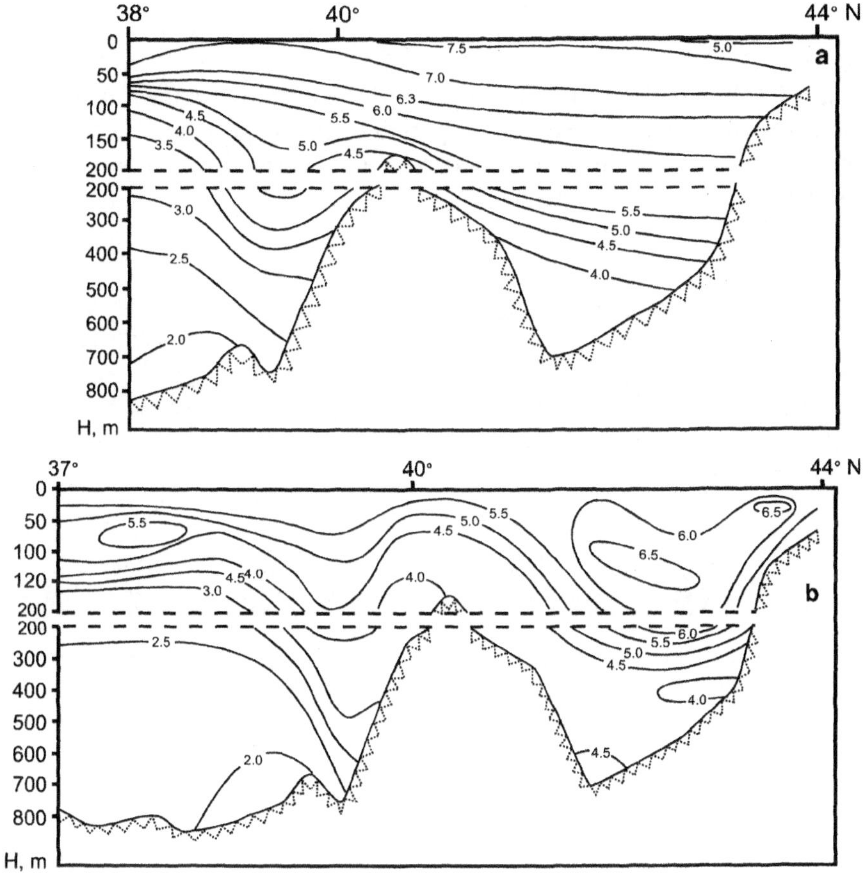

Fig. 7. Distribution of dissolved oxygen (ml/l) at the longitudinal transect: a – in February, b – in August.

In addition to strong salinity variation, temperature parameters are diverse in the Caspian Sea as well, thus promoting biodiversity increase. The greatness of the Caspian more than by 1200 km from North to the South results in the strong difference of seasonal temperatures (Fig. 8). In the Northern Caspian up to the half of water area is covered by ice in winter, while in the Southern Caspian temperature does not fall below +8–9°C even in the coldest months. This kind of spatial distribution of surface water temperature allows occurrence of both cold-water and warm-water hydrobionts here. Besides, strong vertical temperature stratification is observed also in the Caspian Sea. Below termocline water holds low temperatures even in the hottest summer months. In deep sites cold-water thermostatic condition are actually observed without any seasonal changes (Fig. 9). Presence of cold deep waters in the Caspian Sea allows even to Arctic organisms to occupy their upper horizons (Zenkevich, 1963; Kosarev, Yablonskaya, 1994) As well as in the case of salinity, diversity of temperature conditions in the Caspian Sea increases it's biodiversity. It is possible to suggest that today three ecosystem types (cold-water, moderate and warm-water) coexist in the Caspian Sea.

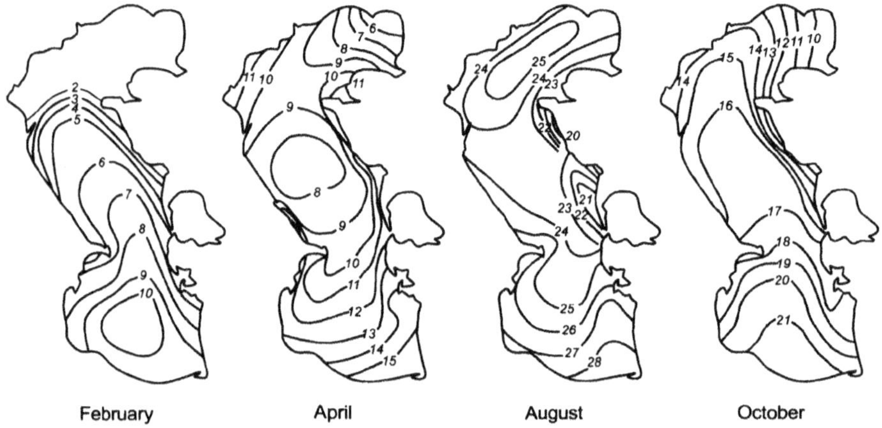

Fig. 8. Mean temperature (°C) on the surface in the Caspian Sea: February, April, August, October.

Production characteristic of various Caspian areas are also rather different, which, as well as in the case of salinity and temperature, results in the biodiversity increase of this reservoir. The Sea parts near the river deltas are characterised by increased productivity due to organic fertilisers brought by river outflow. Due to powerful run-off of Volga and Ural rivers the Northern Caspian is considerably more productive, than the Middle and Southern Caspian are. Some regions along the Middle Caspian Eastern coasts are highly productive as well due to local upwelling. Thus, productivity gradient is observed in the Caspian Sea and as a result hydrobionts living in oligo-, meso- and eutrophic conditions can find a suitable place to live (Karpevich, 1975).

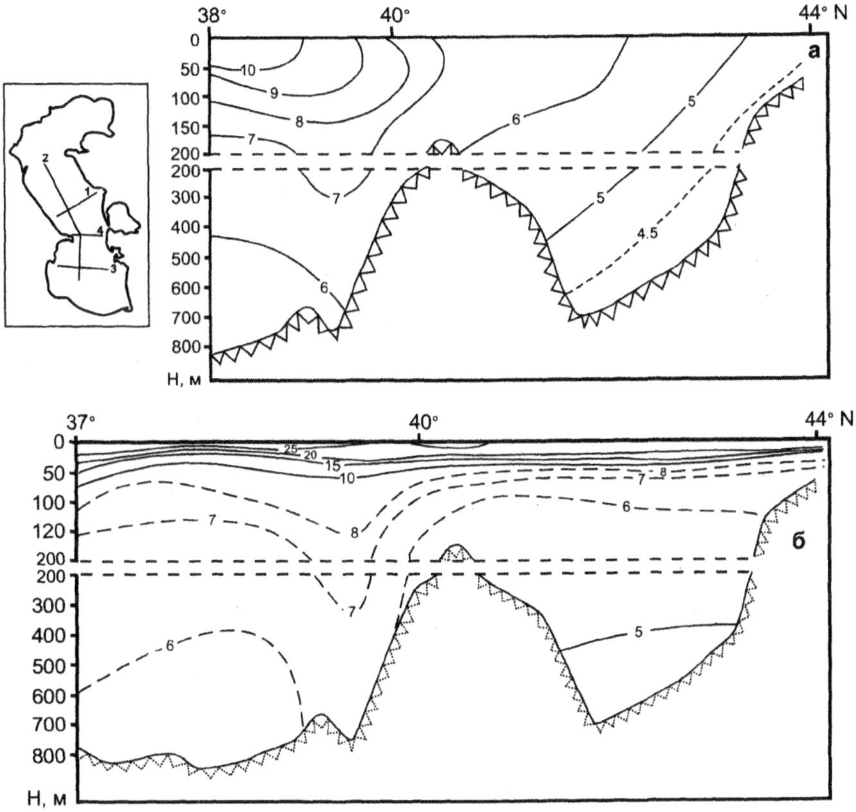

Fig. 9. Water temperature (°C) on the longitudinal transect: a – February, b – August. In the insert are shown the hydrological transects: 1 – Divichi – Kenderli, 2 – Zhiloi – Kuuli, 3 – Kurinsky Kamen – Ogurchinsky, 4 – longitudinal.

Four main components are forming the present fauna and flora of the Caspian Sea: 1, of Caspian origin, 2, of Arctic origin, 3, of Atlantic and Mediterranean origin and 4, of Fresh-water origin (Derjavin, 1912; Knipovich, 1938; Berg, 1928; Zenkevich, 1963) (Fig 10).

According to Zenkevich (1963), the Caspian Sea fauna and flora could not compete with invaders and invading fauna and flora are often killing aborigines.

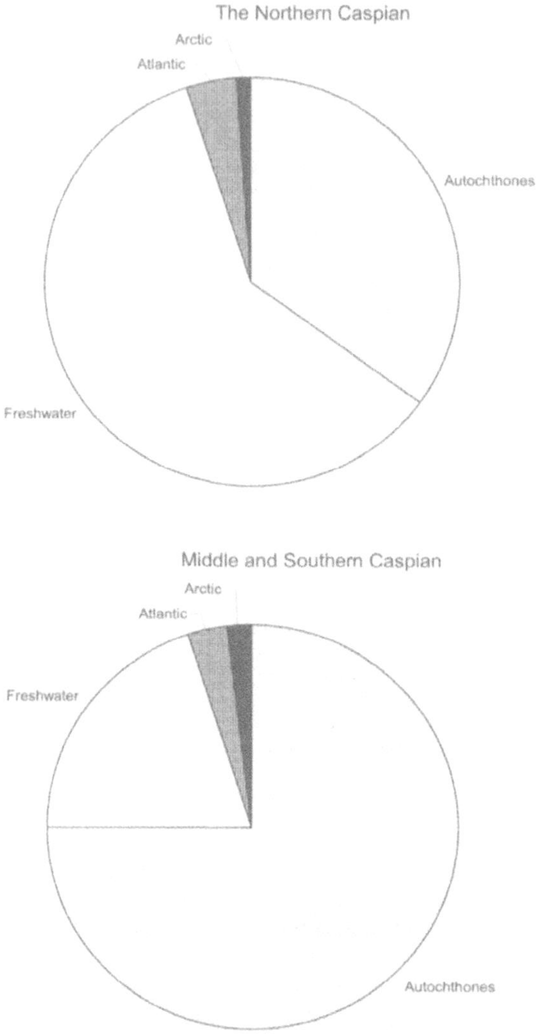

Fig. 10. Autochthones, Freshwater, Atlantic and Arctic elements in the Caspian Sea fauna.

According to data from different sources, the main hydrological features of the Caspian Sea are as following (where L, length, H, width, S, area, V, volume, D, depth):
L_{max} - 1204 km, $H_{Apsheron}$ - 204 km, H_{max} - 566 km, S - 436000 km^2, V - 77000 km^3, D_{max} -1025 m, D_{mean} - 184 m.
For Northern Caspian Sea:

V - 0.94%, S - 27.73%, D_{max} - 10 m, D_{mean} - 6.2 m.
For Middle Caspian Sea:
V - 35.39%, S - 36.63%, D_{max} - 770 m, D_{mean} - 175.6 m.
For Southern Caspian Sea:
V - 63.67%, S - 35.64%, D_{max} - 1025 m, D_{mean} - 325 m.
There is a barrier between Middle and Northern Caspian Sea with depth of no more than 200 m.
Caspian Sea basin is about 3.7 millions km^2 and it is collecting about 355 km^3 of river's water.
Volga is giving mean - 270.83 km^3 or 76.3%.
Kura - 17.22 km^3 or 4.9%.
Ural - 13.17 km^3 or 3.7%.
Terek - 11.31 km3 or 3.2%.
Other small rivers - 42.65 km^3 or 11.9%.
River outflow + Rain fall is giving 451 km^3

Biodiversity of the Caspian Sea is 2.5 times lower than that of the Black Sea, and 5 times lower than that of the Barents Sea (Zenkevich, 1963). Probably the main reason of relatively low species diversity is changeable salinity. For true fresh-water species salinity is too high, but for organisms of marine origin salinity value is too low. So, conditions in the modern Caspian Sea are favourable only for brackish-water species originating from both marine and continental water bodies (Mordukhai-Boltovskoy, 1979; Birstein, 1939). However, in our opinion, in comparison with freshwater lakes, diverse salinity conditions in the Caspian Sea serve to biodiversity increase rather than to its reduction.

Fishes and Crustacea have the largest species number among all the groups of the Caspian Sea fauna (2/3 or 63% off all recent species). The largest species diversity of these groups is determined by their osmoregulation capacities, which allow these organisms live in a wide range of salinities: from fresh water to even more saline one, than ocean water (Zenkevich,1963). Domination of these two groups in the modern Caspian Sea is another evidence that salinity strongly changed in this lake in the past and only species with good osmoregulation capacities could survive and give so good speciation and adaptive radiation. Probably speciation and adaptive radiation of species with poor osmoregulation were suppressed or decreased due to negative influence of changing salinity. Thus, modern biodiversity of the Caspian Sea reflects complicated story of Palaeocaspian transgressions and regressions followed by freshening and salinization.

The first good summary for Caspian Sea fauna and flora was published in 1963 by L. Zenkevitch, who used a lot of data published before (in 1951) by A. Derjavin. According to this two authors 718 species: 62 of Protozoa, 397 of Invertebrata, 79 of Vertebrata and 170 of Parasitic organisms are inhabiting the Caspian Sea. Without Protozoa and Parasitic organisms A. Derjavin and L. Zenkevitch recorded only 476 species of free-living Metazoa. From this species 66% originated from the surrounding Seas (including 46%, which are endemics of the Caspian Sea), 4.4% are of Atlantic and Mediterranean origin and 3% of Arctic origin.

According to A. Derjavin (1951), Mordukhay-Boltovskoy (1960) and L. Zenkevitch (1963) in the Caspian Sea fauna of Arctic origin are presented by: 1 species of Polychaeta, 1 species of Copepoda, 4 species of Mysidacea, 1 species of Isopoda, 4 species of Amphipoda, 2 species of Fishes, 1 species of Mammalia. Atlantic - Mediterranean origin fauna are presented by: 1 species of Turbellaria, 1 species of Coelenterata, 2 species of Polychaeta, 1 species of Copepoda, 2 species of Cirripedia, 3 species of Decapoda, 3 species of Mollusca, 2 species of Bryozoa and 6 species of Fishes. Endemic fauna was richest and diverse. It comprised 4 species of Spongia, 2 species of Coelenterata, 29 species of Turbellaria, 3 species of Nematodes, 2 species of Rotatoria, 2 species of Oligochaeta, 4 species of Polychaeta, 19 species of Cladocera, 3 species of Ostracoda, 23 species of Copepoda, 20 species of Mysidacea, 1 species of Isopoda, 68 species of Amphipoda, 19 species of Cumacea, 1 species of Decapoda, 2 species of Hydracarina, 53 species of Mollusca, 54 species of fishes, и 1 species of mammalia. Large number of species was of freshwater origin. The played significant role in such groups as Rotatoria, Cladocera, Copepoda and Insecta.

Fishes and Crustacea have the largest number of species among different groups of the Caspian Sea fauna.

According to A. Derjavin (1951) and L. Zenkevitch (1963) ichthyofauna included 78 species. Family Petromyzonidae comprised 1 species, Acipenseridae 5, Clupeidae 9, Salmonidae 2, Esocidae 1, Cyprinidae 15, Cobitidae 2, Siluridae 1, Gadida; e 1, Gasterosteidae 1, Syagnathidae1, Atheriiudae 1, Percidae 4, Gobiidae 30, Mugilidae 2, Pleuronectidae 1, Poeciliidae 1 species.

According to the same authors Crustacea included over hundred autochthonous species:
from the family Mysidae: 1 species of *Hemimysis*, 4 species of *Mysis*, 1 species of *Schistomysis*, 10 species of *Paramysis*, 1 species of *Caspiomysis*, 1 species of Katamysis, 1 species of *Diamysis*, 1 species of *Limnomysis*;
from the family Pseudocumatidae: 19 species;
from the other Isopoda: 3 species, including *Mesidothea entomon* of arctic origin
from the family Gammaridae: 60 species;
from the family Corophiidae: 8 species;
from the order Decapoda: 5 species, including 2 aboriginal crayfishes, 2 prones introduced by man and 1 species of crabs that was self-introduced.

The biodiversity of Mollusca is also large. From 57 up to 70 autochthonous species are recorded from the Caspian Sea. Gastropoda are including several species from 4 families: Neritidae – 2 species of *Theodoxus*, Hydrobiidae – 3(?) species, Pyrgulidae – 26 species, Planorbidae – 1 species, that have name *Planorbis eichwaldi* and that is the only one Pulmonata endemic representative in deep Caspian waters, which, strangely enough, have gone through the periods of deep-water anoxy during ancient transgression.
Bivalvia are also including several species from 2 families:
Dreissenidae – 5 species: *Dreissena polymorpha*, *D. rostriformis*, *D. caspica*, *D. grimmi*, *D. andrussovi*. Cardiidae – 13–15 species:
from genera *Adacna* – 4 species: *A. plicata*, *A. vitrea*, *A. minima*, *A. lasviucula*;
from genera *Monodacna* – 2 species: *M. caspia*, *M. edetula*;
from genera *Didacna* – 7–9 species: *D. trigonoides*, *D. pyramidata*, *D. longipes*, *D. barbot-de-marnyi*, *D. baeri*, etc.

4 species of Porifera are known from the Caspian Sea: 2 species from genera *Metschnikovia*, plus *Protoschmidtia flava* and *Amorphina caspia*. Coelenterata *Polypodium hydriforme* is a caviar parasite of sturgeons. Medusa - *Moerisia pallasi* is an endemic species from the Caspian Sea. 3 species of Polychaeta from the family Ampharetidae are known from the Caspian Sea.

According to data published in 1978 by A. Chesunov number of species of free-living Metazoa in the Caspian Sea is larger, 542.

As it was shown above, the Caspian Sea fauna and flora has originated from different sources. There are many invaders to the Caspian Sea from the Arctic: *Limnocalanus grimaldi*, *Mesidothea entomon glacialis*, *Pseudalibrotus caspius*, *Pseudalibrotus platyceras*, *Pontoporeia affinis microphthalma*, *Gammaracanthus loricatus caspius*, *Mysis caspia*, *M. microphthalma*, *M. macrolepis*, *M. amblyops*, *Stenodus leucichthys*, *Salmo trutta*, *Phoca caspia* and *Manayunkia caspia*. How and when all these organisms came to the Caspian Sea is not known exactly. Few hypotheses are available:

The first one proposed by S. Ekman (1916) and G. Sars (1927) claims that it was the result of a direct contact between Caspian Sea and Arctic Ocean. The same idea was given by Humboldt (1915). He said that it was a strait between Caspian Sea and Arctic Ocean. In the scientific literature this strait is used to be named «Humboldt Strait». Today this hypothesis is forgotten due to lack of evidence.

The second hypotheses were put forward by Grimm and Kessler (cit. by Zenkevich, 1963). They said that arctic immigrants came from the North to the Caspian Sea via fresh water streams. Today this hypothesis is supported by some, but very few data.

The third one - was proposed by Guriyanova and Pirozhnikov (cit. by Zenkevich, 1963), who suggested the Kara Sea was a source of immigrants. There is no direct evidence of this idea as well.

The last one was made by Berg. He suggested that there was «Rybnoe Ozero» («Fish's Lake») (Berg, 1928), stretched from Baltic Sea to Ladoga Lake, from Ladoga to Onega Lake, from Onega to Beloe Ozero (White Lake), from Beloe to Sheksna River. This Berg's Lake provided connection between the Caspian Sea and both Arctic and Baltic waters. Today some data are supporting this hypothesis. Some modern scientists call this hypothetical lake «Mologa-Sheksna Lake» (Zenkevich, 1963). They believe that so-called Arctic immigrants came neither from Arctic Ocean directly, no from Kara Sea, but from the Baltic and the White Seas. All invaders from Arctic show little or no sign of speciation, what could be an evidence of recent invasion or genetic conservatism of all Arctic immigrants. Some scientists: Yablokov (personal communication), Dumont (1998) and other are believe that the Caspian seal, *Phoca caspica*, is close if not identical with the Baltic *Phoca hispida*.

Planktonic crustaceans from family Cercopagidae could obviously be considered as northern immigrants as well. According to data of osmoregulation and molecular-biological studies representatives of genus *Cercopagis* is impossible to consider as typical Caspian endemics.

Actually these crustaceans are close to the group of Arctic freshwater immigrants to the Caspian Sea, than to the true autochthonous Caspian fauna. In our opinion, ancestors of modern *Cercopagis* could penetrate into residual reservoirs of Parathethys in the time of Akchagylian or Apsheronian transgressions (2.5-1.1 millions years ago). Thus, modern *Cercopagis* and *Bythotrephes*, obviously, had common ancestor lived in outskirts of Baltic glacial shield. The part of ancestors, apparently, has remained in Plaeoarctic and has given origin to *B. longimanus*, and another has got in the ancient Caspian Sea, when its waters spread far to the north up to the mouth of modern river Kama. Here, in the Akchgilian or Apsheronian Sea, the fast evolution of ancestral forms of *Cercopagis* (accelerated by the cyclomorphosis phenomena, parthenogenetic and gamogenetic cloning, hybridisation etc.) has begun. These specific evolutionary processes have obviously ensured unusually fast (some more than 1 million years) and extensive (up to 13 species) adaptive radiation in *Cercopagis* (Rivier, 1998)

Invasion of freshwater organisms into the Caspian Sea took place several times in a periods when its salinity was lowest. The most ancient invaders among freshwater species are considered to be gastropods. The main ion composition, which differed essentially from that in the ocean, facilitated this process in definite degree.

Atlantic and Mediterranean immigrants came to the Caspian Sea 3 times:

1. The earliest immigrants came during Hvalynian time as early as 50 thousands years B.P. They came naturally via Kuma-Manychenskiy strait between Black and Caspian Seas. They are 7 species: *Zostera nana*, *Cardium edule*, *Fabricia sabella*, *Atherina mochon pontica*, *Syngnathus nigrolineatus*, *Pomatoschistus caucasicus*, *Bowerbankia imbricata*.
2. Next wave of immigrants came in the beginning of the XX th century or a little later. Some of the invasions were natural and some were man-made. Naturally, without man and only by chance (probably together with the water in small wooden boats) invaded 4 species: *Rhizosolenia calcar-avis*, *Mytilaster lineatus*, *Leander squilla*, *Leander adspersus*. *R. calcar-avis has* appeared in the Caspian Sea in 1930, but already in 1936 gave 2/3 of total phytoplankton biomass. Deliberately by man were introduced 5 species: *Mugil auratus*, *Mugil salies*, *Pleuronectes flesus luscus*, *Nereis diversicolor*, *Abra ovata*.
3. Last came immigrants at the middle XX th century. All of them came to the Caspian Sea by Volga-Don canal that was opened in 1954. It was only occasional (probably with boat ballast water or in fouling) invasions. In this period more than 9 algae species and 9 invertebrate

species occupied Caspian. These were algae Acrochaeta parasitica, Ectochaeta leptochaeta, Enteromorpha tubulosa, E. salina, Ectocarpus confervoides v. fluviatilis, Entonema oligosporum, Acrochaetium deviesii, Ceramium diaphanum, Polysiphonia variegata. and invertebrates Moerisia maeotica, Bougainvillia megas, Blackfordia virginica Barentia benedeni, Mercierella enigmatica, Balanus impovisus, Balanus eburneus, Rhithropanopeus harrisii, Conopeum seurati.(Karpevich, 1975). Shortly after invasion *Ceramium diaphanum* became a dominant species in Northern Caspian Sea. Later on some more species invaded Caspian Sea by Volga-Don canal. Among them, 2 species of marine Cladocera: *Pleopis polyphemoides* and *Penilia avirostris* were registered by Mordukhai-Boltovskoi (Mordukhai-Boltovskoi, Rivier, 1987) and Aladin (Harris, Aladin, 1998).

As it was shown earlier, deliberate acclimatization of industrial and food aquatic organisms affected essentially biodiversity of the Caspian Sea. According to data of Karpevich (1975), in 1930-1970 at least 9 species of fishes were introduced successfully into the Caspian Sea (*Pleuronectes flesus luscus*, *Rhombus maeoticus*, *Mugil auratus* and *M. saliens*, *Ctenopharingodon idella*, *Hypophthalmichtys molitrix*, *Aristichthys nobilis*, *Oncorhyncus keta*, *Oncorhynchus gorbuscha*). There were some other species introduced (*Salmo gairdneri*, *Morone saxatilis*, *Mugil so-iuy*), but results of these introductions are still uncertain. Among invertebrate species polychaetes *Nereis diversicolor*, mussels *Abra ovata* and shrimps *Palaemon elegans* (*Leander squilla*) were brought here from the Azov-Black Sea basin and successfully acclimatized.

It is possible to suggest that some more species from Atlantic-Mediterranean are invading Caspian Sea right now and scientists will register this process in the nearest future. Ctenophore that just invaded to the Caspian Sea (Ivanov et al., 2000) could become a reason of huge ecological catastrophe in this lake. This species negatively influenced fishery industry in the Black Sea, basically cutting down stocks of fishes that are preying on zooplankton. Some researchers from Azerbaijan and Turkmenistan has already informed about another Ctenophora (*Beroe ovata*) findings in the Caspian Sea.

Though these data were neither published nor confirmed properly: by photos, figures or caught and fixed individuals. Besides ctenophores one more medusa (*Aurelia aurita*) was recently found. It is not certain whether

Beroe ovata and *Aurelia aurita* are naturalized in the Caspian Sea. What is absolutely certain is catastrophic population explosion of *Mnemiopsis leidyi*. According to personal communication of colleagues from Iran at present every litre of surface caspian water near Iranian coast contains more that 10 g of *Mnemiopsis*. Not only fishery industry but also the whole ecosystem of the Caspian Lake is in big danger due to *Mnemiopsis* invasion (Aladin et al., 2002).

Recently two additional species of Black Sea origin invaded epilimnion of the Caspian Sea. These species are planktonic Copepoda: *Acartia clausi* and *A. tonsa*. Invaders from the Atlantic and Mediterranean will obviously continue to appear in the Caspian Sea until a new balance between aborigines and invaders will appear in the coming years or perhaps decades. It is very difficult to say when this balance will appear because at present the environment in the Caspian Sea is not stable due to changes in climate, anthropogenic pollution and some other important aspects of the environmental impact. Thus changes in the abiotic and biotic components of the ecosystem of the Caspian Sea support invaders.

Last year one more Cladocera (*Moina mongolica*) appeared in central water area of Northern and Middle Caspian Sea. It is not known whether this planktonic organism will naturalize in this lake or not. It is quite possible that resting eggs of this Cladocera were transported by wind from nearby salt lakes. In the Table 1 the complete list of invaders to the Caspian Sea is given.

There are very many immigrants from the Caspian Sea into different basins. According to Birshtein (1939) 44 species of Caspian invertebrates are recorded in the Volga river: 1 species of Isopoda, 26 species of Amphipoda, 10 species of Cumacea, 6 species of Mysidacea, 1 species of Decapoda. 18 species of fishes came to Volga from the Caspian Sea as well. The most famous immigrants from the Caspian Sea into fresh water are: *Cordylophora caspia*, *Polypodium hydriforme*, *Dreissena polymorpha*, *Hypania invalida*, *Hypania kovalewskyi*, species from the genera *Theodoxus* and *Melanopsis*. Nowadays more and more species from the Caspian Sea are conquering adjusted water bodies.

Many of Caspian Sea species are invading Baltic. Hydrozoa without medusa stage, *Cordylophora caspia,* invaded the Baltic Sea last century through canals and later spread to other places. Probably this species was the first invader from the Caspian Sea. Today this hydrozoan is a cosmopolitan species, living nearly everywhere. Not only *Cordylophora caspia* but also *Victorella pavida* became a cosmopolitan species today. Both organisms are living today in coastal waters of North and South America, China, Australia and New Zealand. Recently fish Roundhead goby, some Cladocera from the genera *Cercopagis* and some Mysidae were recorded from the Baltic Sea. From Baltic Sea Mysidae are going into fresh-water and into drinking-water reservoirs in Europe, while *Cercopagis* went to Lake Ontario.

Dreissena polymorpha also came to Europe and America from the Caspian Sea. It has happened in the beginning of XIX century, after the Dnieper-Bug canal was open in 1803. Already in 1824 Dreissena was found in England, then in 1826 in Holland, in 1833 in France, in 1835 in Germany and in 1892 in Spain and Portugal. Recently in the XX century Dreissena came to America and caused many problems as a result of this invasion (Starobogatov, 1994).

One species of Caspian Sea polychaetes, *Hypaniola graizi* from the family Ampharetidae was found near Woods Hall in USA coastal waters.

In our notion the real biodiversity of the Caspian Sea is still very poor investigated for both plants and animals. There are still a lot of new species and subspecies in it, expecting the description. Many true marine animals like: Scyphozoa, Anthozoa, Ctenophora, Gordiacea, Gastrotricha, Kinorhyncha, Sipunculida, Phoronidea, Loricata, Scaphopoda, Tanaidacea, Pantopoda, Tardigrada, Asteroidea, Ophiuroidea, Echinoidea, Holothurioidea, Chaetognatha, Ascidiacea, Appendicularia and Acrania are not known from the Caspian Sea. But some of these organisms are capable to osmoregulation and being most euryhaline can penetrate into the Caspian. Many freshwater and brackish water species are living in the estuaries of rivers flowing into the Caspian Sea. If we include these species to the species list and will find new yet unknown species, the diversity of the Caspian Sea could will appears to be larger than that of Baikal Lake. Zenkevitch (1963) mentioned some 450 species of free-living Caspian Metazoa, Chesunov (1978) pointed out some 550 species, Kasymov (1987) recorded some 950 species. It is possible to imagine that the real number of free-living Metazoa is some 1500-2000 species (Aladin et al., 2001). Speciation in the Caspian Sea has resulted in overall high level of endemism (about 42-46%), which is somewhat lower than that of Baikal Lake (about 54%). The smaller number of endemics in the Caspian Sea is possibly connected with extinction of the Caspian deep-water fauna. There is high probability that after special studies in the Caspian Sea number of found endemic species could increase and reach that of the Baikal Lake. In our opinion, only appearance of deep-water anoxy conditions in the ancient Caspian Sea during transgression has sharply reduced number of Caspian endemics. Only in the Caspian Sea some animal groups which underwent big adaptive radiation show levels of endemism close to 100%. If Caspian deep-water endemics had survived, today the Caspian Sea biodiversity would probably be highest in the world among continental reservoirs.

The complicated story of Caspian Sea natural resources exploitation (both biological and mineral) in 20th century is shown in Annex.

This study was supported by Russian Fund for Basic Research 01-04-49550.

Annex

Chronology of events in XX century influencing exploitation of fish and other biological riches of the Caspian Sea, and events influencing exploitation of its oil-and-gas mineral riches. Composed with reductions and additions on materials published in "The Caspian Bulletin" No. 6 (20) 1999.

Years	Events influencing exploitation of fish and other biological riches	Events influencing exploitation of its oil-and-gas mineral riches
1900		Construction of 26 oil pipelines (total length 277 km) connected Balakhany industrial area with oil refining factories in Baku is completed.
		Azerbaijan had occupied the first place in the world on production of petroleum — 11,500,000 tons per one year.

1901	Fishing in the open Caspian Sea waters has began in Azerbaijan.	Huge fire on Apsheron petroleum deposits lasted five days and night. 96,000 tons of black oil has burned down.
1902	An attempt to introduce flounder (*Platichthys flesus flesus*) in the Caspian Sea was unsuccessful.	
1904	Ichthyological laboratory in Astrakhan for studying zooplankton and benthos of the Caspian Sea northern part was founded.	
	The First is scientific-trade Caspian expedition headed by N. M. Knipovitch.	
1905		On wellsites in Balakhany, Azerbaijan, for the first time at their exploitation, use of compressors was begun.
1906		Putting kerosene pipeline Baku-Batumi (length 882 km, diameter of 203 mm, with 16 pump stations) into operation.
1908		On Cheleken, Turkestan, for the first time from depth 140 m petroleum fountain with the daily debit 24.5 tons is received.
1909		Works on drainage of a gulf Bibi-Eibat about Baku with the purpose to prevent flooding of petroleum deposits by waters of the Caspian Sea have begun.
1910		In the gulf Kara Bogaz Gol taking coastal deposits of mirabilite (Glauber's salt) was began.

1911	Hydrometeorological service of the Caspian Sea consisting of seven hydrometeorological stations and the central station in Petrovsk (Makhachkala) is organised.	The oil production beginning in Kazakhstan on Emba oil field Dossor. For the first time in Surakhany, Azerbaijan, rotary drilling was applied.
	A joint-stock company "Astrakhan refrigerator" — one of the largest firms providing storage and transportation of perishable food was established.	
1912 — 1913	The Second scientific-trade Caspian expedition headed by N. M. Knipovitch. Baku ichthyological laboratory was established.	
1913	The first self-propelled transport refrigerator has begun to serve fish crafts on the Caspian Sea.	
1914 — 1915	Scientific-trade Caspian expedition headed by N. M. Knipovitch continued work. The huge material on hydrology of the Caspian Sea was collected.	
1915		On oil well in Ramany about Baku, Azerbaijan, deep pumps were used for the first time.
1918	Committee on study Kara Bogaz Gol headed by academician N. S. Kurnakov was established under VSNH.	
1919	The Astrakhan Reserve — the first in USSR is established.	
1920		Azerbaijan polytechnical institute — the first in Europe and Asia centre for education of engineers — oil industry workers was established.

1921 The first volume of "Proceedings of the Caspian expedition 1914–1915" and monography of N. M. Knipovitch "Hydrological studies on the Caspian Sea in 1914-1915" were published.

Issuing the books by N. M. Knipovitch "The Caspian Sea and its crafts" in the series "Studies of nature and crafts of Russian seas".

Scientific expedition headed by N. S. Kurnakov was sent to Kara Bogaz Gol. It has made a number of works on studying hydrology, meteorology and hydrochemistry of gulf with the purposes of finding out natrium sulphate production.

1922 Department of safety navigation on the Caspian Sea has established the Baku sea observatory.

1924 Beginning of realisation on the Caspian Sea seasonal (February, May, August, November) hydrological surveys on standard sections (Baku-Krasnovodsk-Ashurade-Astara-Baku), thus execution of the obligations to the International Council for studying the seas were renewed.

Experiments on artificial reproduction of kutum on the river Kumbashinka, Azerbaijan, and in Samur piscicultural station, Dagestan were begun.

Industrial production of natrium sulphate in many sites of gulf Kara Bogaz Gol coast began.

The first in the world petroleum from the Caspian Sea shelf — gulf Bibi-Eibat – is received.

1925 In the system of Caspian Shipping Company in Baku a special Bureau for studying currents of the Caspian Sea is created under guidance of N. N. Struisky (existed till 1929).

Catastrophic frosts (-25°C) in Lenkoran lowland of Azerbaijan. Kirov gulf (nowdays Kyzylagach gulf) has frozen. Death of many water birds (flamingo died in thousand).

Beginning of trade fishery of Caspian sprat.

Beginning of experimental works on mastering sea deposits of petroleum on the Caspian Sea. Under the initiative of S. Kirov in Ilyich bay on artificial island the sea well was layed.

1926	Maximal in XX century Volga run-off— 390.6 km^3 per one year.	
1927	Agreement between Iran and Russia about granting to Russia the right on exploitation of fish resources at the southern coast of the Caspian Sea. On the basis of the Agreement the first Soviet-Iranian fishery company "Shilat" was established (existed till 1953).	
1928		VSNH of USSR has accepted decision about increasing chisel investigation on Cheleken, which results have confirmed the forecast of academician I. M. Gubkin about presence of petroleum stocks in Turkmenistan.
1929	On the basis of Astrakhan Ichthyological Laboratory the Volga-Caspian Scientific Fishery Station was established. Kyzylagach Reserve (in the same gulf of the Caspian sea, Azerbaijan Republic) was founded for protection of wintering aquatic and semi-aqutic birds.	Trust "Karabogazsulphate", carrying out industrial manufacturing (production and processing of Glauber's salt and natrium sulphate from surface brains of gulf) was organized.
1930	Department of United Hydrometeorological Service of Azerbaijan SSR and the Caspian Sea was established. Fishery studies of Caspian herring expedition. Beginning of works on the resettlement of Black Sea mullets — *Mugil auratus* Risso and *M. saliens* Risso in the Caspian Sea. Microscopic plankton alga *Rhizosolenia* penetrated into the Caspian Sea. In 1934–1936 it has received so mass development, that any other representative of Caspian aboriginal phytoplankton flora could not be compared with it.	

1930-1931	Negative result of attempts to introduce kalkan flounder (*Psetta maeotica*) in the Caspian Sea.	
1931-1932	All-Caspian scientific fishery expedition worked (till 1934).	
1932-1934	Unsuccessful attempt of khamsa anchovy (*Engraulis ensrasicolus*) and sultanka (*Mullus barbatus*) acclimatization in the Caspian Sea.	
1933	Session of Acad. Sci. of USSR specially devoted to the problem of Volga-Caspian. It was recognised necessary to carry out developing problem of partial transfer of northern rivers run-off into the basin of Volga and the Caspian Sea. Volume of transfer was determined as 50 km^3.	
	Gasankuli Reserve in Turkmenian SSR was established for protection and studyihg wintering places of aquatic and semi-aquatic birds (includes all shoals of the Southeast Caspian).	
1934	Commission for complex study of the Caspian Sea was established under Acad. Sci. of USSR (CASP). Its task was co-ordination of all works carried out by academic institutions on studying the Caspian Sea and consulting other organisations on these problems. Commission was headed by N. M. Knipovich.	On Artem island, Azerbaijan, sectional drilling, at which some wells are holed from one platform, was carried out for the first time.
	XVII Congress of CPSU(B) — "Congress of the winners". Approval of the Second five years' plan. I. V. Stalin has told: "the task is to begin serious work on organising Trans-Volga region irrigation". After that there was an idea of academician I. G. Alexandrov to isolate gulfs Kara Bogaz Gol and Komsomoletz from the Caspian Sea.	
	Catching mullets in the Caspian sea was allowed, as it has got industrial importance.	

	L.S. Berg has come to conclusion, that in all historical period the Caspian Sea level did not rise even by 5 m above the level of 1925 (-26 m). The first scientific expedition to the ice of the Northern Caspian.	
1935		I. M. Gubkin has put forward and has proved an idea of the sea areas investigation for petroleum. In the area near Artem island, Azerbaijan, the first on the Caspian Sea metal basis for chisel works was build. Development of large petroleum deposit Izberbash (Dagestan) has begun.
1937	Beginning of the State Hydrological Institute works under supervision of B. D. Zaikov on the studying the Caspian Sea level dynamics. Minimal in XX century run-off of Volga — 149.6 km^3 per one year.	
1938	New method of hypophysis injections of sturgeons elaborated in St. Petersburg University by Prof. N.L. Gerbilsky. This method allowed mass reproduction of sturgeons in fish farms.	
1939	XVIII Congress of CPSU(B). Approval of the Fird five years' plan (1938–1942). The problem of the causes of the Caspian Sea level fluctuations received special importance. XVIII Congress has decided: "In the third five years to develop plan of complex reconstruction of the rivers: Volga, Don and Dnieper to prepare measures to maintain the Caspian Sea level and to begin constructing Volga-Don connection".	

	Beginning of extensive experimental researches under the guidance of Prof. L. A. Zenkevich on studying (acclimatization) Azov and Black Sea fauna; the group of species for resettlement from the Azov Sea into Caspian was planned, among them on the first place (61,000 individuals) was polychaete worm - *Nereis diversicolor* Muller фы a perfect forage for sturgeons and other fishes (casually instead of planned for acclimatisation *N. succinea*).	
1940	Work on resettlement of 18,000 individuals of mollusc *Abra ovata* Phil. in the Caspian Sea.	
1941		Azerbaijan has extracted 23,500,000 tons — record for all history of oil-extracting industry in USSR.
1943		Beginning of Makhachkala petroleum deposit, Dagestan, development.
		Start-up of oil refining factory in Krasnovodsk, Turkmenistan, on the basis of evacuated Tuapse factory equipment.
1946	By instructions of USSR Government B. D. Zaikov has carried out fundamental studies (was begun in 1937) of the Caspian Sea level decrease causes (by 1.75 m) for period 1930–1945.	The first oil pipeline in Turkmenia – Kumdag-Vyshka (length 40 km) was put into operation.
1947		Oil field "Petroleum Stones" was discovered on the Caspian Sea in Azerbaijan.

1948	Decision of Council of Ministers of USSR and CC CPSU(B) "About the plan of shelterbelt forests, introduction of grassland crop rotation, construction of ponds and reservoirs for maintenance of high and steady crops in steppe and forest-steppe areas of the European part USSR", so named "Stalin's plan of nature transformation". Volga-Caspian scientific-fishery station was transformed into the branch of VNIRO.	Decision of Council of Ministers of USSR "About measures intensifying petroleum deposits investigation in Turkmenian SSR".
1949		The first sea petroleum from the shelf deposit on the Caspian Sea "Petroleum Stones" was received. Beginning readout of the history of sea petroleum production in Azerbaijan Republic.
1951	B. A. Apollov has drawn conclusion that in the past the level of the Caspian Sea stood below the modern and lower levels in general are peculiar to it, than higher.	
1952	Opening of the Volgodonsk navigable channel in the name of V. I. Lenin (Length 101 km, 9 sluices, total height of rise — -8 m). The most powerful registered water flood on the Caspian sea. As a result in some cities at the coast water has risen by 4.5 м, and between mouths of Volga and Ural the sea has move forward into the continent by 30 km (November 11).	
1959		Discovery of Okarem petroleum and gas deposit, Turkmenia.

1961	Hydrometeorological service of USSR has accepted decision to give characteristics of the Caspian Sea level by its exceeding above the uniform zero of post, which for Caspian is equal -28 m (average weighted sea level for Pleistocene time) relatively the zero Kronstadt tide-gauge ("the Baltic system of heights", 1950).	
1963		The large petroleum deposit Sangachaly-Duvanny, Azerbaijan is discovered.
1964	Industrial catch of sturgeon in the open (Caspian) sea was prohibited.	
	Reconstruction of the Volga-Baltic way, connected the Northwest with the central and southern basins (the Caspian Sea), is completed.	
1965	The Caspian Fishery Research Institute (KaspNIRH) in Astrakhan is established.	
1966	The first tests of airfoil boats on the Caspian Sea. In the West they were named "Caspian Monsters".	
1968	Caspian ornithological station was established (in the system of Astrakhan Reserve).	For the first time on the Caspian Sea the self-elevating drilling rig was used.
	Decision of Council of Ministers of USSR "About measures for Caspian pollution prevention" was accepted.	
1970		The Ministry of Petroleum Industry of USSR has divided the Soviet part of the Caspian Sea on sectors between Azerbaijan, Kazakhstan and Turkmenistan. In a basis of definition the medial line was taken.

1971	Acceptance "Convention (Ramsarian) about water-marsh lands, having the international importance, mainly as inhabiting places of water birds" (Ramsar, Iran) with the purpose of preservation of the world water-marsh lands (Russian Federation has ratified it in 1975).	
1972		On the Caspian Sea there are 1880 steel islands, and also piers of length more than 300 km.
1973	The first in the world nuclear water-electric power station with desalter of productivity 120 thousand m^3 per day was put into operation in Shevchenko city, Kazakhstan.	
1974	Kazakhstan has accepted the decision establishing in the northern part of the Caspian Sea reserved zone. Development of mineral resources in coastal zone was forbidden.	
1975	Decision of Council of Ministers of USSR about the statement of a reserved zone in the northern part of the Caspian Sea.	
	Decision of Council of Ministers of USSR "About measures on maintaining performance of Soviet obligations following from Ramsar Convention".	
	Scientific centre " Caspian" and Caspian Biological Station were established In Azerbaijan.	
1976	Decision of CC CPSU and the Council of Ministers of USSR "About realisation of complex researches for scientific substantiation of volumes and sequence of works on territorial redistribution of water resources".	Astrakhan gas-condensate field (balance taken stocks of free gas made at 01.01.92 – 2,695,000,000,000 m^3, condensate – 425,000,000 tons) was discovered.

	At XXY Congress of CC CPSU the First secretary CC CP of Turkmenistan M. P. Gapurov has raised the question about necessity to regulate the Caspian waters run-off into gulf Kara Bogaz Gol.	
1977	The lowest level (-29.03 m) of the Caspian Sea in XX century. Council of Ministers of USSR has accepted Decision "About additional measures on the protection of the Caspian Sea against pollution". Complete termination of wastewater dump into the rivers of the Caspian Sea basin to 1985 was provided.	For the first time on the Caspian Sea deep-water (84 m) a stationary sea platform for drilling on petroleum was established.
1978	For the Caspian waters economy with the purpose of the sea level maintenance decision about separation of gulf Kara Bogaz Gol was accepted. The state commission of experts of USSR State Planning Committee has rejected the feasibility report on hydroscheme in strait Kara Bogaz Gol and recommended blocking it with solid dam.	Accident on shelf oil field Bakhar, Azerbaijan.
1979	On the Caspian Sea the System of Heights of 1977 year relatively the zero of Kronstadt tide-gauge is entered.	Discovery of the largest in the world Tengiz oil field in Western Kazakhstan at the Caspian Sea coast (taken stocks 781,100,000 tons).
1980	Gulf Kara Bogaz Gol has been blocked with solid dam (level of brine in the gulf by 3.5 m below the Caspian level). Area of the gulf were 9500 km^2, volume of surface brine — 20–22 km^3, average depth — 2.1 m (here and there 3–3.5 m).	
1981	Beginning discussion of the projects on the Caspian sea "rescue" from predicted shallowing.	

1982-1983 "Scheme of complex use and protection of Volga basin water resources" is created. It was based on the concept of transferring part of northern rivers run-off, to 2000 year the putting of canals Volga-Don II (5.5 km^3), Oka-Don (1.4 km^3) and Volga-Ural (17.3 km^3) into full capacity operation was planned. A question on construction of the canal Volga-Chograi simultaneously was put.

1983 Decision of Council of Ministers of USSR "About measures on creation of resort base of general importance at the coast of the Caspian Sea".

1984 The governmental decision is accepted on renewal of sea water drain into gulf Kara Bogaz Gol (volume 2 km^3 per year) constructing temporary water releasing constructions (surface brine to this time has dried up completely, and the gulf has turned to the dry saline lake).

Appearance of first signs of miopatia in Volga-Caspian sturgeon fishes — exfoliation of muscles connected to metabolic disease in muscular tissue.

1985 During 14 months the open flowing of oil well No. 37 of Tengiz oil field, Kazakhstan (there were thrown out into atmosphere more than 4,000,000 tones of petroleum, 2,500,000,000 km^3 of gas with dangerous concentration of hydrosulfuric compounds).

1986 Decision of CC CPSU and the Council of Ministers of USSR "About stopping work on transferring a part of northern and Siberian rivers run off".

	Beginning the edition of book series "The Caspian sea" Acad. Sci. USSR, GKNT USSR, Scientific Council on the complex study of the Caspian Sea problems, IWP Acad. Sci. USSR (first book " Fauna and biological productivity"). During period 1986–1990 6 books were published.	
1987	Disease of Volga-Caspian sturgeons with miopatia has got mass character.	
	The interdepartmental scientific program "Sturgeon" was accepted.	
	Dagestan Reserve for studying and protection of Kizlar gulf of the Caspian Sea and unique natural formation Barkhana-Sarikhum is founded.	
1989	Organisation of the Research and Coordination Centre "Kaspiy", Moscow, Russia.	
1991	By the Decision of Commission of Council of Ministers of RSFSR on extreme situations, Russian Committee for Water Management together with Acad. Sci. USSR and other organisations were charged to prepare "Technical and Economic Report on protection of economic objects and settlements in the coastal strip within the limits of Dagestan ASSR, Kalmyk ASSR and Astrakhan region against flooding in the connection with the increase of the Caspian Sea level".	
1992	Technicoeconomic report "The Caspian Sea".	Company "Kazkhstankaspiyshelf" was established.
	Decree of President of Russian Federation "About measures on protection of sturgeon fishes in the Caspian basin".	

	Decree of President of Russian Federation "About state support of revival of the Russian merchant marine fleet on the Caspian Sea" and special Decision of Government of Russian Federation on its performance.	
	Decree of President of Russian Federation "About measures on protection of the population and decision of problems connected to the rise of the Caspian Sea level".	
	Commission (nongovernmental organization) on water bioresources of the Caspian sea (Azerbaijan, Kazakhstan, Russia) is established.	
1993	Decision of Council of Ministers — Government of Russian Federation "About urgent measures in 1993–95 preventing flooding and underflooding of cities, settlements, industrial and non-productive objects, agricultural grounds and other valuable lands located in the coastal strip of the Caspian Sea".	
1994	Decree of Government of Russian Federation about development and realisation of the Special Federal program "Improvement of ecological conditions on the river to Volga and its tributaries, stopping and prevention degradation of natural complexes of Volga basin", so named program "Revival of Volga".	The first petroleum contract — " Contract of century" on development of oil fields — Chirag, Azeri, Guneshli was signed.
	Interstate Ecological Council of the CIS countries has addressed to the Program UN on environment (UNEP) with the request for rendering assistance in development of the Frame Convention on the protection of sea environment.	
1995	The rise of the Caspian Sea level has reached the maximal mark (-26.61 m) at the end of XX century.	

	Scientific Council on the problem "Complex use and protection of natural-economic resources of the Caspian Sea and its Basin" (Ministery of Science of Russian Federation, Presidium of Russian Acad. Sci., Russian Committee for Water Management) is established.	
1996	United mission of experts of the World Bank, UNEP and UNDP in frameworks of "the Caspian initiative" on the Caspian countries with the purpose of preparation of the concept of international Caspian ecological program (CEP) and creation of the legal tool on protection and steady use of the Caspian Sea.	Azerbaijan has signed petroleum developmental contracts on deposits of the Caspian Sea — Shah-Deniz, Danulduzu and Ashrafi.
	Decision of Government of Russian Federation "About prime measures on 1996–1997 on protection of population and prevention of flooding economic and other objects located at the coast of the Caspian Sea".	
	Between Russia and Islamic Republic of Iran an agreement in the field of fishery was signed.	
1997	In the Appendix II of "Convention on international trade by wild species of fauna and flora, to which the disappearance menaces" sturgeon fishes are included.	Azerbaijan has signed petroleum developmental contracts of deposits of the Caspian Sea Lenkoran-Deniz and Talysh-Deniz, Yalama, Apsheron (former Zeinalabdin Tagiev), Nakhichevan (former D-3).
	In the frameworks of Ministry of Foreign Affairs of Russian Federation the working group on the Caspian Sea is created for preparation of offers about position of Russian Federation concerning the status and regime of the Caspian Sea, exploitation of its alive and mineral resources, and also on the problems of the Caspian petroleum transportation.	

Publication of the first release from the series "Geoecology of the Caspian Sea", generalizing long-term results of scientific researches of Geographical Faculty of Moscow State University.

1998 Agreement between Russian Federation and Republic of Kazakhstan about delimitation of the northern part of the Caspian Sea bottom with the purposes of realisation of sovereign rights on the bowels management.

Decision of Government of Russian Federation "About prolongation on 1998-2000 terms of realisation of prime measures on protection of the population, economic and other objects located at the coast of the Caspian Sea.

Decree of Government of Russian Federation about preparation of the project of special federal program on complex development of territory of Russian Federation, adjoining to the Caspian Sea.

Decision of Government of Russian Federation "About measures on performance of the recommendations of 10th conference of the states which have signed "Convention on international trade by species of wild fauna and flora, being under menace of disappearance" from March 3, 1973 in the relation to the species of sturgeon fishes".

The final variant of the concept, which is in the basis of the Caspian Ecological Program (CEP) is accepted. Beginning of CEP realization. Organization in of the Caspian regional thematic centres (CRTC) the Caspian countries.

Azerbaijan has signed petroleum developmental contracts on deposits of the Caspian Sea: block Kyurdashi, blocks Gobustan and Hyursangi, Gabogly (by land), Inam, Abikh.

New operational company OKIOK, Kazakhstan, is established. Realisation of chisel works on the Kazakhstan shelf of the Caspian Sea is assigned to it.

1999	Putting into operation oil pipeline Baku (Azerbaijan) — Supsa (Georgia) for transportation of "early" Azerbaijan petroleum to the external markets. Drilling of the first oil well on the Kazakh shelf of the Caspian sea (East Kashagan) was began.

References

Aladin N.V., Plotnikov I.S., Filippov A.A. 2002. Invaders in the Caspian Sea. In: Invasive aquatic species of Europe. Distribution, impacts and management. Pp. 351-359.

Berg L. S. 1928. About origin of northern elements in the Caspian fauna. In: Proc. Acad. Sci. USSR. 14, 3. (in Russian)

Birstein Ya. A. 1939. Benthos of the Northern Caspian. In: Zool. Zhurnal. 18, 3. (in Russian)

Chesunov A. V. 1978. New species of free-living nematodes from the Caspian Sea. In: Zool. Zhurnal. 57, 4. Pp. 505-511. (in Russian)

Derjavin A. N. 1912. Caspian elements in the Volga basin fauna. In: Proc. Ichthyol. Lab. of Volga-Caspian Fishery and Seal-hunting. 2, 5. (in Russian)

Derjavin A. N. 1951. Animal world of Azerbajan. Baku: Acad. Sci. AzSSR. (in Russian)

Dumont H. J. 1998. The Caspian Lake: History, biota, structure, and function. In: Limnol. Oceanogr., 43 (1), pp. 44-52.

Ekman S. 1916. Systematische und tiergeographsche Bemerkungen uber einiger glacialmarine Relikte des Kaspischem Meeres. Zool. Anz., 47.

Harris, R. R., N. V. Aladin, 1998. The Ecophysiology of Osmoregulation in Crustacea. Ionic Regulation in Animals: A Tribute to Professor W. T. W. Potts. Springer. Pp. 1-25.

Humboldt A. 1915. Central Asia. (in Russian)

Ivanov V.P., A.M. Kamakin, V.B. Ushintzev, T. Shiganova, O. Zhukova, N. Aladin, S.I. Wilson, G.R. Harbison, H.J. Dumont. 2000. Invasion of the Caspaian Sea by the comb jellyfish Mnemiopsis leidyi (Ctenophora). Biological Invasions 2. Pp. 255-258.

Karpevich A. F. 1975. Theory and practice of aquatic organisms acclimatisation. M.: "Pischevaya promyshlennost". 432 pp. (in Russian)

Kasymov A. G. 1987. The animal kingdom of the Caspian. Elm Publishers, Baku (in Russian)

Knipovich N. M. 1938. Hydrology of seas and brackish waters. M.-L.: "Pischepromizdat". (in Russian)

Kosarev A. N., Tuzhilkin V. S. 1995. Climatic thermohaline fields of the Caspian Sea. M., 91 pp. (in Russian)

Kosarev A. N., Yablonskaya E. A. 1994. The Caspian Sea. SPB. The Hague, 259 pp.

Mordukhai-Boltovskoi F. D. 1960. Caspian Fauna in the Azov and Black Sea Basin. M: Acad. Sci. USSR. (in Russian)

Mordukhai-Boltovskoi F. D., Rivier I. K. 1987. Predatory cladocerans Podonidae, Polyphemidae, Cercopagidae and Leptodoridae of the World Fauna. L.: "Nauka". 184 pp. (in Russian)

Mordukhai-Boltovskoy P. D. 1979. Composition and distribution of Caspian Fauna in light of modern Data. Int. Revue ges. Hydrobiol. 64: pp 1-38.

Rivier, I. K. 1998. The predatory Cladocera (Onychopoda: Podonidae, Polyphemidae, Cercopagidae) and Leptodoridae of the World. In Dumont H., (ed.), Guides to the Identification of the Micro-invertebrates of the Continental Waters of the World 13: 213 pp. Backhuys Publishers, Leiden.

Sars G. O.1927. Notes on the Crustacean fauna of the Caspian Sea. In: A Tribute to Prof. N. M. Knipovich. M.: VNIRO. (in Russian)

Starobogatov Ya. I. 1994. Systematics and palaeontology. In: Dreissena (zebra mussel) Dreissena polymorpha (Pall) (Bivalvia, Dreissenidae). Pp. 18-46. (In Russian)

Zenkevich L. N., 1963. Biology of the USSR seas. M.: "Nauka". (in Russian)

Table 1. Invaders in the Caspian Sea.

Species	Ecol. group	Origin	Naturalization	Region-donor	Way	Years
	Coelenterata					
Blackfordia virginica Mayer	Bf	NA	+	Bl-Az	Sh	1956
Bougaenvillia megas (Kinne)	Bf	NA	+	Bl-Az	Sh	1956
Moerisia maeotica (Ostrumov)	Bf	Bl	+	Bl-Az	Sh	1950s?
Aurelia aurita (L.)*?**	P	Bl	?	Bl-Az	Sh	1999
	Ctenophora					
Mnemiopsis leydii (A. Agassiz)	P	NA	+	Bl-Az	Sh	1999
Beroe ovata Mayer	P	NA	+?-	Bl-Az	Sh	2000
	Turbellaria					
Pentacelum caspium Bekl.	B	Med.	?	?	?	1940-?
	Trematoda					
Apophallus muehlingi (Jagerskiold)	Pa	Bl	+	Bl-Az	H	1980-90s?
Plagiorchis entamiatis Schulz	Pa	?	+	?	H	1960s
Rossicotrema donicum Skrjabin et Lindtrop	Pa	Bl	+	Bl-Az	H	1980-90s?
Nicola scriabini Iwanitzky	Pa	Bl	+	Bl-Az	H	1980-90s?
Haploporus longicollum Wlassenko	Pa	Bl	+	Bl	A+	1930s
Saccocoelium obessum Fischthal et Kuntz	Pa	Bl	+	Bl	A+	1930s
Haplosplanchnus pachymosus Eysenhartd	Pa	Bl	+	Bl	A+	1930s
	Polychaeta					
Nereis diversicolor Müller	B	Bor?	+	Az	A	1940
Mercierella enigmatica Fauvel	Bf	Pac?	+	Bl-Az	Sh	1950s?
Fabricia sabella caspica Zenk.	B	Med				Hvalyn time
Ficopotamus enigmaticus (Fauvel)	B	Med				1960
	Mollusca					
Mytilaster lineatus (Gmelin)	Bf	Med	+	Bl	Rw	1920s
Dreissena bugensis Andrusov	Bf	Az-Bl	+	Bl-Az	Sh	1994
Abra ovata (Philippi)	B	Med	+	Az	A	1940s
Monodacna colorata (Eichwald)	B	Bl	+		Sh	1950s
Hypanis colorata (Eichwald)	B	Med				1959
Cerastoderma rhomboids Lamarck	B	Med				Hvalyn time
C. isthmicum Issel	B	Med				Hvalyn time
Lithogliphus naticoides Pfeifer	B	Az-Bl	+	Bl-Az	Sh	1971
Tenellia adspersa (Nordmann)	Bf	Med	+(?)	Bl-Az	Sh	1989
	Crustacea (Cirripedia)					
Balanus improvisus Darwin	Bf	NA	+	Bl-Az	Sh	1955
B. eburneus Gould	Bf	NA	+	Bl-Az	Sh	1950s

		Crustacea (Cladocera)				
Penilia avirostris Dana	P	Bl	+(?)	Bl-Az	Sh?	?
Pleopis polyphemoides Leukart	P	Bl	+	Bl-Az	Sh	1957
Podon intermedius Liljeborg	P	Med	+	Bl-Az	Sh	1985
Moina mongolica Daday	P	Inland waters	+?	Kazakhstan?	Wind	2000
		Crustacea (Copepoda)				
Acartia clausi Giesb.	P	NA	+	Bl-Az	Sh	1981
A. tonsa Dana	P	NA	+	Bl-Az	Sh	2000
Calanipeda aquaedulcis Kritch.	P	Med				1968
		Crustacea (Amphipoda)				
Corophium volutator (Pallas)	N-B	Bl	+	Bl-Az	?	1950-60s
		Crustacea (Decapoda)				
Palaemon elegans (Rathke)	N-B	Med	+	Bl	A+	1930s
P. adspersus (Rathke)	N-B	Med	+	Bl	A+	1930s
Rhitropanopeus harrisi tridentata (Maitl.)	B	NA	+	Bl-Az	Sh	1950s
		Bryozoa				
Lophopodella carteri (Hyatt)	B, Bf	Ind-Afr	+	?	??	1960s
Conopeum seurati (Canu)	Bf	Med	+	Bl-Az	Sh	1958
Membranipora crustulenta (Pallas)	Bf	Med	+	Bl-Az	Sh	1950s
Bowerbankia imbricata (Adams)	Bf	Med				1959
		Kamptozoa				
Barentsia benedeni (Foetinger)	Bf	Med	+	Bl-Az	Sh	1962
Pisces						
Pleuronectes flesus luscus Pallas	N		-	Bl	A	1902, 1930s
Knipowitschia caucasicus (Kawr.)	N	Med				Hvalyn time
Singnatus nigrolineatus caspius Eichwald	N	Med				Hvalyn time
Atherina mochon caspia Eichwald	N	Med				Hvalyn time
Scomber scombrus L.	N		-	Bl	A	1902
Enguaulis encrasicolus (L.)	N		-	Bl	A	1933-1934
Mullus barbatus (L.)	N		-	Bl	A	1933-1934
Rhombus maeoticus (Pallas)	N		-	Bl	A	1930
Mugil auratus Risso	N		+	Bl	A	1930-1934
Mugil saliens Risso	N		+	Bl	A	1930-1934
Ctenopharingodon idella (Val.)	N		+	China	A	1955-1956 1964-1977
Hypophthalmichtys molitrix (Val.)	N		+	China	A	1964-1977
Aristichthys nobilis (Rich.)	N		+	China	A	1964-1977
Oncorhyncus keta (Walbaum)	N		+	Okh	A	1962-1966
Oncorhyncus gorbuscha(Walbaum)	N		-	Okh	A	1964
Salmo gairdneri Rich. *(Salmo salar?)*	N		-	?	A	1975-1978
Oncorhynchus kisutch (Walbaum)	N		-	?	A	1975-1978

Morone saxatilis (Walbaum)	N	?	USA	A	1972
Stenodus leucichthys nelma (Pallas)	N	-	Ob'	A	1965
Mugil soiuy (Basilewsky)	N	-	Bl	A	1966, 1979

B – benthos; P – plankton; Bf – biofouling; P – parasite; N – nekton; N-B – nektobenthos; SH – shipping; H – with host; A+ – during acclimatization; A – acclimatization; Rw – by railway; Bl-Az – Black Sea – Sea of Azov; Pac – Pacific; Okh – Sea of Okhotsk; NA – Northern America; Med – Mediterranean; Ind-Afr – India-Africa.

Chapter 9

CONTAMINATION OF THE CASPIAN SEA: AN OVERVIEW ON RECENT FINDINGS

Mohammad Reza Sheikholeslami

Caspian Environment Programme, Programme Coordinating Unit. Room 108, 3-d Entrance, Government Building, 40 Uzeir Gadjibekov St., Baku 370016, Azerbaijan

Stephen de Mora

Marine Environment Laboratory, International Atomic Energy Agency, 4 quai Antoine 1er, B.P. 800, MC 98012, Principality of Monaco

Vladimir Vladymyrov

Caspian Environment Programme, Programme Coordinating Unit. Room 108, 3-d Entrance, Government Building, 40 Uzeir Gadjibekov St., Baku 370016, Azerbaijan

INTRODUCTION

The Caspian Sea is the largest inland water body in the world, occupying a deep depression on the boundary of Europe and Asia with a water level at present of approximately 27 m below the level of the world oceans. It lies between 47°13' and 36°34' 35" north latitude and between 46°38' 39" and 54°44' 19" east longitude, surrounded by the five littoral states of Azerbaijan, Russian Federation, Islamic Republic of Iran, Kazakhstan, and Turkmenistan.

J.C.J. Nihoul et al. (eds.), Dying and Dead Seas, 227–251.

The total length of Caspian Sea coast is nearly 7000 km. The length of the Caspian Sea (north-south) is approximately 1200 km. The average breadth of the Caspian from the west to the east is 330 km. This sea is commonly divided into three portions: the Northern, Middle and Southern parts. The Northern part of the sea covers about 80,000 $км^2$ it is relatively shallow, averaging about 5-6 m in depth. The Middle part of the Caspian Sea is a separate depression totaling about 138,000 $км^2$ in area. The bottom is a gently sloped plain with depths of 400-600 m. The average depth of the Middle Caspian is 190 m, and its greatest depth is 788 m. The Southern part of the Caspian Sea, having a total area of about 168,400 $км^2$, is separated from the Middle by the Absheron ridge, which is a continuation of the main Caucasus Ranges. The deepest part of the Caspian Sea (1025m) is in the South Caspian.

The Caspian Sea is a unique system straddling several climate zones from continental to mediterranean due to its north–south orientation. (Kosarev & Yablonskaya, 1994). The Caspian Sea is a special region, with endemism reaching 80% at the species level (Dumont, 1998). However, the biodiversity is relatively low, with the total number of species about 40% of that found in the Black Sea (Karpinsky, 1992). An extensive treatment of biodiversity and biogeography of the Caspian Sea can be found in the summary report by Aladin (2001), which is based on the National Biodiversity reports from the Caspian countries.

About 130 rivers of various sizes drain into the Caspian with an annual input of about 300 km^3. The main rivers are the Volga (80 percent of the total volume of inflow), the Ural (5 percent), the Terek, Sulak, and Samur (total up to 5 percent), the Kura (6 percent), and Iran's small rivers of the Alborz Rangesand others (4-5 percent). Caspian Sea water salinity is about 0.1 psu at the mouth of the Volga and Ural rivers and increases to 12.6-13.5 psu in Middle and Southern parts of the Caspian Sea. Generally, the salinity increases slightly from west to east and with depth. The composition of the salt differs to that of sea water owing to the enrichment of sulphate from runoff (Karpinsky, 1992). Thermohaline circulation in the Caspian Sea ensures that the water remains oxygenated to the bottom (Dumont, 1998). There is also a slight increase in salinity with depth (0.1 to 0.2 psu) observed in all regions of the sea

One of the most important features of the Caspian Sea is its changing water level, a factor that has a significant effect on biodiversity and coastal management in the extensive shallow areas.The highest water level in recent history, -22 m, was reached about 3,800 years ago (Aubrey et al. 1994), and the level may, depending how far back in history one goes, have been as low as -64 m. In early of the last century (up to 1929), the sea level fluctuated around -26.2 m, later decrease to -29.0 m in 1977 (Kosarev and Yablonskaya 1994). This is the lowest level reached during the past 400-500 years. In 1978 a rapid rise began, and the sea level reached –26.42 m by 1995. Since 1995 some regression was observed in the sea level. At present the Caspian Sea level stands around the -27 m mark. The causes of sea level change are mainly natural or related to worldwide anthropogenic effects on the climate.

Water temperature varies considerably with latitude. This difference is greatest (about 10° C) in the winter when temperatures in the north are 0-0.5° C near the ice and 10-11° C in the south. Freezing temperatures are found in the north and in shallow bays along the eastern coast. The water temperature of the west coast is generally 1-2° C higher than along the east coast. In the open sea, the water temperatures are higher than those near the coast by 2-3 ° C in the Middle Caspian and by 3-4° C in the Southern part of the sea (Aubrey et al. 1994).

ROLE OF CONTAMINANTS IN ENVIRONMENTAL QUALITY OF THE CASPIAN SEA

There have been alarming symptoms indicating of an increasing rate of decline in overall environmental quality of the Caspian Sea in recent years. Widespread die-off seals in 2000, Kilka mortality in 2002, rapid and significant drop of certain commercial fish catch, including sturgeon that in recent years landings have decreased dramatically from 30,000 tones in 1985 to only 5,672 tones in 1995, and other similar natural disasters create fear of widespread decline in environmental quality of the Caspian Sea. Decline in environmental quality includes the decline in air, water and sediment quality, damage to ecosystems due to human activities, loss of aesthetic appeal, and related issues (Allchin et al.1997; CEP 2002; World Bank 2002). Various parameters are responsible for decline in environmental quality. Both natural phenomena and human activity can be major sources of pressure on the environment. The major sources of pressure on the environment include natural factors (water level fluctuations, geological conditions such as mud volcanoes, source rocks, and natural oil seeps). Human impact may be caused by oil and gas activities, ports and harbors, shipping, mining, urban activities (such as sewage discharge, solid wastes), industrial and agricultural activities.

Water level fluctuations cause a number of indirect effects, such as, alteration of coastal habitats, massive die-offs of coastal reed beds introducing contaminants from flooded lands that were previously considerable, thereby contributing to the pollution burden of the Caspian Sea (Dumont, 1998)

Other natural factors include geological conditions. The geological characteristics of the Caspian Sea are quite variable (Kosarev & Yablonskaya, 1994; Krilov, 1987). For instance, mud volcanoes dot the landscape and sea bottom along the Absheron ridge. These mud volcanoes are a source of hydrocarbons, methane, certain metals, and other materials. They are one form of natural seep of hydrocarbons to the environment. Mining is another unquantified source.

Chromium mines, for instance, discharge tailings into the Caspian along the upper Ural River, which may reach the Caspian in a relatively short period since there are no impediments on the river. Uranium mining in southern Kazakhstan near Aktau provides another potential source of contaminants to the Caspian, particularly given historical discharge practices. Mines along Iranian rivers discharge various tailings to the rivers. Finally mining in the Caucasus region may create discharges into the main rivers such as the Kura (CEP, 2002).

Oil and gas activities provide another source of inputs of contaminants to the environment. These generally occur due to drilling practices (uses of various types of drilling fluids, or muds), maintenance on rigs, transport of oil, and release of oil and gas from drilling operations This region with huge oil and natural gas reserves has attracted the attention of the international oil gas industry especially since the break-up of the Soviet Union in 1991 (Effimoff, 2000). Most of Azerbaijan's oil resources are located offshore, and perhaps 30-40% of the total potential oil resources of Kazakhstan and Turkmenistan are also offshore. Currently, Iran has no oil or natural gas production in the Caspian region, although has potentially significant reserves and acts as a transit centre for other countries' oil and natural gas exports from the Caspian Sea. Although not "another Middle East," as some analysts believed in the early 1990s, the Caspian Sea region is comparable to the North Sea in its hydrocarbon potential with proven oil reserves estimated at 16-32 billion barrels (Effimoff, 2000). The main sources of pollution are considered to be offshore oil production and land-based sources, notably the Volga River (Karpinsky, 1992). Oil and gas issues are of particular concern, partly due to extensive oil slicks observed in some portions of the Caspian Sea. They simply exist in the region with no obvious source (CEP, 2002). Flooding of former oil wells by rising water levels has been another source of contamination, and one that may get worse in the future. In particular, flooded wells in Azerbaijan, Kazakhstan, and Turkmenistan are known to have released hydrocarbon to the environment. Concern has centered on the Apsheron Peninsula of Azerbaijan where a century of oil production and pipeline construction has left more than 10,000 Hectares of land heavily contaminated. In addition, the industrial complex of Sumgait and the oil refinery and petrochemical plants in Baku are major sources of land-based pollution. These land-based sources, together with offshore oil fields, tanker traffic, and trans-Caspian pipelines, have generated large quantities of toxic waste, run-off, and oil spills to the Caspian Sea (Dumont, 1995; Abilov et al., 1999).

Agriculture also releases chemicals, including fertilizers, pesticides, and insecticides into the environment. Many agrochemicals are persistent organic forms. This agricultural activity extends most intensively along the Iranian coast (where the area is small but the density is intense), southern Azerbaijan, parts of the Russian coast, and parts of Kazakhstan. There is some agriculture in Turkmenistan, but little near the coast. Use of banned pesticides such as DDT is commonly reported in the region, and they appear to be widely available. Recent infestations of locusts in Russian Federation and Kazakhstan resulted in aerial spraying of DDT-based pesticides in these countries (CEP, 2002).

CONTAMINANT DATA SITUATION IN THE CASPIAN REGION

During the preparation of Transboundary Diagnostic Analysis (TDA) for the Caspian Sea, the Caspian Environment Programme (CEP) has conducted a survey on available data and information in the region. The TDA found enormous gaps in data availability and data quality. Unfortunately the contaminant data situation in the region is critical. Historically, few data were acquired by Iran in the Sea itself, and much of the data from the Soviet Union is unavailable, poorly applied the QA/QC and undocumented.

There are some scientific reports on contaminant in sediment and biota. The measurements in sediments such as: Geochemistry and distribution of several metals in Caspian Sea (Krilov, 1987 ;Kosarev and Yablonskaya, 1994), sedimentary records for metals PAHs and PCBs in Volga delta (Winkels et al., 1998), Pollution discharge from upper chatchment region in Armenia, (Domont, 1995), Industrial and waste site and oil field of Azerbaidjan (Domont, 1995 and 1998; Abilov et al., 1999), Flux of organochlorinated pestcides and metals in Russian rivers (Zhulidov *et al.*, 2000 and 1998). There are some measurements on biota such as: Metal levels in seals (Anan et al., 2000; Watanabe et al., 1999), Organic pollutants in sturgeon (Södergren et al., 1978), DDT in Caspian sturgeon and its breakdown (Ballschmiter *et al.*, 1983), Chlorinated pesticides in fishes and stergeon (Moore *et al.*, 2003; Vetter *et al.*, 1995), Chlorinated hydrocarbons in Caspian seals (Hall *et al.*, 1999; Kajiwara *et al.*, 2002; Watanabe *et al.*, 1999).

The baseline studies of oil and gas companies in the Caspian Sea is another sources of contaminant data, but these data are limited to a certain area applying different methodology, therefore the basin-wide assessment of contaminant level would not be possible using these data and information.

In order to complete the TDA, particularly assessment of Persistent Toxic Substances, the CEP conducted a Basin-Wide Contaminant Screening Programme (At Sea Training Programme "ASTP") that covered the sediments of the entire Caspian Sea except the Turkmenistan sector. The results of this investigation are enumerated in the report by de Mora and Sheikholeslami (2002) and summarized in the TDA report (CEP, 2002). On the basis of ASTP the first basin wide contaminant assessment, particularly on Persistent Toxic Substances was made in the history of the Caspian Sea.

Another recent study carried out was CEP's Ecotoxicology Project "ECOTOX" (World Bank, 2002). The overall objectives of this study were; to determine the geographical pattern of contamination in the Caspian Sea, identify problem areas and sources of contamination, and analyze the passage of contaminants through the food chain to the top predator species in the system (i.e. seals, sturgeon, and predatory bony fish). The results of this study are summarized in its final Workshop and incorporated in TDA report, (World Bank 2002).

All available data along with other studies results were managed and integrated in Data and Information Management system (DIM) that was located in the CEP Programme Coordinating Unit (PCU) in Baku. The following provides an overview on contaminant assessment that is based on major findings of ASTP.

ASTP CONTAMINANT SCREENING PROGRAMME

In ASTP Cruise a total of 105 surface sediment samples (3-4 cm) were collected in the coastal zone of the Caspian Sea using a Van Veen grab. The focus was primarily on sediments from shallow depths, although material from as deep as 100 m was also obtained. Sample collection was accomplished during four separate oceanographic cruises during October 2000 to October 2001(Fig.1).

The samples from the Russian sector were analysed at the Russian laboratory "Typhoon" Center for Environmental Chemistry. All other samples were analyzed at the International Atomic Energy Agency Marine Environmental Studies Laboratory in Monaco. The samples were analysed for measurements of grain size distribution, together with the inorganic and organic contaminants including: Metals, Hydrocarbons, Polyaromatic Hydrocarbons, Chlorinated Pesticides and Polychlorinated Biphenyls (PCBs). Other parameters include grain size distribution, Total Organic Carbon (TOC), Carbonate content, Extracted Organic Matter (EOM) and Lipid content. The analysis of the sediment samples in Typhoon laboratory mostly covered the same parameters, but with slight differences.

MAJOR FINDINGS

A summary of organic and inorganic contaminants data in ASTP along with some global respective data presented in Table 1 to 5 and the following provides a brief review of the findings:

Sediment nature: Surface sediments in the North Caspian Sea were generally characterized by low percentage (13 ± 14%) of fine material (< 62 μm) and low content of total organic carbon (TOC: 0.68 ± 0.67). In the South Caspian Sea, grain size was consistently finer, averaging 43± 18% and TOC ranged from 0.3 to 2.4 % with an average of 1.1± 0.5 % (Fig.1). No correlation of TOC with the sediment fine fraction was found. In general, TOC tends to be higher in the deep sediments.

Petroleum Hydrocarbons (PHs): The total hydrocarbon concentration ranges from 29 to 1820μg g^{-1}, with the highest amounts in Azerbaijan and the lowest in the

North Caspian Sea. Whereas total hydrocarbon concentrations higher than 500 μg g^{-1} are generally indicative of significant pollution, values below 10 μg g^{-1} are considered unpolluted sediments (Fig. 2).

Total Polyaromatic Hydrocarbons PAHs: The overall trend behaving that of the PH data with highest PAHs concentrations observed in Azerbaijan. The range of values was from 280 to 3000 ng g^{-1}. A hot spot at south of Baku Bay was evident; the content never exceeded the NOAA Sediment Quality Guideline value for Effects Range Low (ERL) of 4000 ng g^{-1} dry weight (Fig. 3).

DDT-related compounds: Noting that the NOAA ERL is 1600 pg g^{-1}, clearly DDT – type compounds exceed this quality standard at a number of locations in Azerbaijan and Iran. The maximum value (13400 pg g^{-1}) observed at Kura River mouth. The lowest concentrations were found in the North Caspian Sea, particularly in the north-eastern shallow area. Relatively high proportions of DDT compared to the breakdown products are evident at locations in all coastal stations, indicative of recent DDT influxes and, by implication, ongoing DDT usage throughout the region (Fig 4).

Lindane and other Hexachlorocyclohexanes (HCHs): This pesticide consists of a mixture of eight steric isomers, including the well-known gamma isomer lindane, which is the main insecticidal component. The lowest values were found in the Iranian and Kazakhstan sectors, several stations in the Russian coastal zone had elevated levels. The maximum concentration 609 pg g^{-1}, exceeded the ISQG value of 320 pg g^{-1}. Lindane degrades rapidly in the environment to compare with DDT. Moreover, upon ingestion by animals, it tends to be rapidly mobilised to water-soluble chlorophenols and chlorobenzenes (Kennish 1996). Therefore, high levels of lindane in the environment reflect the widespread and continuous use of this compound in the Russian sector (Fig 5). Regarding the other HCHs, the highest concentrations were found in Azerbaijan. Very low concentrations were measured in the north-eastern sector that is the shallow coastal zone of Kazakhstan.

Other Chlorinated Pesticides: Several other organochlorinated pesticides were measured and their distributions in the sediments as mentioned in-above. Although the concentrations of other pesticides are not high as for DDT or Lindane, and are generally not of concern, three distinct distribution trends evidently reflect different usage patterns. Higher values for HCB, cis- and trans-Chlordane, Aldrin and Endrin were observed in the Azerbaijan sector. In the Russian sector Methoxychlor, Heptachlor and Heptachlor Epoxide were significantly higher than elsewhere in the Caspian Sea. In the Iranian sector, high levels were generally observed for Endrin and α-Endosulfan. However, national differences were observed with Endosulfan sulfate being high in the eastern region and Aldrin relatively higher in the western part of the Iranian sector.

Polychlorinated Biphenyls (PCBs): The highest values were observed in the Russian sector of the Caspian Sea. However, it should be noted that the Σ PCBs concentrations in this study are relatively low by global standards. Moreover, the NOAA ERL value for Σ PCBs, 23 ng g^{-1}, was never exceeded (Fig 6).

Goldberg Index: With respect to organochlorinated compounds, the Goldberg Index is defined as the ratio of Σ PCBs to Σ DDTs. A small value indicates that the organochlorinated substances are predominantly derived from agricultural sources rather than industrial ones. In Kazakhstan, the very low content of both Σ PCBs and Σ DDTs renders the index meaningless. Elsewhere, distinct differences are evident for the Russian sector compared to the coastal regions of Azerbaijan and Iran. In the north, the high Goldberg index reflects the relative importance of industrially derived PCBs. In contrast, the values were quite low throughout Azerbaijan and Iran, highlighting the overall importance of agrochemical DDTs, especially in the vicinity of the Kura River.

Metal Contaminants:

Arsenic (As): The NOAA ERL value (8.2 μg g^{-1}) is exceeded throughout Iran and Azerbaijan. A couple of sites in the northern Caspian Sea show elevated levels.

Barium (Ba): There are several sites in the central Caspian region with anomalous high levels relative to other locations. The highest concentration, 1250 μg g^{-1} found in Kazakhstan, exceeds those found in the coastal region of Azerbaijan. Barium shows no relationship to Al. As barium is used in drilling mud, the elevated levels may reflect such a source.

Cadmium (Cd): Although levels are higher in the central and southern part of the Caspian Sea compared to the northern sector, concentrations never exceed the NOAA ERL value of 1.2 μg g^{-1}.

Chromium (Cr): The distribution of chromium shows much higher concentrations in the central and southern regions compared to the north Caspian Sea. The Cr content exceeds the NOAA ERL value, 81 μg g^{-1}, at most locations in Azerbaijan and Iran. In the north, the influence of the Ural River as a source of Cr is obvious. Overall, the elevated concentrations of Cr stem from its high natural background in the region. The Caspian region is mineral-rich and several countries, most notably Kazakhstan, are important producers of chromium. The results showing well correlation between Cr and Al, thus the elevated concentrations in Azerbaijan and Iran reflect the deposition of the clay mineral fractions in the sedimentary material. Cr does not accumulate in the northern sector where the sediments are largely coarse-grained.

Copper (Cu): While the concentrations of copper much lower in the north Caspian Sea, but the Cu content exceeded the NOAA ERL, 34 μg g^{-1}, value at several locations in Azerbaijan and Iran. A hot spot is also evident in Kazakhstan. The distribution suggests that the Kura River is an important source of Cu locally. This may be derived from either mining or agricultural activities in the catchment's area.

Lead (Pb): Like cadmium, the lead concentrations are not very high for the sites investigated in the Caspian Sea. Although the maximum concentration (28.6 μg g^{1}) is found just south of Baku Bay, but the levels never exceeded the NOAA ERL value of 47 μg g^{-1}. There is a strong relationship between Pb and Al, thus the concentration of Pb is determined largely by the amount of fine-grained material in the sediments. This explains the tendency for metals, such as Cd and Pb, to have somewhat higher concentrations in Azerbaijan and Iran compared with Russia and Kazakhstan.

Mercury (Hg): The mercury content is low in the northern sector having sediments that are relatively coarse or composed mostly of carbonates. Mercury concentrations are notably high at a number of sites in Azerbaijan, where Hg content exceeds the NOAA ERL value of 0.15μg g^{-1}. In particular, the sediments to the south of Baku Bay are polluted. The overall trend is representative of industrial inputs (Fig 7).

Nickel (Ni): Distribution of Ni is similar to that of chromium, reflecting the high natural background due to mineralization. Exceptionally, Ni displays very high concentrations in sediments throughout the central and southern Caspian Sea. The NOAA ERL (21 μg g^{-1}) was always exceeded and Ni concentrations were higher than the ERM (52 μg g^{-1}) values at several sites. The highest concentrations were found near the mouth of the Kura River, but the Ural River influence is also evident. The elevated content reflects a high natural background, but could be augmented by mining activities.

Silver (Ag): The Baku Bay samples appeared to contain a slightly elevated concentration of Ag, a known indicator of an anthropogenic source. However, the levels were not statistically different from those at other locations in Azerbaijan. The maximum concentration observed was in Iran. Overall, no particular pattern was obvious and the levels never exceeded the NOAA ERL value of 1 μg g^{-1}.

Uranium (U): There have been claims of significantly elevated concentrations of certain naturally occurring radionuclides, especially uranium, in the Caspian Sea. A report from ISAR (Initiative for Social Action and Renewal in Eurasia) in 2001 indicated that the bottom sediments in the Caspian Sea contain levels of uranium five to seven times higher than those in other seas, due mostly to the complex migration patterns of naturally occurring radioactive nuclides. The report also indicated that there were anthropogenic sources of uranium in the Caspian Sea from a uranium enrichment plant in Aktau, Kazakhstan. In study presented here, the highest concentrations of uranium were observed at two deep sites in the central eastern Caspian. Concentrations were 11.1 and 6.2 μg g^{-1}, respectively. Otherwise, all sediments contained <5 μg g^{-1} uranium (consistent with crustal abundance) and many of the sediments showed <1 μg g^{-1} U (Fig 8).

Vanadium (V): The distribution of Vanadium closely behaves that of aluminium. Accordingly, relatively high concentrations are found along the coast of Azerbaijan and Iran where fine-grained material has been deposited.
Zinc (Zn): Generally the overall pattern of distribution is like that of lead and cadmium, again due to the grain size influence. However, one important hot spot is observed in Iran at Sefidroud mouth amounting 148 $\mu g\ g^{-1}$, exceeded the Canadian ISQG value of 124 $\mu g\ g^{-1}$.

CONCLUSION

Overall, local source strengths and the propensity of fine-grained material to accumulate influence the distribution of contaminants in sediments of the Caspian Sea. The north Caspian Sea is a shallow water environment with quite coarse sediments. They tend to have low aluminium concentrations, but often have high carbonate content.
Petroleum hydrocarbon (PH) concentrations are quite high by global standards at some locations, notably to the south of Baku Bay. The recent oil input in the northern part of Azerbaijan sector is obvious. Total PAH concentrations never exceed the NOAA Sediment Quality Guideline value for Effects Range Low (ERL) of 4000 $ng\ g^{-1}$ dry weight. PAHs tend to be derived predominantly from oil or combustion products, with the later notable in the Russian sector. Natural sources, namely in situ biological activity is only important at a few limited sites in Iran.
Several organochlorinated pollutants concentrations were invariably very low in Kazakhstan. DDT-related compounds exhibited concentrations higher than NOAA ERL values at numerous locations in the coastal zone of Azerbaijan and Iran, but were quite low in the Russian sector. However, Lindane concentrations exceeded the Canadian sediment quality guideline value in the Russian sector. Similarly, the PCBs content was higher in the Russian sector than elsewhere, but in this case did not surpass the NOAA ERL of 23 $ng\ g^{-1}$ dry weight. As concerns other organochlorinated pesticides, sources strengths (i.e. local usage) in the different regions varied considerably. The Goldberg Index reflects the relative importance of industrial and agricultural sources of organochlorinated compounds. The Goldberg Index reinforced the observation of the relative importance of agricultural sources in both Azerbaijan and Iran, in contrast to industrial sources in Russia.
As indicated above, the metal concentrations are strongly correlated to the aluminium concentration, a good proxy for terrigenous material and the amount of fine-grained material present. The exception to this is barium, for which some anomalous high concentrations are probably from drilling muds. Several metals (As Cr, Ni) exhibit concentrations sufficiently high to exceed sediment quality guidelines. Such metals undoubtedly have a high natural background in this mineral-rich region. However, anthropogenic activities, notably mining, may have

further enhanced the metal burdens in the sediments of the Caspian Sea. This might explain apparent hot spots for copper and zinc in Azerbaijan and Iran. Uranium levels are generally low (< 3 $\mu g\ g^{-1}$), except for a couple of sites in the central eastern Caspian Sea where the concentration reaches 11.1 $\mu g\ g^{-1}$. Several metals (Ag, Cd, and Pb) have relatively low levels that pose no environmental concerns.
In many cases findings of ASTP study confirm the findings in biota from ECOTOX study. In ECTOX, the Sturgeons have been measured with high concentration of DDT metabolites, PCBs, HCH as well as some heavy metals (Zinc, Copper, Cadmium, and Lead). The Gobies, since they live in the sediments where pollutants tend to accumulate, show some spatial variability appears to reflect directly the sediment and water quality. Seals have been observed to have high concentrations of heavy metals and organics for the past several decades. Recent data support the previous findings in seals indicating high in PCBs, DDT metabolite, HCH chlordane, HCB and certain heavy metals.

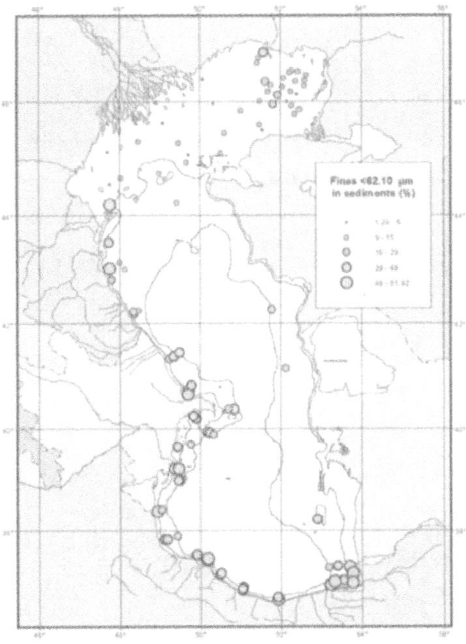

Fig.1 Sampling stations and sediment grain size distribution

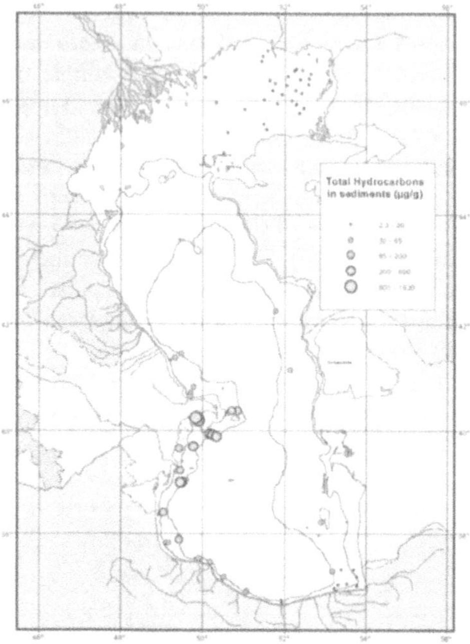

Fig.2 Total Hydrocarbons distribution

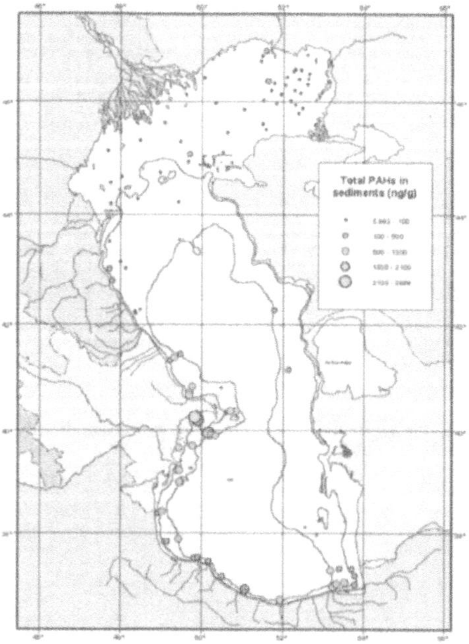

Fig.3 Total PAHs distribution

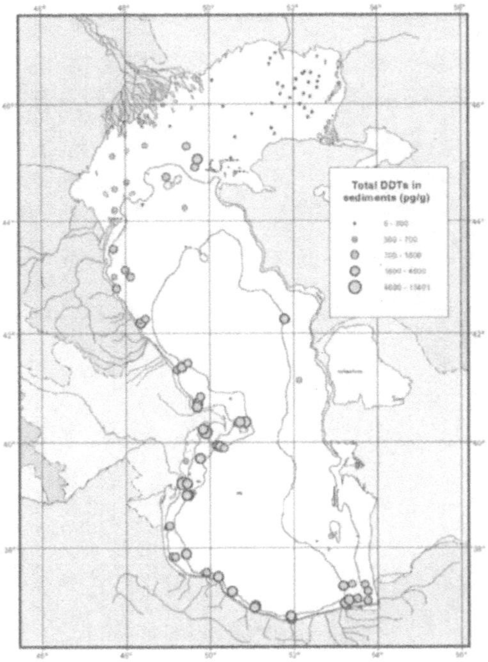

Fig.4 Total DDTs distribution

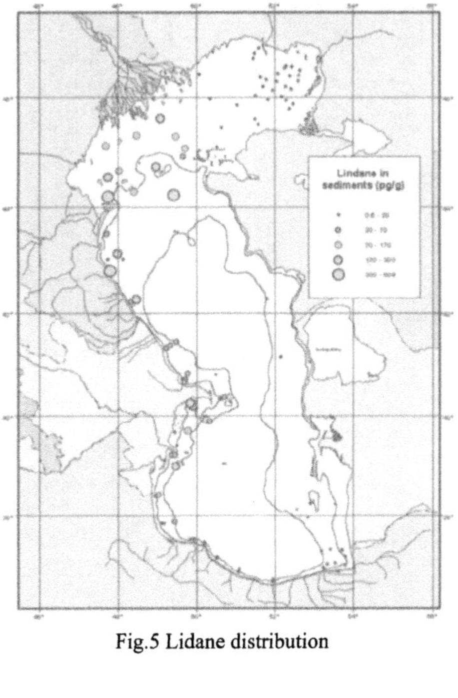

Fig.5 Lidane distribution

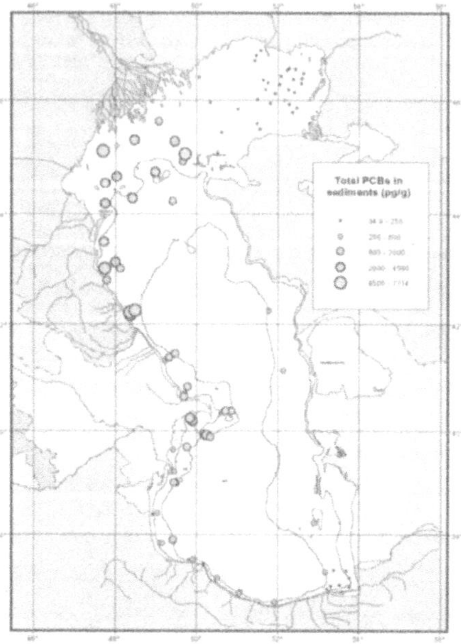

Fig.6 Total PCBs distribution

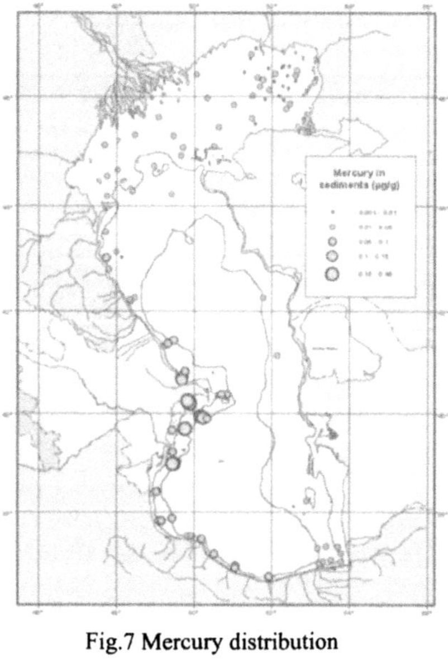

Fig.7 Mercury distribution

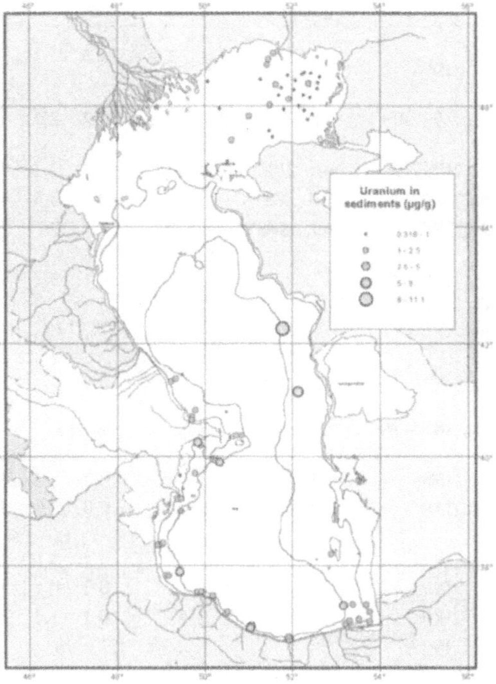

Fig. 8 Uranium distribution

Table1. Comparison of World-wide concentrations of total hydrocarbons in sediments with ASTP findings in the Caspian Sea ($\mu g\ g^{-1}$ dry weight)

Area	Survey Year	Total Hydrocarbons	Oil equivalent	*n*-Alkanes	References
Offshore, Gulf	1994		5.4…92		Al-Lihaibi and Ghazi, 1997
Saudi Arabia, Persian Gulf	1991-93	11…6900	5.3…26000	0.2…28	Readman et al., 1996
Kuwait, Persian Gulf	1992-93	40…240	72…1400	0.3…2.2	Readman et al., 1996
Xiamen Harbour, China	1993	3.1…33*		0.4…3.4	Hong et al., 1995
Victoria Harbour, Hong Kong	1992	60…646*		3.1…20	Hong et al., 1995
Mississipi-Alabama Continental Shelf, USA	1987-89			0.1…3.2	Kennicut et al., 1995
Eastern Harbour, Alexandria, Egypt	-			7…143	Aboul-Kassim and Simoneit, 1995
Western Coast, Taiwan	1990	869…10300*			Jeng and Han, 1994
Rhone River, France, Mediterranean Sea	1985-86	25…170		2…12	Bouloubassi and Saliot, 1993
Kuwait, Persian Gulf	1991	28	13	0.2	Fowler et al., 1991
Saudi Arabia, Persian Gulf	1991	19…671	5…1400	0.9…23	Fowler et al., 1991
Bahrain, Persian Gulf	1991	23…41	3…14	0.3…2.6	Fowler et al., 1991
UAE, Persian Gulf	1991	16	5…7	0.3…0.5	Fowler et al., 199
Oman, Gulf	1991	6…22	1…12	0.1…1.2	Fowler et al., 1991
Danube River	1992			1…40	Equipe Cousteau, 1993
Bosphorus, Black sea, Turkey	1995	12…76	6.5…340	1.3…2.6	Readman et al., 1999
Sochi, Black Sea, Russia	1995	7.6…170	52…680	0.7…3.4	Readman et al., 1999
Odessa, Black Sea, Ukraine	1995	110…310	220…1300	1.4…1.6	Readman et al., 1999
Coastline, Black Sea, Ukraine	1995	2.1…6.6	3.2…42	0.1…0.6	Readman et al., 1999
Danube Coastline, Black Sea, Ukraine	1995	49…220	66…1750	1.2…2.1	Readman et al., 1999
Caspian Sea, Azerbaijan (recent ASTP findings)	**2000**	**20…1820**	**25 … 2000**	**1.1…17**	**De Mora et al, 2002**
Caspian Sea, Russia (recent ASTP findings)	**2000**			**0.02...0.27**	**De Mora et al, 2002**
Caspian Sea, Iran (recent ASTP findings)	**2001**	**8.5 … 167**	**8.7 … 210**	**0.45… 3.5**	**De Mora et al, 2002**
Caspian Sea, Kazakhstan (recent ASTP findings)	**2001**	**2.3 … 41**	**0 … 32**	**0.69 ... 1.9**	**De Mora et al, 2002**

Table 2.Comparison of World-wide concentrations of Polyaromatic Hydrocarbons (PAHs) in sediments with recent findings of ASTP in Caspian Sea

Area	Survey Year	Concentrations ($ng\ g^{-1}$ dry weight)	References
France, Mediterranean Sea	1996	36...6900 (18 cong)	Baumart et al., 1998
Spain, Mediterranean Sea	1996	1.2...8400 (18 cong)	Baumart et al., 1998
Majorca, Mediterranean Sea	1996	0.3...100 (18 cong)	Baumart et al., 1998
Guanabara Bay, Rio de Janeiro, Brazil	1995	1570...18440 (23 cong)	Lima, 1997
North-West Coast, Mediterranean Sea	1991	86.5...48090 (14 cong)	Benlahcen et al., 1997
North Western, Persian Gulf	1991-93	< 20...4740 (13 cong)	Readman et al., 1996
San Quintin Bay, Mexico	1992	N.D....< 50 (44 cong)	Galindo et al., 1996
Xiamen Harbour, China	1993	70...33000 (9 cong)	Hong et al., 1995
Victoria Harbour, Hong Kong	1992	350...3450 (9 cong)	Hong et al., 1995
Baltic Sea	1993	9.5...1871 (15 cong)	Witt, 1995
Saratosa Bay, Florida, USA	-	17...26771 (11 cong)	Sherblom et al., 1995
Western Coast, Australia	1991	1.0 ...3200 (11 cong)	Burt and Ebell, 1995
Italy Coast, Adriatic Sea	1990	27...527 (9 cong)	Guzzella and DePaolis, 1994
Rhone River, France, Mediterranean Sea	1985-86	1070...6330 (15 cong)	Bouloubassi and Saliot, 1993
Lake Burley Griffin, Australia	1989	80...538 (8 cong)	Leeming and Maher, 1992
Tabasco State Continental Shelf, Mexico	1989	454...3120 (15 cong)	Botello et al., 1991
Danube River	1992	< 10...3700 (4 cong)	Equipe Cousteau, 1993
Bosphorus, Black sea, Turkey	1995	13.8...168 (17 cong)	Readman et al., 1999
Sochi, Black Sea, Russia	1995	57.7...360 (17 cong)	Readman et al., 1999
Odessa, Black Sea, Ukraine	1995	66.5...632 (17 cong)	Readman et al., 1999
Coastline, Black Sea, Ukraine	1995	7.2...126 (17 cong)	Readman et al., 1999
Danube Coastline, Black Sea, Ukraine	1995	30.3...604 (17 cong)	Readman et al., 1999
Gulf of Trieste, Italy	1996	30...600 (22 cong)	Notar et al., 2001
Black Sea	1988	12...2400 (28 cong)	Wakeham, 1996
White Sea, Russia	1994	13...208 (27 cong)	Savinov et al., 2000
Caspian Sea, Azerbaijan (recent ASTP findings)	**2000**	**320 ... 3109 (37 cong)**	**De Mora et al, 2002**
Caspian Sea, Russia (recent ASTP findings)	**2000**	**1339 ... 7714**	**De Mora et al, 2002**
Caspian Sea, Iran (recent ASTP findings)	**2001**	**72 ... 954**	**De Mora et al, 2002**
Caspian Sea, Kazakhstan (recent ASTP findings)	**2001**	**35 ... 681**	**De Mora et al, 2002**

Table 3. *Comparison of World-wide concentration of DDTs and HCHs in sediments and recent findings of ASTP in Caspian Sea*

Area	Survey Year	Σ DDT ($ng\ g^{-1}$)	Σ HCH ($ng\ g^{-1}$)	References
South Western Coast, Baltic Sea	1993	< 0.04...88	< 0.04...1.2	Dannenberg et al., 1997
Vanuatu and Tonga, South Pacific Islands	1991	< 0.1...1027	< 0.1...0.3	Harrison et al., 1996
San Quintin Bay, Mexico	1992	< 2...15	n.d....< 1	Galindo et al., 1996
Lake Baikal, Russia	1992	0.01...2.7	0.02...0.1	Iwata et al., 1995
Xiamen Harbour, China	1993	4.5...311	0.1...1.1	Hong et al., 1995
Victoria Harbour, Hong Kong	1992	1.4...97	< 0.1...9.4	Hong et al., 1995
Western Coast, Australia	1991	1...22	-	Burt and Ebell, 1995
Saratosa Bay, Florida, USA	-	< 1...69	-	Sherblom et al., 1995
Gulf of Alaka, Bering Sea, Chukchi Sea	1990	0.01...0.2	0.04...0.3	Iwata et al., 1994 (b)
San Francisco Estuary, USA	1992	< 0.1...9	-	Pereira et al., 1994
Cities, India	1989	8...450	0.6...38	Iwata et al., 1994 (a)
Cities, Thailand	1990	4.8...170	0.5...3.1	Iwata et al., 1994 (a)
Cities, Vietnam	1990	0.4...790	0.4...12	Iwata et al., 1994 (a)
Cities, Indonesia	1991	3.4...42	0.04...0.1	Iwata et al., 1994 (a)
Cities, Papua New Guinea	1990	4.7...130	0.2...0.3	Iwata et al., 1994 (a)
Cities, Solomon Islands	1990	9.3...750	< 0.3...2.2	Iwata et al., 1994 (a)
Cities, Japan	1990	2.5...12	4.5...6.2	Iwata et al., 1994 (a)
Cities, Taiwan	1990	0.4...11	0.3...0.8	Iwata et al., 1994 (a)
Cities, Australia	1990	0.08...1700	0.02...17	Iwata et al., 1994 (a)
Danube River	1992	< 0.04...41	0.03...6.4	Equipe Cousteau, 1993
Bosphorus, Black sea, Turkey	1995	0.2...7.2	0.08...1.1	Readman et al., 1999
Sochi, Black Sea, Russia	1995	3.3...12	0.3...0.8	Readman et al., 1999
Odessa, Black Sea, Ukraine	1995	35...65	1.3...2.3	Readman et al., 1999
Coastline, Black Sea, Ukraine	1995	0.06...0.6	0.02...0.2	Readman et al., 1999
Danube Coastline, Black Sea, Ukraine	1995	9.2...43	1.3...2	Readman et al., 1999
Romania Coastline, Black Sea	1995	0.6...72	0.2...40	Readman et al., 1999
Coast of North Vietnam	1997	6.2...10.4	0.1...6.5	Nhan et al., 1999
Coastal Lagoon, Nicaragua	1995	n.d....270	n.d....1.38	Carvalho et al., 1999
Caspian Sea, Azerbaijan (recent ASTP findings)	**2000**	**0.56...13.4**	**0.196...3.46**	**De Mora et al, 2002**
Caspian Sea, Russia (recent ASTP findings)	**2000**	**0.006...1.86**		**De Mora et al, 2002**
Caspian Sea, Iran (recent ASTP findings)	**2001**	**0.057..3.897**	**0.029..0.589**	**De Mora et al, 2002**
Caspian Sea, Kazakhstan (recent ASTP findings)	**2001**	**0.011...1.89**	**0.004...0.26**	**De Mora et al, 2002**

Table 4: comparison of World-wide concentration of Polychlorinated Biphenyls (PCBs) in sediments with recent ASTP findings in Caspian Sea

Area	Survey Year	Concentrations (ng g^{-1} dry weight)	References
South Western Coast, Baltic Sea	1995	< 0.1…16 (23 cong)	Dannenberg et al., 1997
South Western Coast, Baltic Sea	1995	0.1…11 (23 cong)	Dannenberg et al., 1997
Humber Estuary, UK	1991-93	n.d.…84 (7 cong)	Tyler and Millward, 1996
Irish Sea, UK	1993	0.2…42 (21 cong)	Thompson et al., 1996
San Quintin Bay, Mexico	1992	< 10	Galindo et al., 1996
Gulf of Bothnia, Baltic Sea	1991-92	2…14 (68 cong)	Van Bavel et al., 1995
Lake Baikal, Russia	1992	0.08…6.1 (sum of Kaneclors)	Iwata et al., 1995
Xiamen Harbour, China	1993	0.05…7.2 (Supelco PCB mixt)	Hong et al., 1995
Victoria Harbour, Hong Kong	1992	3.2…81 (Supelco PCB mixt)	Hong et al., 1995
Western Coast, Australia	1991	< 10	Burt and Ebell, 1995
Firth of Clyde, Scotland	1989	0.5…500 (7 cong)	Kelly and Campbell, 1995
Gulf of Alaska, Bering Sea, Chukchi Sea	1990	0.1…2 (36 cong)	Iwata et al., 1994 (b)
San Francisco Estuary, USA	1992	0.1…8.1 (Sum of 4, 5 and 6 Cl)	Pereira et al., 1994
Cities, India	1989	4.8…1000 (Sum of Kaneclors)	Iwata et al., 1994 (a)
Cities, Thailand	1990	11…520 (Sum of Kaneclors)	Iwata et al., 1994 (a)
Cities, Vietnam	1990	0.2…630 (Sum of Kaneclors)	Iwata et al., 1994 (a)
Cities, Indonesia	1991	5.9…220 (Sum of Kaneclors)	Iwata et al., 1994 (a)
Cities, Papua New Guinea	1990	3.3…54 (Sum of Kaneclors)	Iwata et al., 1994 (a)
Cities, Solomon Islands	1990	1.1…5.0 (Sum of Kaneclors)	Iwata et al., 1994 (a)
Cities, Japan	1990	63…240 (Sum of Kaneclors)	Iwata et al., 1994 (a)
Cities, Taiwan	1990	2.3…230 (Sum of Kaneclors)	Iwata et al., 1994 (a)
Cities, Australia	1990	0.5…790 (Sum of Kaneclors)	Iwata et al., 1994 (a)
Danube River	1992	0.02…85 (Ar54+Ar60)	Equipe Cousteau, 1993
Bosphorus, Black sea, Turkey	1995	0.4…4.4 (13 cong)	Readman et al., 1999
Sochi, Black Sea, Russia	1995	0.3…4.7 (13 cong)	Readman et al., 1999
Odessa, Black Sea, Ukraine	1995	5.7…6.8 (13 cong)	Readman et al., 1999
Coastline, Black Sea, Ukraine	1995	n.d.…0.4 (13 cong)	Readman et al., 1999
Danube Coastline, Black Sea, Ukraine	1995	1.4…2.7 (13 cong)	Readman et al., 1999
Romania Coastline, Black Sea	1995	0.1…24 (9 cong)	Readman et al., 1999
Coast of North Vietnam	1997	0.47…28.1 (Ar 1254)	Nhan et al., 1999
Coastal Lagoon, Nicaragua	1995	n.d.…50 (Ar54+Ar60)	Carvalho et al., 1999
Caspian Sea, Azerbaijan (recent ASTP findings)	**2000**	**0.296…3.5 (Ar54+Ar60)**	De Mora et al, 2002
Caspian Sea, Russia (recent ASTP findings)	**2000**	**1.75…10.6 (Ar54)**	De Mora et al, 2002
Caspian Sea, Iran (recent ASTP findings)	**2001**	**0.05…0.995**	De Mora et al, 2002
Caspian Sea, Kazakhstan (recent ASTP findings)	**2001**	**0.032…0.09**	De Mora et al, 2002

Table 5: Range of elemental concentrations (µg g⁻¹) in Caspian Sea sediments in recent findings of ASTP

	Azerbaijan		Iran		Kazakhstan		Russia	
	Min	**Max**	**Min**	**Max**	**Min**	**Max**	**Min**	**Max**
Ag	0.031	0.117	0.026	0.171	0.003	0.061	0.005	0.032
Al	50010	81289	37549	77512	1069	45142	2557	38290
As	8.87	22.6	6.97	20.1	2.13	20.2	0.42	6.71
Ba	314	1076	200	679	75.2	1250	69.7	669
Ca			60730	191938	21522	343571		
Cd	0.076	0.185	0.098	0.244	0.008	0.252	0.022	0.099
Co	11.5	18.1	6.91	24.2	0.73	12.1	1.33	7.62
Cr	56.4	100	59.6	128	1.9	103	2.08	69.3
Cu	14.5	57.6	13.2	50.9	1.2	49.5	2.54	21.9
Fe	29317	43469	22231	44035	1943	27995	1600	9680
Hg	0.047	0.450	0.021	0.092	0.001	0.040	0.009	0.068
Li	27.6	63.6	24	52.5	1.23	32.3	3	35.3
Mg			13933	22353	701	14516		
Mn	543	971	470	1111	45.1	630	90	455
Ni	34.5	68	29.4	67.8	1.8	54.8	5.42	34.2
Pb	12.2	28.6	11.3	24.6	1.43	14.6	0.69	8.03
Sb	0.434	0.879	0.368	1.35	0.072	1.54	0.3	0.3
Se	0.1	0.803	0.1	0.555	1.67	1.67	0.1	0.1
Sn	0.95	2.51	1.32	4.78	0.01	1.38	10	10
Sr	269	1057	362	1707	220	3175	356	6240
U	0.88	2.96	1.53	4.46	0.32	11.10		
V	73.9	136	76.5	145	5.62	81.2	7.25	84.5
Zn	51.1	110	55.9	146	1.04	59.9	2.77	52.9

REFERENCES

Abilov, F.A., Orudjev, A.G., Lange, R. (1999). Optimization of oil-containing wastewater treatment processes. Desalination 124, 225-229.

Aboul-Kassim T.A.T. and Simoneit B.R. (1995). Petroleum hydrocarbon fingerprinting a sediment transport assessed by molecular biomarker and multivariate statistical analyses in the eastern harbor of Alexandria, Egypt. Marine Pollution Bulletin, 30, pp. 63-73.

Aladin, N.V., 2001. Biodiversity of the Caspian Sea. Caspian Environment Programme, Baku, AZ.

Allchin, C., Barrett, T., Duck, C., Eybatov, T., Forsyth, M., Kennedy, S., and Wilson, S. 1997. Surveys
of Caspian seals in the Apsheron Peninsula region and residue and pathology analyses of dead seal tissues. pp. 101-108 in H. Dumont, S. Wilson, and B. Wazniewicz (eds.), Proceedings from the firstBio-network workshop, Bordeaux, November 1997. Caspian Environment Program, World Bank.

Al-Lihaibi S.S. and Ghazi S.J. (1997). Hydrocarbon distributions in sediments of the open area of the Arabian Gulf following the 1991 Gulf war oil spill. Marine Pollution Bulletin, 34, pp. 941-948.

Anan, Y., Kunito, T., Ikemoto, T., Kubota, R., Watanabe, I., Tanabe, S., Miyazaki, N. & Petrov, E. A. (2002). Elevated concentrations of trace elements in Caspian seals (Phoca caspica) found stranded during the mass mortality events in 2000, *Archives of Environmental Contamination and Toxicology 42*, 354-362.

Aubrey, D.G., V.A. Ivanov, T.A. Glushko, and J.D. Winston, 1994. The North Caspian Basin: Environmental Satus and Oil and Gas Operational Issues. Mobil Oil Corporation, Dallas, TX, USA, 311 pp. + Appendices.

Ballschmiter, K., Buchert, H., Scholz, C., Zell, M. (1983). Baseline studies of the global pollution: VI. The pattern of pollution by chlorinated hydrocarbons in the Caspian Sea, Fresenius' Zeitschrift für Analytische Chemie 316, 242-246.

Baumard P, Budzinski H, Mchin Q., Garrigues P, Burgeot T and Bellocq J (1998). Origin and bioavailability of PAHs in the Mediterranean Sea from mussel and sediment records. Estuarine, Coastal and Shelf Science, 47, pp. 77-90.

Baumard, P., Budzinski, H., Mchin, Q., Garrigues, P., Burgeot, T. and Bellocq, J. (1998). Origin and bioavailability of PAHs in the Mediterranean Sea from mussel and sediment records. Estuarine, Coastal and Shelf Science 47, 77-90.

Benlahcen K.T., Chaoui A., Budzinski H., Bellocq J. and Garrigues P. H. (1997). Distribution and sources of polycyclic aromatic hydrocarbons in some Mediterranean coastal sediments. Marine Pollution Bulletin, 34, pp. 298-305.

Botello A.V., Gonzalez C and Diaz G. (1991). Pollution by petroleum hydrocarbons in sediments from continental Shelf of Tabasco State, Mexico. Bulletin of Environmental Contamination and Toxicology, 47, pp. 565-571.

Bouloubassi I. and Saliot A. (1993). Dissolved, particulate and sedimentary naturally derived polycyclic aromatic hydrocarbons in a coastal environment- geochemical significance. Marine Chemistry, 42, pp. 565-571.

Burt J.S. and Ebell G.F. (1995). Organic Pollutants in mussels and sediments of the coastal waters off Perth, Western Australia. Marine Pollution Bulletin, 30, pp. 723-732.

Carvalho F.P., Montenegro-Guillen S., Villeneuve J.-P., Cattini C., Bartocci J., Lacayo M., Cruz A. (1999). Chlorinated hydrocarbons in coastal lagoons of the Pacific coast of Nicaragua. Archives of Environmental Contamination and Toxicology, 36, pp. 132-139.

Caspian Environmental Program. http://www.neapsd.kz/kaspi/eng/text/nature/sea/sea.htm

CEP. (2002). Trasboundary Diagnostic Analysis final report Vol.1 & 2

de Mora, S. and M.R. Sheikholeslami, 2002. ASTP: Contaminant Screening Programme, Final Report: Interpretation of Caspian Sea Sediment Data. Caspian Environment Programme, CRTC ERACL, Baku, AZ, 27 pp. + figures and tables.

Dannenberg D., Andersson R and Rappe C. (1997). Levels and patterns of polychlorinated dibenzo-p-dioxins, dibenzofurans and biphenyls in surface sediments from the western Baltic sea (Arkona Basin) and the Oder river estuarine system. Marine Pollution Bulletin, 34, pp. 1016-1024.

Dumont, H. J. (1995). Ecocide in the Caspian Sea, *Nature 377*, 673-674.

Dumont, H. J. (1998). The Caspian Lake: History, biota, structure, and function, *Limnology and Oceanography 43*, 44-52.

Effimoff, I. (2000). The oil and gas resource base of the Caspian region. Journal of Petroleum Science and Engineering 28, 157-159.

Equipe Cousteau (1993). Concentration of chemical pollutants in sediments and mussels along the Danube. In: The Danube…for whom and for what? Equipe Cousteau's final report, Paris, pp. 104-126.

Fowler S.W., Readman J.W., Oregioni B., Villeneuve J.-P. and McKay K. (1993). Petroleum hydrocarbons and trace metals in nearshore Gulf sediments and biota before and after the 1991 war: an assessment of temporal and spatial trends. Marine Pollution Bulletin, 27, pp. 171-182.

Fowler, S.W., J.W. Readman, B. Oregioni, J.-P. Villeneuve and K.McKay. Petroleum Hydrocarbons and Trace Metals in Nearshore Gulf Sediments and Biota Before and After the 1991 War: An Assessment of Temporal and Spatial Trends. Marine Pollution Bulletin, Volume 27 (1993) 171-182.

Galindo E.A.G., Celaya J.A.V., Munoz G.F. and Sericano J.L. (1996). Organic contaminants in sediments from San Quintin Bay, Baja California, Mexico. Marine Pollution Bulletin, 32, pp. 378-381.

Guzzella L. and DePaolis A. (1994). Polycyclic aromatic hydrocarbons in sediments of the Adriatic Sea. Marine Pollution Bulletin, 28, pp. 159-165.

Hall, A. J., Duck, C. D., Law, R. J., Allchin, C. R., Wilson, S., Eybator, T. (1999). Organochlorine contaminants in Caspian and harbour seal blubber. Environmental Pollution 106, 203-212.

Harrison N., Gangaya P. and Morrison R.J. (1996). Organochlorines in the coastal marine environment of Vanuatu and Tonga. Marine Pollution Bulletin, 32, pp. 575-579.

Hong H., Xu L., Zhang L., Chen J.C., Wong Y.S. and Wan T.S.M. (1995). Environmental fate and chemistry of organic pollutants in the sediment of Xiamen harbor and Victoria harbor. Marine Pollution Bulletin, 31, pp. 229-236.

Iwata H., Tanabe S., Aramoto M., Sakai N. and Tatsukawa R. (1994b). Persistent organochlorine residues in sediments from the Chukchi Sea, Bering Sea and Gulf of Alaska. Marine Pollution Bulletin, 28, pp. 746-753.

Iwata H., Tanabe S., Sakai N., Nishimura A. and Tatsukawa R. (1994a). Geographical distribution of persistent organochlorines in air, water and sediments from Asia and Oceania, and their implications for global redistribution from lower latitudes. Environmental Pollution, 85, pp. 15-33.

Iwata H., Tanabe S., Ueda K. and Tatsukawa R. (1995). Persistents organochlorine residues in air, water, sediments and soils from Lake Baikal region, Russia. Environmental Science and Technology, 29, pp. 792-801.

Jeng W.L. and Han B.C. (1994). Sedimentary coprostanol in kaohsiung harbor and the Tan-Shui estuary, Taiwan. Marine Pollution Bulletin, 28, pp. 494-499.

Kajiwara, N., Niimi, S., Watanabe, M., Ito, Y., Takahashi, S., Tanabe, S., Khuraskin, L. S., Miyazaki, N. (2002). Organochlorine and organotin compounds in Caspian seals (*Phoca caspica*) collected during an unusual mortality event in the Caspian Sea in 2000, Environmental Pollution 117, 391-402.

Karpinsky, M. G. (1992). Aspects of the Caspian Sea benthic ecosystem, *Marine Pollution Bulletin 24*, 384-389

Kelly A.G. and Campbell L.A. (1995). Persistent organochlorine contaminants in the Firth of Clyde in relation to sewage sludge input. Marine Environmental research, 41,

Kennish M.J. (1996) *Practical Handbook of Estuarine and Marine Pollution,* CRC Press, New Jersey, pp. 524.

Kosarev, A. N. & Yablonskaya, E. A. (1994). *The Caspian Sea*, 259 pp. SPB Academic Publishing, The Hague.

Krilov, H. A. (1987). *Caspian Sea. Geology and Oil-Gas Presence*, pp. 296. Nauka, Moscow.

Leeming R. and Maher W. (1992). Sources of polycyclic aromatic hydrocarbons in lake burley griffin, Australia. Organic Geochemistry, 18, pp. 647-655.

Lima A.L.C. (1997). Geocronologia de hidrocarbonetos poliaromaticos (PAHs): Estudo de caso de Guanabara. M. Sc. Thesis, PUC, Rio de Janeiro, Brazil, 107 p.

Moore, M.J., Mitrofanov, I.V., Valentini, S.S., Volkov, V.V., Kurbskiy, A.V., Zhimbey, E.N., Eglinton, L.B. & Stegeman, J.J. (2003). Cytochrome P4501A expression, chemical contaminants and histopathology in roach, goby and sturgeon and chemical contaminants in sediments from the Caspian Sea, Lake Balkhash and the Ily River Delta, Kazakhstan. *Marine Pollution Bulletin 46*, 107-119.

Nhan D.D., Am N.M., Carvalho F.P., Villeneuve J.-P., Cattini C. (1999). Organochlorine pesticides and PCBs along the coast of north Vietnam. The Science of the Total Environment, 237/238, pp. 363-371.

Notar M., Leskovsek H. and Faganeli J. (2001). Composition, distribution and sources of polycyclic aromatic hydrocarbons in sediments of the Gulf of Trieste, northern Adriatic Sea. Marine Pollution Bulletin, 42, No 1, pp. 36-44.

Pereira W.E., Hostettler F.D., Cashman J.R. and Nishioka R.S. (1994). Occurrence and distribution of organochlorine compounds in sediment and livers of striped bass (*morone saxatilis*) from the San Francisco Bay delta estuary. Marine Pollution Bulletin, 28, pp. 434-441.

Readman J.W., Bartocci J., Tolosa I., Fowler S.W., Oregioni B. and Abdulraheem M.Y. (1996). Recovery of the coastal marine environment in the Gulf following the 1991 war related iol spills. Marine Pollution Bulletin, 32, pp. 493-498.

Readman J.W., Fillmann G., Tolosa I., Bartocci J., Villeneuve J.-P., Cattini C.,Coquery M., Azemard S., Horvat M., Wade T., Daurova Y., Topping G. and Mee L.D. (1999). The Black Sea: A comprehensive survey of contamination. Black Sea Environmental Series. Vol. 10, edited by L.D. Mee and G. Topping. UN Publications. New York. Pp. 171-252.

Readman J.W., Mantoura R.F.C. and Rhead M.M. (1987) A record of polycyclic aromatic hydrocarbons (PAH) pollution obtained from accreting sediments of the Tamar estuary, U.K.: Evidence for non-equilibrium behaviour of PAH. Sci. Total Environ., 66: 73-94.

Savinov V.M., Savinova T.N., Carroll J., Matishov G.G., Dahle S. and Naes K. (2000). Polycyclic aromatic hydrocarbons (PAHs) in sediments of the White Sea, Russia. Marine Pollution Bulletin, 40, No 10, pp. 807-818.

Sherblom P.M., Kelly D and Pierce R.H. (1995). Baseline survey of pesticide and PAH concentrations from Sarasota bay, Florida, USA. Marine Pollution Bulletin, 30, pp. 568-673.

Södergren, A., Djirsarai, R., Gharibzadeh, M. & Moinpour, A. (1978). Organochlorine residues in aquatic environments in Iran, 1974. *Pesticides Monitoring Journal 12*, 81-86

Thompson A., Allen J.R., Dodoo D., Hunter J., Hawkins S.J. and Wolff G.A. (1996). Distribution of chlorinated biphenyls in mussels and sediments from Great Britain and the Irish Sea coast. Marine Pollution Bulletin, 32, pp. 232-237.

Tyler A.O. and Millward G.E. (1996). Distribution and partitioning of polychlorinated dibenzo-p-dioxins, polychlorinated dibenzofurans and polychlorinated biphenyls in the Humber estuary, UK. Marine Pollution Bulletin, 32, pp. 397-403.

VanBavel B., Naf C., Bergqvist P.A., Broman D., Lundgren K., Papakosta O., Rolff C., Strandberg B., Zebuhr Y., Zook D. and Rappe C. (1995). Levels of PCBs in the aquatic environment of the Gulf of Bothnia: benthic species and sediments. Marine Pollution Bulletin, 32, 210-218.

Vetter, W., Natzeck, C., Luckas, B., Heidemann, G., Kiabi, B. & Karami, M. (1995). Chlorinated hydrocarbons in the blubber of a seal (*Phoca caspica*) from the Caspian Sea. *Chemosphere 30*, 1685-1696.

Wakeham S.G. (1996). Aliphatic and polycyclic aromatic hydrocarbons in Black Sea sediments. Marine Chemistry, 53, 187-205.

Watanabe, M., Tanabe, S., Tatsukawa, R., Amano, M., Miyazaki, N., Petrov, E.A. & Khuraskin, S.L. (1999). Contamination Levels and Specific Accumulation of Persistent Organochlorines in Caspian Seal (*Phoca caspica*) from the Caspian Sea, Russia. *Archives of Environmental Contamination and Toxicology 37*, 396-407.

Winkels, H.J., Kroonenberg, S.B., Lychagin, M.Y., Marin, G., Rusakov, G V. & Kasimov, N.S. (1998). Geochronology of priority pollutants in sedimentation zones of the Volga and Danube delta in comparison with the Rhine delta. *Applied Geochemistry 13*, 581-591.

Witt G. (1995). Polycyclic aromatic hydrocarbons in water and sediment of the Baltic Sea. Marine Pollution Bulletin, 31, 237-248.

World Bank. (2002). Ecotoxicological Study: Investigation into toxic contaminant Accumulation and related Pathology in the Caspian Sturgeon, Seal and Bony Fish (ECOTOX Study). Final Workshop Report

Zhulidov, A.V., Headley, J.V., Pavlov, D.F., Robarts, R.D., Korotova, L.G., Fadeev, V.F., Zhulidova, O.V., Volovik, Y. & Khlobystov, V. (1998). Distribution of organochlorine insecticides in rivers of the Russian Federation. *Journal of Environmental Quality 27*, 1356-1366.

Zhulidov, A.V., Headley, J.V., Pavlov, D.F., Robarts, R.D., Korotova, L.G., Vinnikov, Y.Y. & Zhulidova, O.V. (2000). Riverine fluxes of the persistent organochlorine pesticides hexachlorocyclohexane and DDT in the Russian Federation. *Chemosphere 41*, 829-841.

Chapter 10

1,000-YEAR ENVIRONMENTAL HISTORY OF LAKE ISSYK-KUL

Santiago Giralt and Ramon Julià

Institute of Earth Sciences "Jaume Almera" (CSIC), Lluís Solé i Sabarís, s/n, E-08028 Barcelona, Spain (sgiralt@ija.csic.es)

Jean Klerkx

International Bureau for Environmental Studies (IBES) Rue Audrey Hepburn 9/13, B-1090 Bruxelles, Belgium

Santiago Riera

Department of Archaeology, Faculty of History, University of Barcelona, Baldiri i Reixach s/n, E-08028 Barcelona, Spain.

Suzanne Leroy

Department of Geography and Earth Sciences, Brunel University (West London), Uxbridge UB8 3PH, UK.

Teresa Buchaca and Jordi Catalan

Centre d'Estudis Avançats de Blanes (CEAB- CSIC), Accès a la cala Sant Francesc 14, E-17300 Blanes, Spain.

Marc De Batist

Renard Centre of Marine Geology, Geological Institute, University of Gent, Krijgslaan 281 S.8, B-9000 Gent, Belgium.

J.C.J. Nihoul et al. (eds.), Dying and Dead Seas, 253–285.

Christian Beck

Laboratoire de Géodynamique des Chaines Alpines, U.M.R. C.N.R.S. 5025, Université de Savoie, Campus Savoie-Technolac, 73376 Le Bourget-du-Lac Cedex, France.

Vladimir Bobrov, Vselodov Gavshin, Ivan Kalugin and Feodorov Sukhorukov

United Institute of Geology, Geophysics and Mineralogy, Acad. Koptyug Prospekt, 3, RU-630090 Novosibirsk, Russia.

Matthias Brennwald, Rolf Kipfer and Frank Peeters

Swiss Federal Institute of Technology (ETH) and Swiss FederalInstitute of Environmental Science and Technology (EAWAG) CH-8600 Dübendorf, Switzerland.

Salvatore Lombardi

Department of Earth Sciences, University of Rome 'La Sapienza', P.le A. Moro, 5, Rome 00185, Italy

Vladimir Matychenkov and Vladimir Romanovsky

Institute of Water Problems and Hydropower. National Academy of Sciences of the Kyrgyz Republic, 720033 Bishkek, Republic of Kyrgyzstan

Victor Podsetchine

Pirkanmaa Regional Environment Centre, P.O.Box 297, FIN-33101 Tampere, Finland

Nunzia Voltattorni.

Istituto Nazionale di Geofisica e Vulcanologia Sezione Roma 1, Via di Vigna Murata 605, 00143 Roma, Italy.

1. INTRODUCTION

Lake Issyk-Kul constitutes one of the most important economic resources in the Republic of Kyrgyzstan, with more than 100 recreational centers along its shore. Some 370,000 holidaymakers visit the lake annually, and this number is expected to increase in the near future given the growing interest in natural environments (Romanovsky, 1990; Savvaitova and Petr,

1992). Thus, a fuller understanding of the past and present evolution of this ecosystem is essential for promoting and sustaining this natural habitat.
At present the lake is faced with two main hazards:

1) **A decrease in the instrumental water level record by more than 3 m since 1927** (Fig. 1). Up to now, the precise reasons for the fall in the present water level remain unknown although a number of factors have been identified such as climate (a long-term decrease in regional moisture), tectonic activity (a deepening of the bottom of the lake) and anthropogenic activities (a progressive increase in water consumption for agricultural purposes) (Romanovsky, 2002).

2) **The progressive pollution of the lake.** This is attributed to the increase in the use of fertilizers given the expansion of agricultural activities on the shore of the lake (eutrophy of the water) and to the rise in the population occupying its shores. The later factor includes the input of non-treated waters from the surrounding villages to the lake as well as waste from the local industry (oil products, phenols and suspended particles). Particular cases of contamination include a cyanide spill in 1998 and the influx of particles into the lake from former uranium mining wastes on the southern shores since the 1950s.

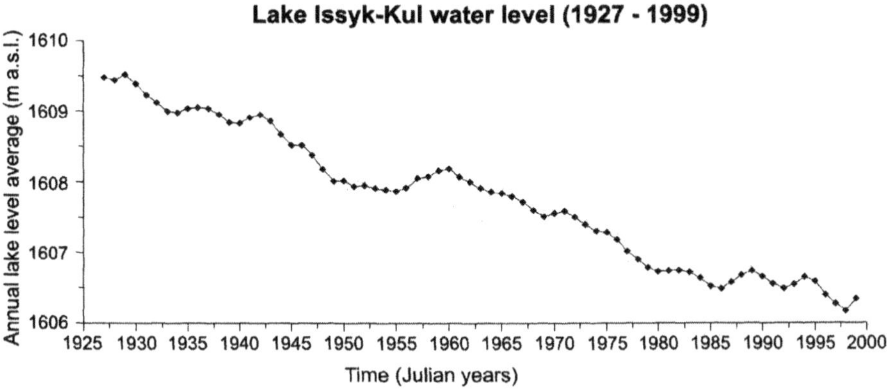

Figure 1. Instrumental water level oscillations of Lake Issyk-Kul for the period 1927 - 1999 (Modified from Romanovsky, 2002).

This has led to the drastic decline in fishing (from 1129 Tm in 1965 to 160 Tm in 1985), the degradation of the water quality and the incorporation of toxic chemical compounds in the food web with deleterious consequences for the population of this area (Savvaitova and Petr, 1992).

In an attempt to determine the present-day environmental status a large multidisciplinary study was conducted. Thus, physical parameters of the water were measured, together with current measurements. An equation of state was established to develop a conceptual model for deep-water mixing (Hofer et al., 2002). Rates of deep-water exchange were estimated from measurements of transient tracers. One-dimensional box-model was developed to determine the long-term deep-water exchange and to quantify the mass fluxes of chemicals. A hydrodynamic model, which illustrated the transport and mixing today and in the near future, was built. Geochemical profiles in the water column were carried out to determine the geochemical cycle, to identify the places of input, and also to define the possible areas of contaminant concentration. The amount of radioactive contamination was evaluated (see the web page of the EU project APELIK for further details http://www.uiggm.nsc.ru/issyk-kul).

The recent environmental evolution of Lake Issyk-Kul was also studied to establish historical trends and past environmental changes. A thorough knowledge of the sedimentary processes and environments in the lake is essential for the selection of the most suitable coring sites. To this end, an extensive high-resolution seismic profiling as well as a coring campaign were carried out (De Batist et al., 2002). The main biogeochemical processes controlling the formation and precipitation of the different endogenic mineral phases (especially those related to the microbial biscuits of vaterite (Giralt et al., 2001)) were also characterized. Finally, the environmental evolution of Lake Issyk-Kul for the last 1,000 years was determined and is presented below.

2. SITE DESCRIPTION

Lake Issyk-Kul (180 km long and 50 km wide, at about 1,600 m a.s.l.,) is a large, closed, slightly saline (about 6 g/l of TDS), tectonically active intra-mountainous lake, located between the Kungey (northern part) and Terskey (southern part) Ranges in the Tien Shan mountains (Republic of Kyrgyzstan, Central Asia). These ranges reach altitudes of 5,000 – 6,000 m a.s.l. (Fig.2).

The catchment area of the lake is relatively small (250 km long and 100 km wide) when compared with the size of the lake (Zabirov and Korotaev, 1978).

The lake is almond shaped. The littoral platforms, mainly located in the eastern and western sides of the lake, slightly dip up to water depths of 200 m, transiting abruptly to a central deep, rectangular and flat shaped area 668 m deep (Fig.2). The littoral platforms of the northern and the southern shores are almost non-existent and they strongly dip up to this flat deep area. These steep slopes contain deep canyons which transfer the sedimentary particles from the shores to the flat deep area (De Batist et al., 2002).

The water chemistry of the lake is dominated by Na^+ and K^+ among the cations and by Cl^- among the anions (Karmanchuk, 2002). The alkalinity ranges from 310 mg/l to 330 mg/l and the calcium content is about 115 mg/l. These slightly saline waters are oversaturated with respect to calcite, monohydrocalcite and vaterite, minerals found in the sediments (Giralt et al., 2001). The lake is monomictic and is considered to be oligotrophic to ultra-oligotrophic (2 - 3.8 μg/l of phosphorous) (Aladin and Plotnikov, 1993). The pH values range between 8.75, at the surface of the lake, and nearly 8 at the bottom. The lake surface temperature in January is not less than 2 - 3°C, and in July it reaches 19 – 20 °C. At depths of more than 100 m, the water temperature remains constant all the year at 3.5 - 4.5 °C. High values of dissolved oxygen at the bottom of the lake (6.5 – 7.5 mg/l) demonstrates that the lake is well mixed (Hofer et al., 2002).

The lake water circulation is cyclonic with strong horizontal water currents. Drift buoy measurements conducted during the expedition of March 2001 indicate the presence of water currents with horizontal velocities in the order of 0.5 m/s down to at least 350 m depth. The deep-water renewal in Lake Issyk-Kul is rapid and the residence time of the deep water of the lake seems to be less than 10 yr (Hofer et al., 2002). The mechanism driving the deep-water renewal seems to be the differential cooling. Differential cooling leads to dense cold water in the shallow eastern regions of the lake. This dense bottom water propagates along the channels towards the deep open-water region where it sinks as a density plume to a depth corresponding to the water density in the plume (Peeters et al., 2003).

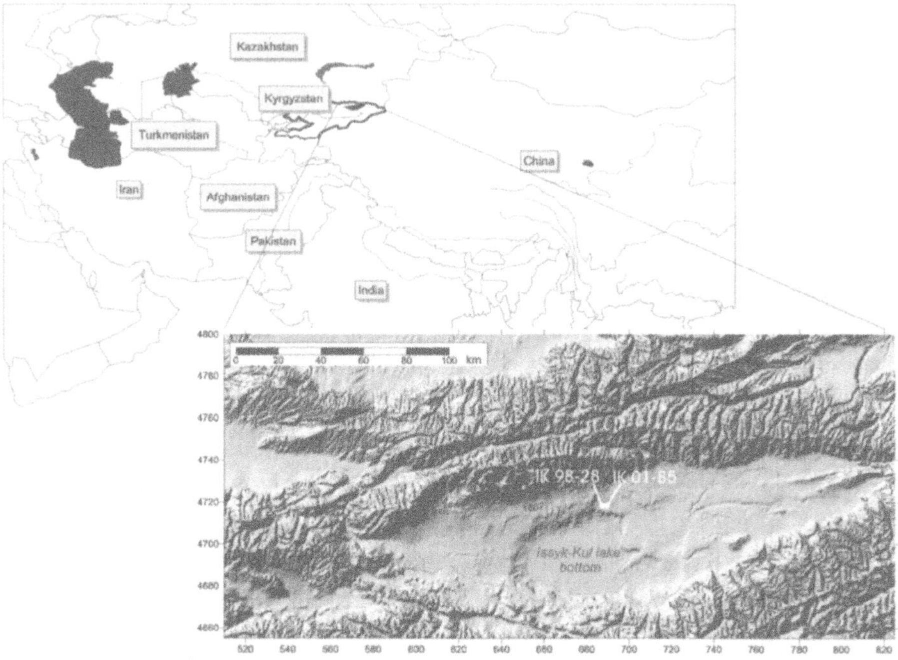

Figure 2. Geographical location of Lake Issyk-Kul (top). Digital Elevation Model (DEM) of Lake Issyk-Kul and its catchment area (Modified from Delvaux, 2001). Location of the cores IK 98-28 and IK 01-85.

The lake is fed by 102 streams and rivers, but the Jergueland and Tyup rivers are the main rivers contributing to it. All these streams and rivers are, in turn, fed by ice caps located in the surrounding mountains with an altitute exceeding 3,300 m a.s.l. This induces an annual lake level oscillation of about 20 cm. The lake level rises from February to September because of thawing glaciers and falls from September to the following February owing to the accumulation of snow in the ice caps (Giralt et al., 2002).

In the Central Asia region airflow from the west predominates in the middle and upper troposphere and is one of the main sources of moisture (Aizen et al., 2001). The low pressure over Europe during the prevalence of Meridional Atmospheric Circulation Pattern favours the increasing frequency of synoptic processes associated with precipitation, e.g. influxes of air masses from west to east bringing moisture from the Mediterranean Sea (Aizen et al. 2001). Thus, the humid air masses are primarily carried by mid-latitude cyclonic air currents from the Atlantic Ocean through the

Mediterranean Sea in spring and autumn although secondary cycloniccenters are also found over the Caspian Sea (Alpert et al., 1990; Cullen and DeMenocal, 2000). The climate in the lake catchment area is continental but highly variable (from warm, temperate, and dry in the western part to slightly moist in the eastern part). The annual precipitation averages 250 mm and the annual evaporation from the surface of the lake is approximately 700 mm.

The vegetation around lake Issyk-Kul is arranged in altitudinal belts (Walter and Box, 1983; Issyk-Kol Biosphere, 2001; Chemonics International Inc., 2001; Giralt et al., 2002). The western part of the lake basin is surrounded by semi-desert vegetation that is characterized by *Ephedra* and Chenopodiaceae (*Salsola, Suaeda*). The dry steppe (*Artemisia* and Poaceae) is widespread at the bottom of the mountain slopes and along the shores of the eastern basin.
These areas have been transformed by agricultural use. *Hippophaë rhamnoides* thickets grow along the river courses. The riparian ecosystem also contains *Populus, Salix* and *Betula.*

Only a very restricted deciduous belt of mostly fruit trees is present on the wetter west-facing slopes: *Juglans,* a series of Rosaceous fruit trees and pistachios. The tree belt in the south is dominated by *Juniperus* sp. and in the north by *Picea shrenkiana.* Abies sp. forests grow in the far north and west. The alpine and sub-alpine meadows and pastures (with sedges) are characterised by Poaceae and Lamiaceae amongst a rich diversity of other herbs.

The mountains are affected by considerable deforestation despite recent protection laws (Ministry of the National Environment Protection of the Kyrgyz Republic and UNEP/GRID Arendal, 2000). Excessive grazing has resulted in the loss of meadows. Wild and human-induced fires constitute a problem.

3. MATERIALS AND METHODS

Extensive seismic and coring campaigns were carried out to select the most suitable place for conducting the palaeoenvironmental reconstructions (see De Batist et al. (2002) for further details).

The palaeoenvironmental reconstruction was done using two sediment cores (IK-98-28 and IK-01-85) both located in the northern part of the lake

(Fig. 2) since this area apparently has not been disturbed by tectonic phenomena. Both cores were retrieved using a gravity corer. Core IK-98-28 was 1.80 m long whereas core IK-01-85 was 1.50 m long.

Both cores were split longitudinally, and one half was used for lithological description, digitalization with a CCD camera, and sampling. Core IK-98-28 was subsampled for granulometry, x-ray diffraction, thin sections, charcoal, pollen and other palynological remains whereas core IK-01-85 was subsampled for x-ray diffraction, pigments, spherical carbonaceous particles and charcoal remains.
The grain size analyses were conducted every 3 millimeters using a Malvern Instruments MasterSizer/E with a 2 mW He - Ne laser (633 nm wavelength). X-ray diffractions were done with an automatic Brucker D - 5005 x - ray diffractometer: Cu radiation ($K\alpha = 1.5405$), 40 kV, 30 mA and graphite monochromator. The mineralogical composition of core IK-98-28 was obtained every 5 cm whereas analyses were performed every 1.5 cm for core IK-01-85.

The thin sections from sediment samples were obtained after freeze-drying and balsam hardening and 100 thin sections were studied. Thin sections were digitalized using a CCD camera, and images were for several measurements employing Scion Image software, following the technique of Francus (1998).

The uppermost 38 cm of the core IK-01-85 were inspected for charcoal remains and carbonaceous particles. The samples were treated according to Yang et al. (2001). They were digested with HNO_3 at 80 °C for 2 h to remove the reactive organic material as well as the carbonate particles, and with HF also at 80 °C for 2 h to remove the siliciclastic minerals (Rose, 1994). Lycopodium tablets were added after the chemical treatment in order to quantify carbonaceous particle concentrations (Stockmarr, 1971). The quantification of particles was carried out under the light microscope. Electron microscope observations were carried out to characterize the different particle morphologies, and the attached microprobe to this microscope was used to determine their chemical composition.

Three main categories of carbonaceous particles were distinguished according to their morphologies: Spheroidal Carbonaceous Particles (SCP), unperforated-like spheres and semispheres and plant remains (ligneous and herbaceous) (Fig. 5). In addition, plant particles were subdivided into two groups according to their size (>50 μm, 20 - 50 μm).

The frozen sediment samples were freeze-dried prior to pigment extraction.

Pigments were extracted from c. 1 g freeze-dried sediment in 4 ml 90% acetone with a probe sonicator (50W, 2 min). The extract was filtered through Whatman ANODISC 0.1 µm (ø = 25 mm) and analyzed by HPLC. The HPLC system was equipped with a Waters 600E solvent delivery system, an autosampler Waters 717 set at 4 °C, a C18 column (dimensions: 250 x 4.6 mm, particle size: 5µm, Spherisorb ODS1 Waters) and a Waters 996 photodiode array detector. The detector was set at 440 and 660 nm for carotenoid and phorbin peak integration, respectively. Analytical separation was achieved by linear gradient (1.2 ml/min, constant flow) using a modification of the system described by Wright et al. (1991). After sample injection (100 µl of extract), pigments were eluted by linear gradient from 100% solvent A (51:36:13 methanol:acetonitrile:MilliQ water, v/v/v 0.3 M ammonium acetate) to 75% A in 5 min followed by an isocratic hold for 5 min at 75% A and to 100% solvent B (70:30 ethyl acetate:acetonitrile, v/v) in 20 min. The solvent composition was returned to initial conditions over a 5 min gradient, followed by 5 min of system equilibration before injection of the next sample.

Pigments were identified by checking them against a library of pigment spectra obtained from several extracts of pure cultures of algae from the Culture Collection of Algae and Protozoa (CCAP) of The Windermere Laboratory (UK). Chl-a and Chl-b standards were obtained from Sigma Chemical Co. Ltd. (UK). The extinction coefficients used for calculations were those compiled by Davies (1976), Rowan (1989) and Jeffrey et al. (1997).

Twenty-three 5 mm-thick samples for palynological studies were taken from the uppermost 41 cm of core IK 98-28 every two centimeters. The c. 1 ml samples were treated with the following method: sodium pyrophosphate, HCl, HF, HCl, and finally sieving on 250 and 10 µm meshes. The slides were mounted in glycerol. An estimation of the concentration was obtained by the initial addition of Lycopodium tablets to a known volume of sediment.
The sum for percentages was usually larger than 400 terrestrial pollen grains. The same sum was used for spores, pollen from aquatic plants, colonies of the green algae Botryococcus, fungal spores and other microfossils. The concentration units are number of pollen and spores per ml of wet sediment. The diagrams (Figures 8 and 9) were plotted using the software psimpoll 4.10.

Zonation of different proxies such as mineralogy, pigments and pollen was established by the CONISS program of TILIA package, using the Edwards & Cavalli-Sforza chord distance method (Grimm, 1987). The same zonation was applied to the concentration diagram. The palynological indicators of human activity were derived from the categories proposed by Behre (1981) for Europe.

In the palynological sequences located in the limits between forest and steppic environments AP is an excellent indicator of climate change, which is usually attributed to variations in moisture and to the continentality degree (Yafeng, 1993). The forest in the Lake Issyk-Kul area is located in the northern face of the Tien Shan Range whereas the southern face and the plains are dominated by steppe and semi-desert vegetation. Some authors have proposed moisture and continentality index based on several pollen taxons in an attempt to determine climate fluctuations using pollen sequences. Some studies on the present-day pollen deposition have shown that Chenopodiaceae is related to desert conditions whereas Artemisia indicates steppe environments. Since the pollinization mechanism of both vegetal species is similar, the Artemisia/Chenopodiaceae index can be used as a moisture index (Rhodes et al., 1996; Giralt et al., 2002).

The relatively high salinity of the water of Lake Issyk-Kul favours the extension of Chenopodiaceae on the shores of the lake during low lake level stands. Hence, the fluctuations of this vegetal taxon are not only triggered by moisture oscillations. The Artemisia/Ephedra index was used as an indicator of the regional moisture variations given that Ephedra is a palynological taxon indicating arid conditions (Gasse and Van Campo, 1994).

The chronology of the Lake Issyk-Kul sequences was based on ^{210}Pb, ^{137}Cs and on AMS ^{14}C data derived from Ricketts et al. (2001). ^{210}Pb was measured in two cores (IK 98-28 and IK 01-85) whereas ^{137}Cs only was measured in IK 01-85 (Fig. 9). IK 01-85 was subsampled in 5 mm-thick slices between 0 and 6 cm depth and 10 mm thick from 6 to 9 cm depth. IK 98-28 subsampling was more spaced. A low background γ-spectrometer equipped with NaI(Tl) scintillation detectors was used.

4. RESULTS

4.1 Lithology

Both cores showed the same lithological units, although their relative thicknesses were not the same. From the bottom to the top of the cores two lithological units can be differentiated:

1) **Unit 2**: This unit ranges from the 15 cm depth of core IK 98-28 and from the 13 cm of core IK 01-85 to the bottom of both cores. It is mainly composed of a light – dark color alternation of silty-clays. From 86 cm to 136 cm of core IK 98-28 and from 107 cm to the bottom of core IK 01-85 the textural composition of this alternation is slightly coarser (Giralt et al., 2002). Two types of color alternation are present: 1.- a millimetric alternation (1 - 2 mm thick) of light and dark gray lamina with clear boundaries. 2.- a centimetric alternation (1 - 2 cm thick) of light and dark gray layers. The boundaries are sharp at the top of the dark lamina and transitional at the bottom.

2) **Unit 1**: This unit is found in the uppermost 15 cm depth of the core IK 98-28 and in the 13 cm depth of the core IK 01-85 and is formed by massive light brownish silty-clays.

4.1.1 Microlithofacies

According to the thin sections, three microfacies were defined (see Giralt et al. (2002) for further details).

Light centimetric lamina: The first microlithofacies is composed of massive micritic carbonate with some terrigenous grains floating in this matrix. These grains are mainly rounded and homogeneous in size (with a mean size of 5 μm) and are mainly composed of quartz, hornblende and some elongated phyllosilicates, such as biotite, mainly arranged parallel to the surface of sedimentation. The presence of fragments of gastropods and diatoms is also noticeable.

Dark centimetric lamina: The second microlithofacies is constituted by massive micritic carbonate with large fragments of plant remains, mostly partially pyritized and some of them with a framboidal texture, and charcoal, as well as, black masses, which could be interpreted as sulfide-

and/or manganese- rich accumulation. These accumulations are laterally discontinuous, and their borders are diffuse.

Peletoidal lamina: The third microlithofacies is mainly composed of light massive micritic carbonate, rich in fecal pellet aggregates, with terrigenous minerals floating in this matrix, and large quantities of ostracods (complete shelves and fragments) and diatoms. In this case, the elongated particles are arranged without any preferential orientation.

4.1.2 Facies distribution

Petrological and SEM observations suggest that the main part of the carbonates composing these sediments is endogenic. The facies of these cores seem to correspond to distal shelf sedimentation, affected by muddy contribution (evidenced by the terrigenous grains) related to the Jergueland and Tyup rivers, as well as to the aeolian input. These microlithofacies are arranged forming couples of light (first microlithofacies) and dark (second microlithofacies) lamina. Only in the coarser textural alternation of the two cores are the light lamina represented by the peletoidal microlithofacies.

The origin of the millimetric alternation seems to be associated with the annual water level oscillation due to the melting of the ice cap that occurs in spring and to the ice accumulation in autumn. On the other hand, the centimetric alternation of these microfacies seems to be related to early diagenetic redox mobilization of iron associated with sulfate reduction and organic matter decay (see Giralt et al. (2002) for further details).

4.2 Mineralogy

The mineralogical composition of the two uppermost meters of the Lake Issyk-Kul infill is formed by two main fractions: an endogenic one (calcite, magnesian calcite, monohydrocalcite and palygorskite) and a terrigenous one (quartz, illite, clinochlorite, microcline, albite and riebeckite) (Giralt et al., 2002). Monohydrocalcite is the most abundant mineralogical species in the endogenic fraction whereas illite is the most common within the terrigenous fraction (Fig. 3).

From bottom to top, and on the basis of the carbonate species dominance, four zones were defined:

1) **Zone D**: this is only found at the base of core IK 98-28 and it ranges from 180 cm to 147 cm depth. This zone is dominated by magnesian calcite with percentages that range between 30 and 55% of the total weight. Illite exhibits the highest percentages.

2) **Zone C**: this zone is located in core IK 98-28 from 147 cm to 97 cm depth and in core IK 01-85 from 147 cm to 121 cm depth. Calcite is the main carbonate in this zone and its percentages range from 30 to 70%. In both cores, the terrigenous fraction displays the lowest percentages.

3) **Zone B**: this zone is located between 97 cm and 83 cm depth in core IK 98-28 and between 121 cm and 107 cm depth in core IK 01-85. The main carbonate in this zone is magnesian calcite with percentages that range between 50 and 70%.

4) **Zone A**: this zone corresponds to the upper 83 cm of core IK 98-28 and to the upper 107 cm of core IK 01-85 and is mainly composed of monohydrocalcite and of minor percentages of calcite. The percentages of the first carbonate range between 30 and 50% whereas those of the second carbonate are roughly constant (at about 15%).

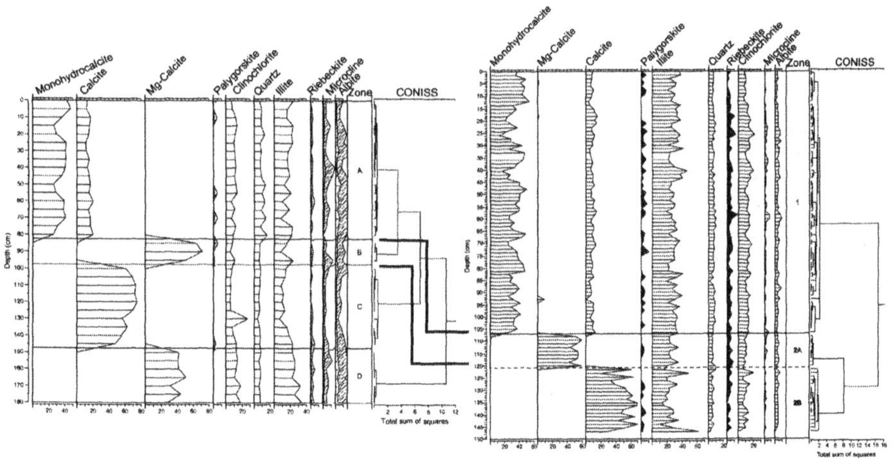

Figure 3. Mineralogical composition of core IK 98-28 (left) and of core IK 01-85 (right) and their mineralogical correlation.

In spite of the different sampling resolution applied in the two cores (a sample every 5 cm in the case of core 98-28 and a sample every 1.5 cm in the case of core 01-85), their mineralogical content shows the same oscillations (Fig. 4 left).

This correlation has allowed defining homologous points between both cores (Fig. 4 left). The monohydrocalcite increases at 84 cm and at 47 cm in core IK 98-28 are located at 107 cm and at 65 cm in core IK 01-85. The differences in core depths between both cores were interpreted as different sedimentation rates. The depths of core IK 98-28 were adjusted to core IK 01-85 using the equation determined by the homologous points (Fig. 4 right) given that nearly all the analyses were performed in this second core. This adjustment was performed in order to facilitate the comparison of the different proxies studied.

4.3 Carbonaceous particles

According to the chemical composition, the sphere and the semisphere particles could be attributed to a combustion process, but their source is unclear. However, they do not preserve vegetal structures (Fig. 5).

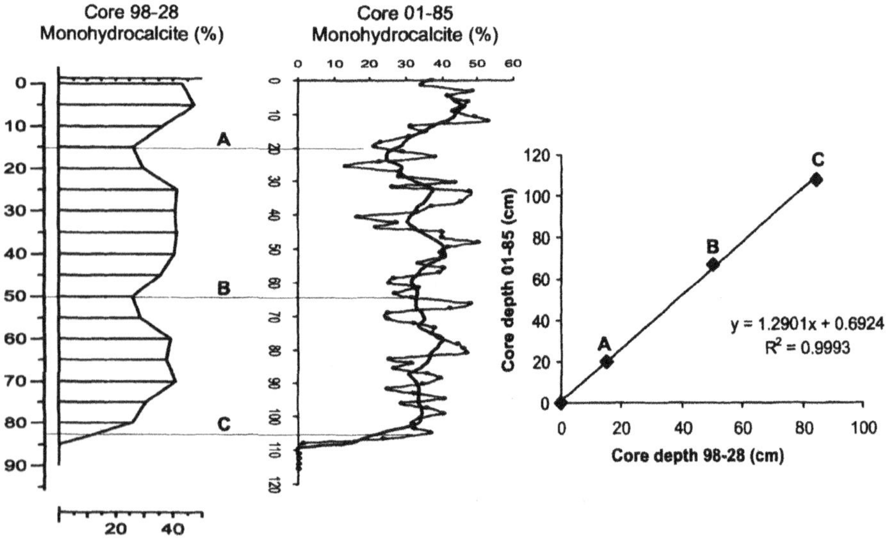

Figure 4. Correlation of the monohydrocalcite percentages of cores IK 98-28 and IK 01-85 (left). They show the same oscillations. Mathematical equation that relates the core depths of core 98-28 with those of core 01-85 (right).

Total Carbonaceous Particles (TCP), spheres and semispheres as well as charcoal concentrations show similar patterns, suggesting that their origin and diffusion are related. The maximum carbonaceous particle concentration occurs between 22 and 15 cm depth, with a peak at 20 cm. Secondary peaks occur at 32, 17 and 11 - 14 cm depth (Fig. 6).

Plant remains are more abundant than SCP, spheres and semispheres. Relative values of non-plant particles increase at 13 - 22, 11 and 2 cm depth. During the periods with low carbonaceous particle concentrations, plant remains are much more dominant (Fig. 6).

Inside the group of plant particles, a progressive increase in burned herbaceous particles occurs, mainly after 24 cm depth coeval with the highest peak in carbonaceous particle concentration. Commonly, concentration of herbaceous particles is higher when total particle concentration is low.

Plant particles ranging from 20 to 50 μm are dominant along all the sequence with values higher than 80%. Concentrations of plant particles larger than 50 μm increase in the uppermost 22 cm (Fig. 6).

4.4 Pigments

Pigment extracts contained a mixture of carotenoids and phorbins including some derivatives. Identified carotenoids were of two types, the carotenes and the xanthophylls. The carotenes were represented by ββ-carotene, which was not detected in high amounts in the sediment record despite being present in all algal groups together with chlorophyll-a (Chl-a). On the other hand, the xanthophylls made up the majority of the carotenoids found and their diversity lead to their utility as taxonomic markers. Xanthophylls identified were marker pigments of Cryptophyta, Chlorophyta, Cyanobacteria and zooplanktonic invertebrates. Xanthophylls used as marker pigments of specific classes of algae are presented in Fig. 7, which also shows their taxonomic affinities.

Phorbins were less important as marker pigments because most of them gave redundant information when xanthophylls were present. Identified a-phorbins included Chl-a together with two of their derivatives, pheophytin-a and Chl-a epimer. Pheophytin-b is a Chl-b derivative the only representative b-phorbin identified. Chl-a is the most ubiquitous pigment and it has traditionally been used as a general indicator of phytoplankton production whereas Chl-b is associated with Chlorophyta and higher plants.

In addition, Bpheophytin-a, a derivative product of bacteriochlorophyll-a, was identified with no associated xanthophyll and was used as a marker pigment substitute for specific phototrophic bacteria.

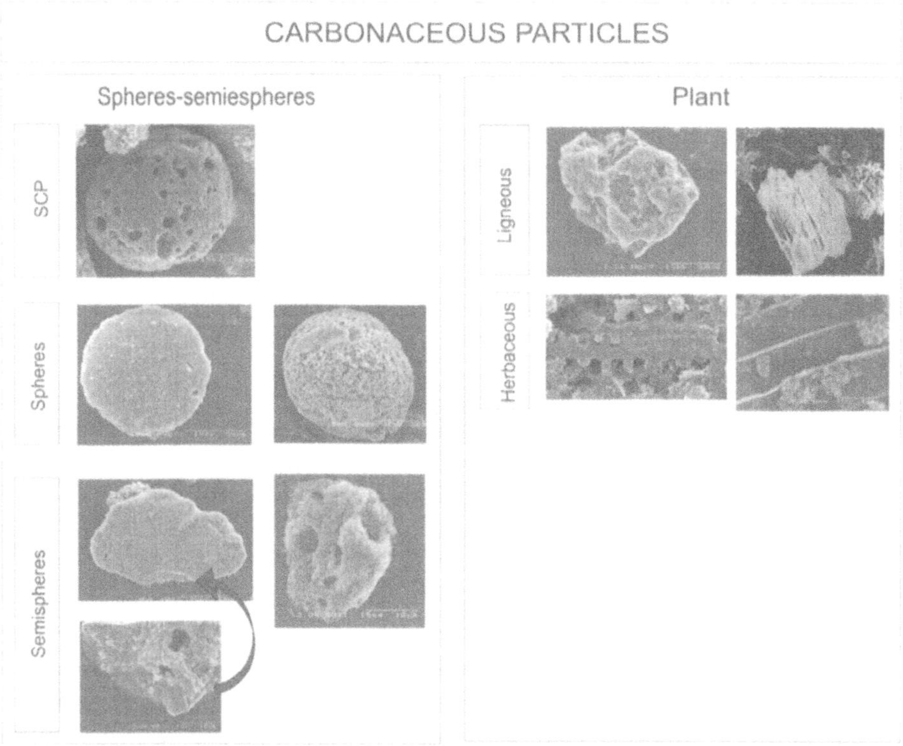

Figure 5. Different morphological combustion particles found in Lake Issyk-Kul sediments.

Pigment concentration was in general very low along the record despite the fact that the samples integrated long time because of the high sediment compaction.

On the basis of pigment composition and from the bottom to the top of the sequence three zones were defined. The lowest pigment concentration was found in zone 3 (between 58 and 25 cm depth in Figure 7) and the highest in zone 2 (between 25 and 7 cm depth) with a maximum close to 15 cm depth. The sum of the Phorbins in Figure 7 behaves as a general phytoplankton production indicator since the main contributors to that sum in Lake Issyk-Kul sediments were a-phorbins. If we assume a constant sedimentation rate of terrigenous material along the sequence, the main productive period, evidenced by an increase in the Phorbin sum concentration, was situated in zone 2.

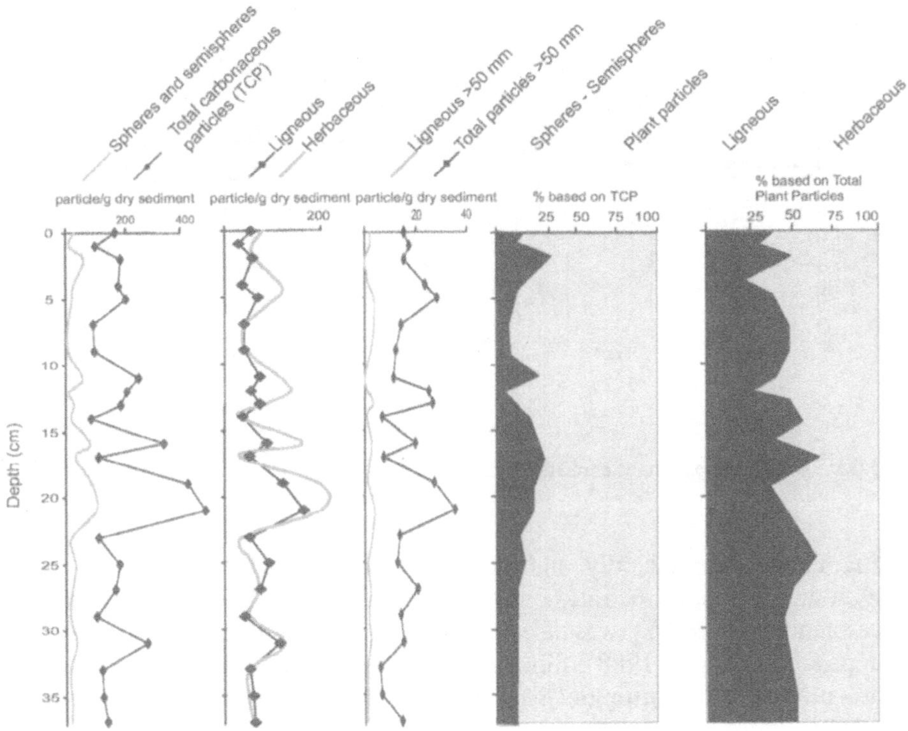

Figure 6. Different Spherical Carbonaceous Particles (SCP) and other plant remains of core 01-85. Number of particles expressed as total per gram of dry sediment (left) and expressed in percentages (right).

Echinenone, the main Cyanobacteria marker pigment identified, was present between 24 and 9 cm depth conciding with the main productive period. Situated at the maximum of this period appeared oscillaxanthin, which is a pigment highly specific to only two genera in the family Oscillatoriaceae (Cyanobacteria). The rise in Oscillaxanthin concentration has been used to date the onset of anthropogenic disturbances (Swain, 1985) and its presence suggests a change in the trophic status. The distribution of Cryptophyta and Chlorophyta marker pigments was wider, and extended above and below this layer. The zooplankton marker pigment astaxanthin appeared before the main productive period at the beginning of zone 2 whereas Bpheophytin-a was present in only one sample of the uppermost part after the main productive period. The presence of a phototrophic bacteria marker pigment suggests the existence of anoxic layers in the sediment surface.

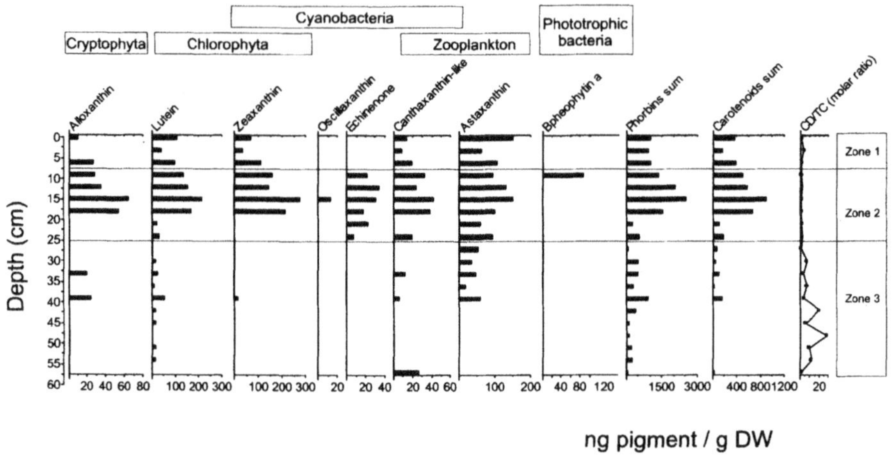

Figure 7. Pigment profile of core IK 01-85.

Carotenoids degrade only slightly more rapidly than chlorophyll derivatives with the result that the ratio between total chlorophyll derivatives and total carotenoids (CD/TC) was used as an indicator of the amount originally deposited (Sanger, 1988). Since relatively more carotenoids than phorbins are produced in eutrophic lakes (Swain, 1985), the marked decrease in CD/TC values above 40 cm depth can be attributed to a change in the overall algal community composition of the lake suggesting, a quick eutrophication of the water.

4.5 Palynology

The diagram is largely dominated by Non-Arboreal Pollen (NAP) (Fig. 8 top, middle and bottom). *Artemisia* pollen percentages are mainly around 60 %.

Next in importance are Amaranthaceae-Chenopodiaceae (15 %) and Poaceae (8 %). In general, a dry steppe belt (*Artemisia* and Poaceae) dominates the sequence, with strong influences from the nearby semi-desert. The *Abies* forests do not seem to be recorded at all in the sedimentary sequence. It seems that pollen grains of *Quercus* are mostly the result of long-distance aerial transport, which is frequent in mountain settings especially with low biomass.

According to the cluster analysis and from the bottom to the top of the sequence six zones were defined.

4.5.1 Palynological zone 6 (63 to 56 cm)

This zone is characterized by maximum values of Asteraceae Liguliflorae (1-2 %) and Cyperaceae (4-6 %), and a minimum of *Ephedra distachya* type percentages. Poaceae > 37 µm are present with 1.5 % (Fig. 9).

The large Poaceae pollen grains may have come from a variety of sources: wild or cultivated. Their small peak at 34 cm is not accompanied by other indicators of agriculture (such as *Centaurea cyanus*, *Polygonum aviculare*, *Rumex acetosella*, *Fagopyrum* and Brassicaceae, occasionally present in the diagram), inviting caution as to the reconstruction of agricultural activities at this depth (Andersen, 1978).

4.5.2 Palynological zone 5 (56 to 33 cm)

This is a fairly stable zone in the percentage diagrams, whereas in the diagram of concentrations there is a maximum in the lower part (63,114 g/ml) followed by a decrease in the upper part. This is especially reflected by maximal concentrations of *Artemisia*, Amaranthaceae-Chenopodiaceae and Poaceae in the lower part of this pollen zone. *Picea* percentages are the lowest in this zone (2 %). This zone displays an increase in *Ephedra distachya* type, a bell-shaped curve of Caryophyllaceae at the center, and gently decreasing values of Cyperaceae. Amaranthaceae-Chenopodiaceae (19 %) and *Typha-Sparganium* (1.2%) have their maximum development in this zone. The values of *Botryococcus* show a progressive increase.

4.5.3 Palynological zones 4 and 3 (33 to 28 cm)

These two zones are discussed together given that they are made up of only one sample. They illustrate a brief period of rapid change. Zones 4 and 3 constitute the limit between the lower part of the diagram, rich in pollen, and the upper part, which is poorer. In addition, zone 3 is the poorest of the whole diagram (10,531 g/ml). Zone 3 contains a brief maximum of *Picea* in the percentage diagrams, which scarcely appears in the concentration diagram (Fig. 9). *Artemisia* drops by 23 %. The concentration diagram confirms this drop. This zone shows a continuous curve of fungal spores.

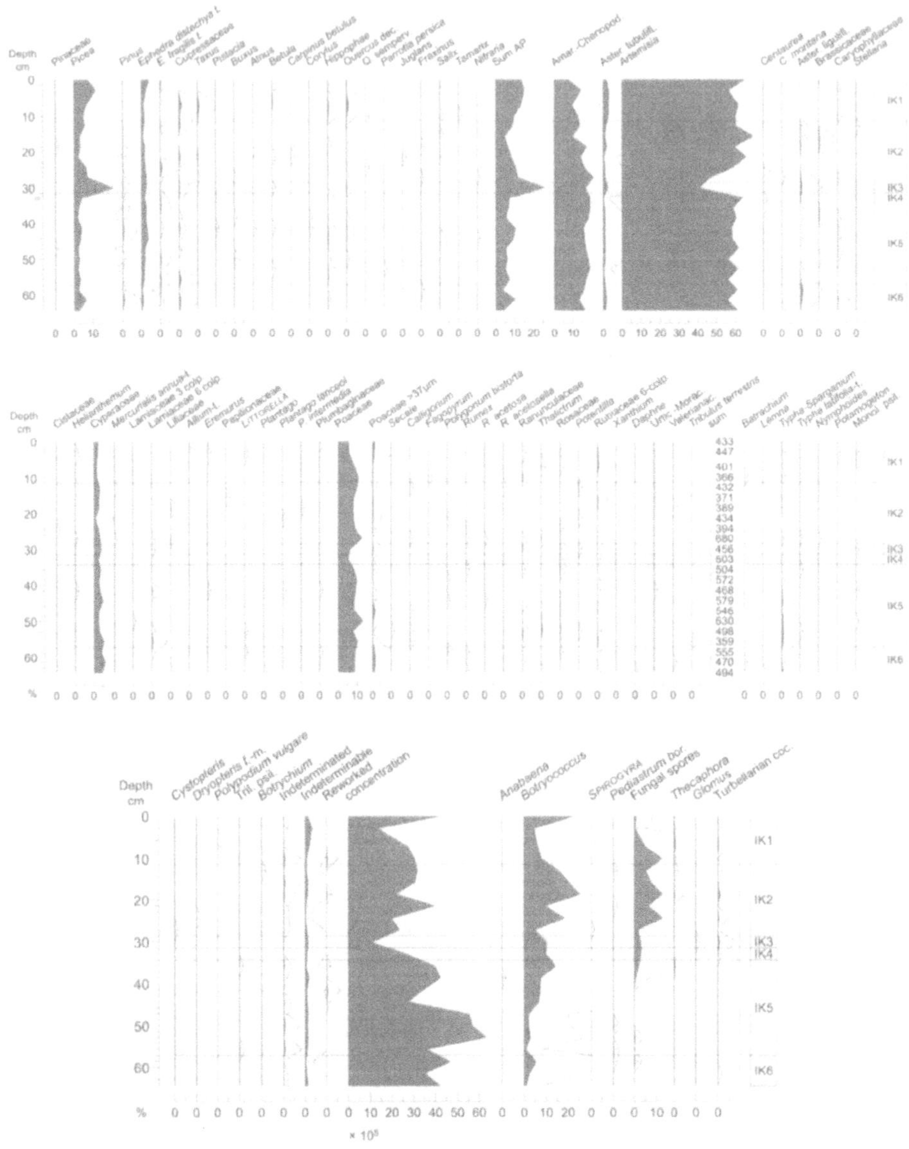

Figure 8. Arboreal and shrub pollen (AP) (top), Non Arboreal pollen diagrams (NAP) (middle) and other biological remains (bottom) from Lake Issyk-Kul. Curve exaggerated ten times. Black dot for values smaller than 0.05 %. "Sum AP" = sum of all arboreal pollen in percentages; "sum" number of terrestrial pollen grains counted; this sum is used to calculate all percentages. Concentration in grains per ml of wet sediment.

4.5.4 Palynological zone 2 (28 to 12 cm)

In this zone the percentage values of *Picea* start by decreasing and end by re-increasing. *Pinus* peters out towards the end of this zone. Amaranthaceae-Chenopodiaceae percentages progressively decrease whereas those of *Artemisia* reach their maximum. Rosaceae values do not return to previous percentages. The zone displays a maximal representation of fungal spores and *Botryococcus.*

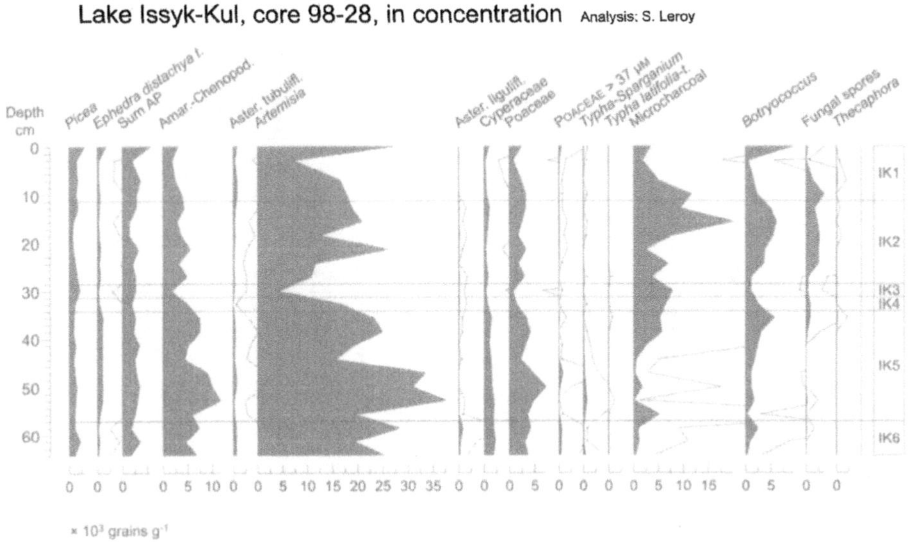

Figure 9. Concentration pollen diagram. Palynological diagram in concentration of selected taxa in grains per ml of wet sediment. Ten times exaggeration curve. Black dot for values smaller than 100 g.ml-1.

4.5.5 Palynological zone 1 (12 cm to top)

The Arboreal Pollen (AP) values increase in this zone owing to *Picea* and the two *Ephedra* taxa. The two *Ephedra* taxa reach their maximum percentages in the uppermost sample. The absence of *Pinus* pollen grains in this zone is striking. Cupressaceae (*Juniperus* most probably) percentages are relatively high (1.7 %) in the lower part of this zone but they disappear towards the top of the sequence. As regards *Picea* at 3 cm the percentage peak is visible but not the concentration peak. Maximum values are reached for Asteraceae Tubuliflorae percentages. Excluding the lowest sample of this pollen zone Rubiaceae (possibly *Galium*) have a continuous representation. In the concentration diagram, *Artemisia* values sharply increase in the uppermost sample. The percentages of Amaranthaceae-Chenopodiaceae keep decreasing throughout this zone. The percentages of *Botryococcus* are now lower except for the uppermost sample. There is a steady decline of microcharcoals and fungal spores. This zone however shows a maximum of *Thecaphora* fungal spores.

4.6 Chronology

The ^{210}Pb and ^{137}Cs concentration profiles of both cores are shown in Fig.10.

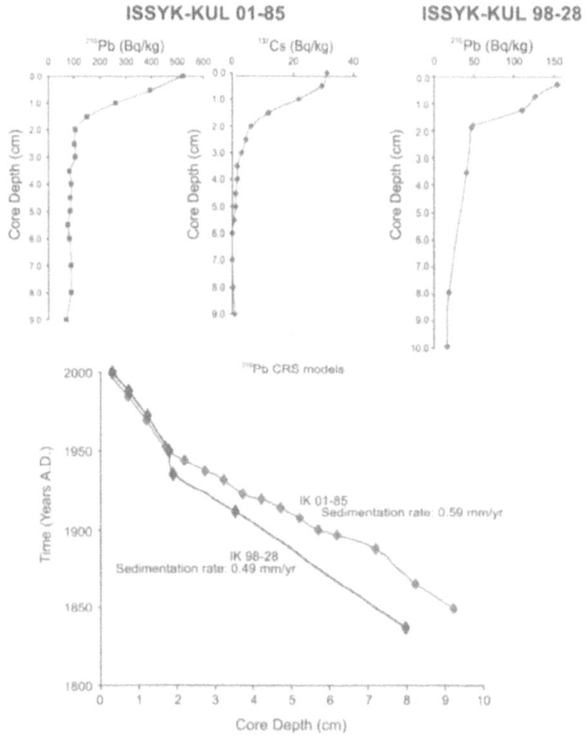

Figure 10. ^{210}Pb and ^{137}Cs profiles of the cores 98-28 (top left) and 01-85 (top right). CSR model calculated for both cores (bottom).

The highest value of ^{210}Pb (155 Bq/kg of dry sediment) for core IK 98-28 is found in the uppermost-analyzed sample (0.29 cm). The ^{210}Pb values abruptly decrease downcore to 50 Bq/kg at 1.9 cm of core depth. From this depth down to 10 cm of core depth, the ^{210}Pb values slightly decrease to 16 Bq/kg, this value being the lowest one. The same ^{210}Pb pattern is observed for core IK 01-85. The highest ^{210}Pb values are at the top of the core (523 Bq/kg) progressively decreasing downcore to 108 Bq/kg at 2 cm of core depth. From this core depth to the lowest analyzed sample the ^{210}Pb values slightly decrease almost to 72 Bq/kg. On the other hand, the ^{137}Cs profile of this core shows an increase in this artificial radionuclide between 3 and 2 cm depth. This increase was attributed to the 1950s on account of nuclear weapons testing (Cambray et al., 1989) .

5. DISCUSSION

5.1 Chronology

The lowest ^{210}Pb activity values measured at 9 cm in core IK 98-28 and at 8 cm in core IK 01-85 were attributed to supported ^{210}Pb. The activities measured between 8 and 2 cm are quite similar, with minor oscillations. These oscillations could be ascribed to small changes in sedimentation rates. Owing to changes in sedimentation rates, a Constant Rate of Supply (CRS) model was used to establish a chronological model for the uppermost 10 cm (Appleby and Oldfield, 1978; Ugur et al., 2003) (Fig. 10 bottom).

The results of the CRS models were the same for the uppermost 1.5 cm (1954) with a sedimentation rate of 0.32 mm/yr. Moreover, the results of core IK 01-85 resemble the ^{137}Cs increase in the 1950s. Between 2 and 10 cm depth, the sedimentation rate of core IK 01-85 was higher than in core IK 98-28. Sedimentation rates were 0.59 mm/yr for the first core and 0.49 mm/yr for core IK 98-28.

These sedimentation rates agreed with those that were already available based on a set of 12 AMS ^{14}C dates (Ricketts et al., 2001). Sedimentation rates derived from these radiocarbon dates were 0.56 mm/yr for core 11 P and 0.47 mm/yr for core 10 P (Fig. 11).

The age of the samples located above 10 cm of core depth were calculated using the CRS model for each sequence. Chronology was obtained for both cores below 10 cm depth, using the sedimentation rates obtained in each core. Similar ages were obtained by applying the previous calculated sedimentation rates (0.59 mm/yr for IK 01-85 and 0.49 mm/yr for core IK 98-28) to the samples located below 10 cm up to the level of appearance of monohydrocalcite. In core IK 98-28, the monohydrocalcite appeared at 84 cm depth corresponding to 289 AD whereas in core IK 01-85 this mineral appeared at 107 cm depth corresponding to 189 AD. The difference in age of both cores (100 years) would only represent an error of 5%.

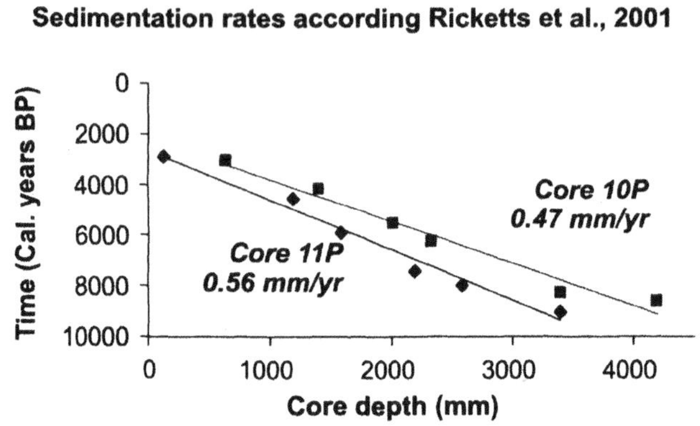

Figure 11. Sedimentation rates of two cores from Lake Issyk-Kul calculated using AMS ^{14}C dates (Modified from Ricketts et al., 2001).

5.2 Origin of the sediment

The sediments of cores IK 98-28 and IK 01-85 could have two main origins:

5.2.1 Terrigenous fraction

The terrigenous fraction is mainly composed of siliciclastic minerals (quartz, illite, albite, microcline, riebeckite and clinochlorite).
The mountain ranges of the Lake Issyk-Kul basin are mainly made up of crystalline basement rocks (granites and granodiorites) of Archaean to Middle Paleozoic age, covered by volcano-sedimentary strata of Devonian-Carboniferous age (Atlas Kyrgyzskaja, 1987). The main source of terrigenous material entering in the lake is due to the erosion of these rocks by glacial and fluvial networks.

On the other hand, the grain size analysis and the electron microscope observations indicated that the mean grain size ranged between 6 and 10 □m. These grain sizes have also been described as belonging to loess deposits (Ding et al., 2002). Moreover, the mineralogical composition of these loess deposits is dominated by siliciclastic particles (clays and feldspars). Recent works have shown that the Asian dust storm particles have a distinctive chemical signature with predominant calcium, silica and aluminum (Ma et al., 2001), indicating that they are siliciclastic minerals (Sun, 2002). This distinctive chemical signature has also been found in the Inilchek glacier located very close to Lake Issyk-Kul (Kreutz et al., 2001).

The Siberian and the Mongolian High-pressure cells over Central Asia are responsible for the remobilization of fine dust mainly from the Taklamakan desert, and for transporting it, through the westerlies over the Tien Shan to the Japan Sea (Sun, 2002). Historical Chinese records of dust rain, mainly generated by these dust storms, show that these storms were more frequent during cold and/or dry periods (Deer, 1984). Moreover, these dust storm periods coincided with the main eastern Asia nomadic migrations (Fang and Liu, 1992).

Thus, the terrigenous fraction is mainly composed of a fine mixture of aeolian and riverine sedimentary particles. The Jergueland and Tyup rivers are the main sediment contributors to Lake Issyk-Kul. Both cores are located in the distal part of deltaic deposits, and their sediments represent the fine-grained suspended load sediments from the water column, mainly controlled by the current regime, the thermal stratification characteristics of the water column and the suspended sediment load of the inflowing rivers (De Batist et al., 2002).

5.2.2 Endogenic fraction

The endogenic fraction is mainly composed of calcium carbonates (monohydrocalcite, calcite and magnesian calcite) and clays (palygorskite). Microbial biscuits of vaterite have been found only in the uppermost sediment of the shallow parts of the lake (Giralt et al., 2001).

The precipitation of monohydrocalcite and of vaterite has been usually associated with biological processes, especially with bacterial communities, among other factors (Rasmussen et al., 1996; Léveillé et al., 2000; Giralt et al., 2001). In fact, large communities of microbialites located at 100 m of water depth have been described in Lake Issyk-Kul and it has been proposed that they are the main promoters of the precipitation of monohydrocalcite (Rasmussen et al., 1996). A number of works have also demonstrated that the presence of phosphorous in the water (p.e. sodium hexametaphosphate or triphosphate) inhibits the spontaneous precipitation of anhydrous carbonates such as calcite even if these are oversaturated, and stabilizes the hydrated forms (Hull and Turnbull, 1973; Stoffers and Fischbeck, 1974; Clarkson et al., 1992). Whatever the exact mechanism triggering the monohydrocalcite precipitation, it seems evident that this mineral is an indicator of the productivity of the lake.

Over a longer timescale, the productivity of Lake Issyk-Kul seems to be controlled by the nutrient availability and by solar activity (solar irradiance, among other parameters). In fact, the nutrient input will also be related to the water availability (more rainfall, more vegetal cover and biomass but also more runoff, and thus, more nutrient input). Thus, the primary productivity of the lake (or in other words, the monohydrocalcite percentages) would be a direct function of the solar activity.

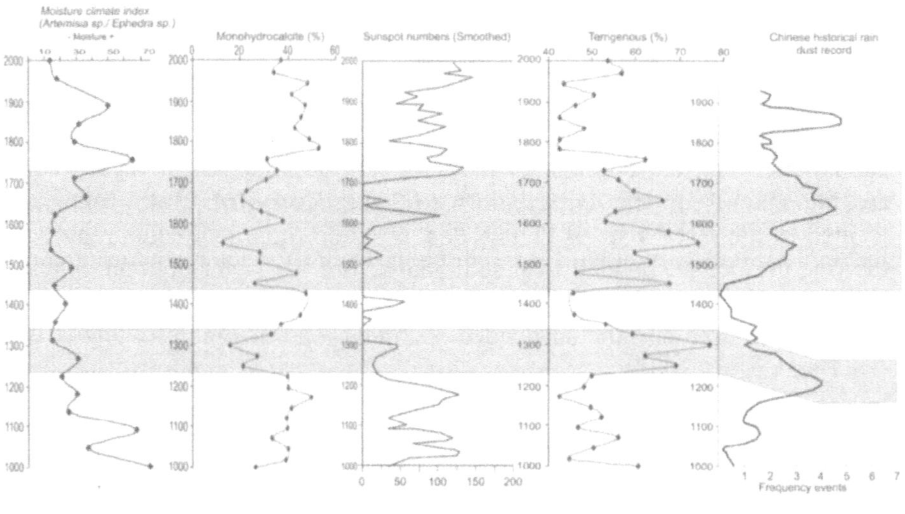

Figure 12. Comparison of the Moisture Climate index, the monohydrocalcite percentages, the reconstructed sunspot number (Modified from Rigozo et al., 2001), the terrigenous content and the reconstructed Chinese rain dust record (Modified from Deer, 1984).

The comparison of the monohydrocalcite percentages with the reconstructed sunspot numbers for the last millennium (Rigozo et al., 2001) shows a good agreement (Fig. 12). In spite of the general low water availability and cold conditions during the period AD 1,200 - 1,710 and given the low number of sunspots, small fluctuations in the water availability (such as those that took place at about AD 1,400 and at about AD 1,700) led to a considerable increase in the lake productivity. Thus the long-term evolution of the water availability (high from at least AD 1,000 to 1,200, low from 1,200 to 1,710 and high again from 1,170 to the present-day) is not recorded by the lake. On the other hand, this long-term evolution is evidenced by the vegetal cover. The moisture climate index (established using the Artemisia sp / Ephedra sp ratio) clearly shows these three water availability phases, but the short-term oscillations are less evident.

5.3 Environmental periods

According to the chronological model, the last 1,000 years are recorded in the upper 48 cm of the sedimentary infill of Lake Issyk-Kul. From the bottom to the top of the sequence, and on the basis of the oscillations of the analyzed proxydata 7 environmental periods were defined (Fig. 13).

5.3.1 Period VII (AD 1,000 - 1,180)

During this environmental period, Lake Issyk-Kul was characterized by high water levels (high monohydrocalcite percentages) and by oligotrophic conditions (observed in the pigment index Chlorophyll derivatives / Total carotenoids) which increased upwards (progressive augmentation of the oligotrophy index). The regional environmental setting was characterized by a progressive decrease in the moisture (decrease in the pollen moisture climate index), provoking a slight fall in the arboreal biomass.

5.3.2 Period VI (AD 1,180 - 1,308)

From AD 1,180 to 1,308, the Lake Issyk-Kul water levels were reconstructed as progressively declining with its minimum at the beginning of the 14th century. On the other hand it was during this period when the lake reached its maximum oligotrophic conditions (low monohydrocalcite percentages and highest values in the oligotrophy index). The regional moisture continued its progressive and fluctuating decrease. The beginning of the 14th century recorded the driest conditions, as was suggested by a minimum in the moisture climate index. These driest conditions narrowed the spruce forest belt. The increase in the percentages of Typha-Sparganium suggest the formation of shallower areas in the lake and marshes related to a drop in the water level. The maximum expansion of marshy vegetation and hence the lower lake levels occurred in the first half of the 13th century. Historical documents point out that the palace of Timur was constructed in these shallow areas at the end of the this century.

5.3.3 Period V (AD 1,308 - 1,562)

During this period of time the lake water level rose slightly (and the dilution of the pollen concentration indicates higher sedimentation rates), although the regional moisture continued to be low (as indicate by the moisture climate index), favouring a semi-desert vegetation similar to periods VII and VI. This rise in the water level led to an growth of primary productivity, mainly because of an increase in soil erosion (higher

Botryococcus values) allowing nutrient input (with the subsequent decrease in the oligotrophy index). Towards the 16th century the lake level diminished, reaching its minimum level at about 1,560 AD. Grazing activities near the lake shores commenced owing to the increase in fungal spores.

5.3.4 Period IV (AD 1,562 - 1,681)

From 1,560 AD there was a progressive increase in the regional moisture (steppe vegetation invades the semi-desert environment), which triggered a progressive rise in the water level (increase of monohydrocalcite percentages). On the other hand, this higher water availability favored soil erosion (evidenced by the high occurrence in fungal spores), and thus, the nutrient input to the lake, with a progressive growth of primary productivity of this lacustrine ecosystem (reduction of the pigment productivity index and enlargement of *Botryococcus* percentages). During this period anthropogenic pressure (grazing practices) was maximum. Deforestation due to wildfires reached its maximum values (highest carbonaceous particle values), contributing to soil erosion and to fertilization of the lake.

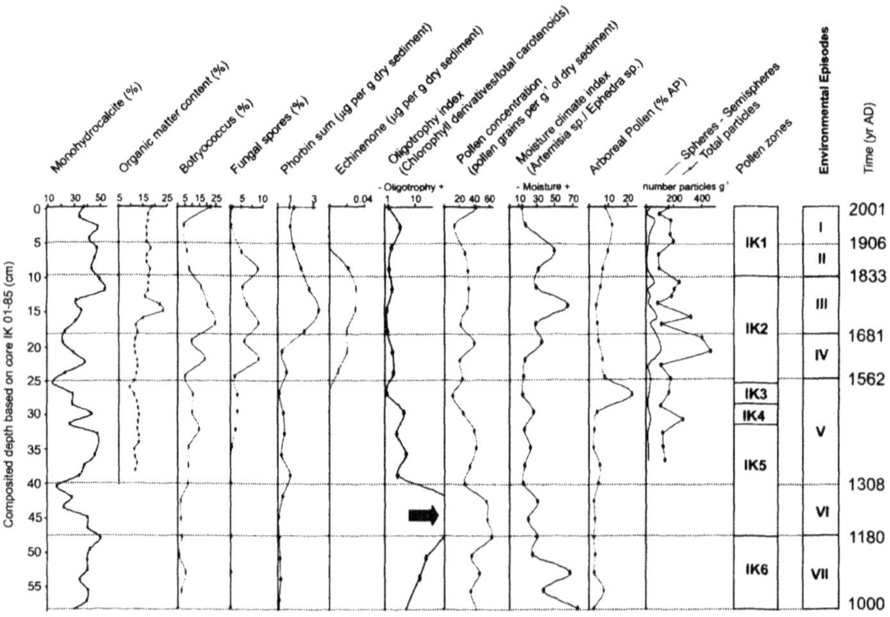

Figure 13. Environmental fluctuations occurred in Lake Issyk-Kul during the last 1,000 years.

5.3.5 Period III (AD 1,681 - 1,833)

In this period the lake reached its maximum primary productivity, which is reflected by the pigments but also by the total organic content. Moreover, the water level rose, reaching its highest level at the beginning of the 19th century, mainly because of the higher values of the regional moisture (maximum values of the moisture climate index and the increase of monohydrocalcite percentages). Historical data also demonstrate the existence of this lake level maximum (Romanovsky, 2002). The grazing activities were maximal. In fact, the incidence of wildfires seems to decrease favoring forest regeneration.

5.3.6 Period II (AD 1,833 - 1,906)

At the beginning of the 19th century there was a slight decline in the anthropogenic pressure, mainly due to the decrease in grazing activities, which resulted in fewer wildfires. The improvement in the climate favored the expansion of forests. After the primary productivity and the water level maxima recorded in Period II there was a progressive decline in both parameters (decrease in the productivity index and in the *Botryococcus* percentages, respectively), indicating a reduction in the fertilization of the lake and soil erosion.

5.3.7 Period I (AD 1,906 - 2,001)

During the last 100 years, the water level and primary productivity continued their decline despite renewed activity in recent years. This fall in the water level, which has also been recorded in the instrumental water record, has been attributed to the expansion of the irrigated agricultural activities on the shores of the lake. Futhermore, a decrease in regional moisture has contributed to this. The reforestation by Scotch Pines, Iberian larches and walnut-trees (since 1948, Kyrgyz-Swiss Forestry Support program, 2003) does not seem to show in the palynological record.

6. ACKNOWLEDGEMENTS

We would like to thank George von Knorring for improving the final version of the manuscript. This work has received finantial support from the Inco-Copernicus program 'Assessment and prognosis of environmental changes in Lake Issyk-Kul (Kyrgyzstan) - APELIK' (Contract number : ICA2-CT-2000-10003).

7. REFERENCES

Aizen EM, Aizen VB, Melack JM, Nakamura T, Ohta T (2001) Precipitation and atmospheric circulation patterns at mid-latitudes of Asia. International J. Clim. 21: 535-556.

Aladin NV, Plotnikov IS (1993) Large saline lakes of former USSR: a summary review: Hydrobiologia 267: 1 - 12.

Alpert P, Neeman BU, Shay-El Y (1990) Climatological analysis of Mediterranean cyclones using ECMWF data. Tellus 42A: 65-77.

Andersen ST (1978) Identification of wild grass and cereal pollen. Damm. Geol. Unders. Arbog. 1978: 68-92.

Appleby PG, Oldfield F (1978) The calculation of 210Pb dates assuming a constant rate of supply of unsupported 210Pb to the sediment. Catena 5: 1 - 8.

Atlas Kyrgyzskaja (1987) Moscow, 157 pp.

Behre KE (1981) The interpretation of anthropogenic indicators in pollen diagrams. Pollen et spores 23: 225 - 245.

Cambray RS, Playford K, Lewis GNJ, Carpenter RC (1989) Radioactive fallout in air and rain: results to the end of 1987. Herwell, AERE-R, 13226.

Chemonics International Inc. Biodiversity Assessment for Kyrgyzstan. USAID Central Asian Republics Mission, Almaty, Kazakhstan (2001) www.biofor.com/documents/Kyrgyzstan.pdf

Clarkson JR, Price TJ, Adams CJ (1992) Role of metastable phases in the spontaneous precipitation of calcium carbonate. J. Chem. Soc. Faraday Trans. 88: 243 - 249.

Cullen HM, DeMenocal PB (2000) North Atlantic influence on Tigris-Euphrates streamflow. Int. J. Clim. 20: 853-863.

De Batist M, Imbo Y, Vermeesch P, Klerkx J, Giralt S, Delvaux D, Lignier V, Beck C, Kalugin I, Abdrakhmatov KE (2002) Bathymetry and sedimentary environments of Lake Isssyk-Kul, Kyrgyz republic (Central Asia): a large, high-altitude, tectonic lake. In: J. Klerkx. and B. Imanackunov (Editors), Lake Issyk-Kul: its natural environment. NATO Science Series, Kluwer Academic Publishers, Netherlands, Earth and Environmental Sciences 13: 101 - 123.

Davies BH (1976) Carotenoids. Academic Press, London, 36 pp.

Deer Z (1984) Synoptic-climatic studies of dust fall in China since historic times. Scientia Sinica (Series B) 27: 825 - 836.

Delvaux D (2001) DTM of Issyk-Kul. Internal Report and CD-Rom. Unpublished.

Ding ZL, Ranov V, Yan SL, Finaev A, Han JM, Wang GA (2002) The loess record in southern Tajikistan and correlation with the Chinese loess. Earth and Planet. Sci. Let. 200: 387-400.

Francus P (1998) An image-analysis technique to measure grain-size variation in thin sections of soft clastic sediments. Sed. Geol. 121: 289-298.

Frang JQ, Liu G (1992) Relationship between climatic change and the nomadic southward migrations in eastern Asia during historical times. Clim. Change 22: 151 - 169.

Fritz SC (1996) Paleolimnological records of climatic change in North America. Limnol. and Oceanogr. 41: 882 - 889.

Gasse F, Van Campo E (1994) Abrupt post-glacial climate events in West Asia and North Africa monsoon domains, Earth and Planet. Sci. Let. 126: 435-456.

Giralt S, Julià R, Klerkx J (2001) Microbial biscuits of vaterite in Lake Issyk-Kul (Republic of Kyrgyzstan). J. Sed. Res. 71: 430 - 435.

Giralt S, Riera S, Klerkx J, Julià R, Lignier V, Beck C, De Batist M, Kalugin I (2002) Recent paleoenvironmental evolution of Lake Issyk-Kul. In: J. Klerkx. and B. Imanackunov (Editors) Lake Issyk-Kul: its natural environment. NATO Science Series, Kluwer Academic Publishers, Netherlands, Earth and Environmental Sciences 13: 125 - 145.

Grimm E (1987) A Fortran 77 Program for Stratigraphically Constrained Cluster Analysis by the Method of Incremental Sum of Squares. Computers & Geosciences 13: 13-35.

Hofer M, Peeters F, Aeschbach-Hertig W, Brennwald M, Holocher J, Livingstone DM (2002) Rapid deep-water renewal in Lake Issyk-Kul (Kyrgyzstan) indicated by transient tracers. Limnol. and Oceanogr. 47: 1210 - 1216.

Hull H, Turnbull AJ (1973) A thermodynamical study of monohydrocalcite. Geochim. Cosmochim. Acta 37: 163 - 170.

Issyk-Kol Biosphere (2001) www.issyk-kol-br.com.kg

Jeffrey SW, Mantoura RFC, Wright SW (1997) Phytoplankton pigments in oceanography: guidelines to modern methods, UNESCO Publishing, Paris, 661 pp.

Karmanchuk AS (2002) Water chemistry and ecology of Lake Issyk-Kul. In: J. Klerkx. and B. Imanackunov (Editors) Lake Issyk-Kul: its natural environment. NATO Science Series, Kluwer Academic Publishers, Netherlands, Earth and Environmental Sciences 13: 13 - 26.

Kreutz K, Aizen V, Cecil D, Wake C (2001) Dust deposition and isotopic composition of precipitation recorded in a shallow ice core, Inilchek glacier, central Tien Shan. J. Glaciol. 47: 549 - 554.

Kyrgyz-Swiss Forestry Support program (2003)
http://www.forest.kg/ForestrySector/for_type.htm

Léveillé RJ, Fyfe WS, Longstaffe FJ (2000) Geomicrobiology of carbonate - silicate microbialites from Hawaiian basaltic sea caves. Chem. Geol. 169: 339 - 355.

Ma CJ, Kasahara M, Höller R, Kamiya T (2001) Characteristics of single particles sampled in Japan during the Asian dust-storm period. Atm. Environ. 35: 2707-2714.

Ministry of the National Environment Protection of the Kyrgyz Republic and UNEP/GRID Arendal (2000) The State of the Environment of Kyrgyz Republic.
http://www.grida.no/enrin/htmls/kyrghiz/soe2/index.htm

Peeters F, Finger D, Hofer M, Brennwald M, Livingstone DM, Kipfer R (2003) Deep-water renewal in Lake Issyk-Kul driven by differential cooling. Limnol. and Oceanogr. 48: 1419-1431.

Rasmussen KA, Romanovsky VV, MacIntyre IG (1996) Late Quaternary coastal microbialites and beachrocks of Lake Issyk-Kul, Kyrgyzstan: geologic, hydrographic, and climatic significance (abstract). Geological Society of America, Annual Meeting, Abstracts with Programs, 28: A-304.

Rhodes TE, Gasse F, Ruifern L, Fontes JC, Keqin W, Bertrand P, Gibert E, Mélières F, Tucholka P, Zhixiang W, Zhi-Yuan C (1996) Late Pleistocene Holocene lacustrine record from Lake Manas, Zunggar (northern Xinjiang, Western China). Palaeogeogr., Palaeoclim., Palaeoecol. 120: 105 - 121.

Ricketts R, Johnson TC, Brown ET, Rasmussen KA, Romanovsky VV (2001) The Holocene paleolimnology of Lake Issyk-Kul, Kyrgyzstan: trace element and stable isotope composition of ostracodes. Palaeogeogr., Palaeoclim., Palaeoecol. 176: 207 - 227.

Rigozo NR, Echer E, Vieira LEA, Nordemann DJR (2001) Reconstruction of Wolf sunspot numbers on the basis of spectral characteristics and estimates of associated radio flux and solar wind parameters for the last millennium. Solar Physics 203: 179 - 191.

Romanovsky VV (1990) Lake Issyk-Kul - a natural system. Ilim., Frunze (in Russian).

Romanovsky VV (2002) Water level variations and water balance of Lake Issyk-Kul. In: J. Klerkx. and B. Imanackunov (Editors). Lake Issyk-Kul: its natural environment. NATO Science Series, Kluwer Academic Publishers, Netherlands, Earth and Environmental Sciences 13: 45 - 58.

Rose NL (1994) A note on further refinements to a procedure for the extraction of carbonaceous fly-ash particles from sediments. J. Paleolim. 11: 201-204.

Rowan KS (1989) Photosynthetic Pigments of Algae. Cambridge Univ. Press, Cambridge, 334 pp.

Sanger JE (1988) Fossil pigments in paleoecology and paleolimnology. Palaeogeogr., Palaeoclim., Palaeoecol. 62: 343-359.

Sapozhnikov DG, Viselkina MA (1960) Recent sediments of the Lake Issyk-Kul and its bays, Moscow, Publishing House Akademii Nauk (in Russian).

Savvaitova K, Petr T (1992) Lake Issyk-Kul, Kirgizia. Int. J. Salt Lake Res. 1: 21 - 46.

Sevastyanov DV (1991) Historical-archaeological data on the changes of shores and lake level fluctuations. In The history of Sevan, Issyk-Kul, Balkhash, Zaisan and Aral lakes. Leningrad, Nauka.

Stockmarr J (1971) Tablets with spores used in absolute pollen analysis. Pollen et Spores 13: 615-621.

Stoffers P, Fischbeck R (1974) Monohydrocalcite in the sediments of Lake Kivu (East Africa). Sedimentology 21: 163 - 170.

Sun J (2002) Provenance of loess material and formation of loess deposits on the Chinese Loess Plateau. Earth and Planet. Sci. Let. 203: 845-859.

Swain EB (1985) Measurement and interpretation of sedimentary pigments. Freshwater Biology 15: 53-75.

Ugur A, Miquel JC, Fowler SW, Appleby P (2003) Radiometric dating of sediment cores from a hydrothermal vent zone off Milos Island in the Aegean Sea. The Sci. of the Total Environ. 307: 203 214.

Walter H, Box EO (1983) The orobiomes of Middle Asia. In: N.E. West (Editor) Ecosystems of the world: temperate deserts and semi-deserts. Elsevier scientific publishing company, Amsterdam, 161 - 191.

Wright SW, Jeffrey SW, Mantoura RFC, Llewellyn CA, Bjornland T, Repeta D, Welschmeyer N (1991) Improved HPLC method for the analysis of chlorophylls and carotenoids from marine phytoplankton. Mar. Ecol. Prog. Ser. 77: 183-196.

Yafeng S, Zhaozheng K, Sumin W, Lingyu T, Fubao W, Tandong Y, Xitao Z, Peiyuan Z, Shaohua S (1993) Mid-Holocene climates and environments in China, Global and Planet. Change 7: 219-233.

Yang HR, Battarbee R (2001) Dating of recent catchment peats using spheroidal carbonaceous particle (SCP) concentration profile with particular reference to Lochnagar, Scotland. The Holocene 11: 593-597.

Zabirov RD, Korotaev VN (1978) Location and morphometry of the lake. Lake Issyk-Kul. Frunze, 12-20 (In Russian).

Chapter 11

THE DEAD SEA AS A DYING LAKE

Ittai Gavrieli[1] and Aharon Oren[2]

[1]*Geological Survey of Israel, 30 Malkhei Yisrael St., 95501 Jerusalem, Israel*
[2]*The Institute of Life Sciences, and the Moshe Shilo Minerva Center for Marine Biogeochemistry, The Hebrew University of Jerusalem, 91904 Jerusalem, Israel*

> Where Sodom and Gomorrah reared their domes and towers, that solemn sea now floods the plain, in whose bitter waters no living thing exists - over whose waveless surface the blistering air hangs motionless and dead - about whose borders nothing grows but weeds ...
>
> (Mark Twain, The Innocents Abroad or The New Pilgrims Progress. 1869).

1. SUMMARY

Anthropogenic intervention has severely disturbed the Dead Sea as an ecosystem. The water level has been decreasing at a rate of nearly one meter per year during the last decade. In 1979 the water column overturned, thereby ending centuries-long stratification. Since then the lake is mostly holomictic, with annual built up of stratification in spring and its destruction in late autumn. The negative water balance of the lake results in a gradual increase in its salinity. The lake is saturated with respect to NaCl, and halite is precipitating to the bottom. The main components of the Dead Sea biota are the unicellular green alga *Dunaliella* and several red halophilic Archaea.

J.C.J. Nihoul et al. (eds.), Dying and Dead Seas, 287–305.

Massive microbial development is possible only when the upper water layers become diluted with more than 10% fresh water and phosphate is available. Dense microbial blooms occurred in 1980 and in 1992. In both cases the archaeal community imparted a reddish color to the lake. Today the lake is virtually devoid of microbial life.

Because of the hygroscopic nature of its solutes, the Dead Sea can never fully dry out and die. A steady state between inflow and evaporation is expected in 200-400 years at water levels some 100-150 m below present level, depending on the volumes of future inflows. A proposal for the construction of a "Peace Conduit" between the Gulf of Eilat/Aqaba and the Dead Sea is currently being investigated. This water carrier is intended to stabilize the water level. The difference in elevation of over 400 m will enable to use it for seawater desalination by reverse osmosis. However the project will greatly change the characteristics of the Dead Sea and therefore must be considered very carefully.

2. INTRODUCTION

The Dead Sea is a hypersaline terminal desert lake located on the border between Israel and Jordan (Fig. 1). Geologically, the lake is situated within the large Dead Sea basin which is one of the pull aparts that formed along the Dead Sea rift (Quennell, 1959; Garfunkel and Ben Avraham 1996). The lake's deepest point (-730 m) is the deepest terrestrial spot on Earth. At the time of writing (mid 2003), the lake's shoreline was located at about 416 m below main sea level. The lake shore is therewith by far the lowest exposed surface on Earth.

The modern Dead Sea evolved in the early Holocene after a major decline in the water level of Lake Lisan, the late Pleistocene precursor of the Dead Sea. Since then the water level of the Dead Sea fluctuated around -400 m (Ken-Tor et al., 2003), which is the elevation of the sill dividing between the shallow southern basin of the lake and the much deeper northern basin. Higher water levels were attained during rainy periods when the lake extended into the southern basin and the surface water was diluted. Lower levels reflect dry periods, with negative water balance and large area shrinkage, including the drying out of the southern basin. The smaller surface area and higher salinity resulted in a drastic decrease in evaporation which served to buffer further lake level drop.

In the 20th century, human intervention in the water balance of the lake has resulted in rapid decline in the water level. In 1977 the southern basin dried out but water level continues to decline at a high rate of about 1 m/yr. This decline is accompanied by other undesired changes in the Dead Sea:

development of hundreds of sinkholes around the shore of the lake (Abelson et al., 2003), exposure of large mudflats, de-watering and sediment shrinkage which lead to local ground sinking (Baer et al., 2002) and rapid geomorphological changes, which result in damages in the surrounding infrastructure, mainly to roads and bridges. Accordingly, the Dead Sea area, which has a major economic, touristic and environmental potential is in fact being abandoned, and future planning becomes difficult. The Dead Sea is thus a prime example of a "dying" lake.

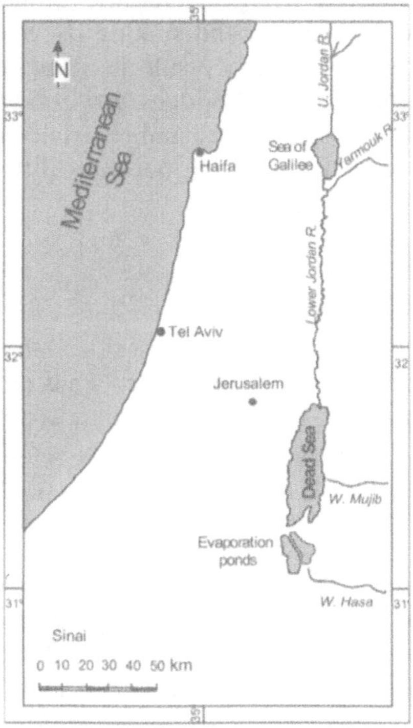

Figure 1: The Dead Sea: location map

The modern Dead Sea evolved in the early Holocene after a major decline in the water level of Lake Lisan, the late Pleistocene precursor of the Dead Sea. Since then the water level of the Dead Sea fluctuated around -400 m (Ken-Tor et al., 2003), which is the elevation of the sill dividing between the shallow southern basin of the lake and the much deeper northern basin. Higher water levels were attained during rainy periods when the lake extended into the southern basin and the surface water was diluted. Lower levels reflect dry periods, with negative water balance and large area shrinkage, including the drying out of the southern basin. The smaller surface area and higher salinity resulted in a drastic decrease in evaporation which served to buffer further lake level drop.

In the 20th century, human intervention in the water balance of the lake has resulted in rapid decline in the water level. In 1977 the southern basin dried out but water level continues to decline at a high rate of about 1 m/yr. This decline is accompanied by other undesired changes in the Dead Sea: development of hundreds of sinkholes around the shore of the lake (Abelson et al., 2003), exposure of large mudflats, de-watering and sediment shrinkage which lead to local ground sinking (Baer et al., 2002) and rapid geomorphological changes, which result in damages in the surrounding infrastructure, mainly to roads and bridges. Accordingly, the Dead Sea area, which has a major economic, touristic and environmental potential is in fact being abandoned, and future planning becomes difficult. The Dead Sea is thus a prime example of a "dying" lake.

3. DEAD SEA WATER LEVEL

During the 20th century, the Dead Sea level has dropped by more than 20 meters (Fig. 2). As a result, in 1977 the shallow southern basin of the Dead Sea completely dried (Fig. 3) and presently the lake, whose level continues to decline by about 1 m/yr, is limited to the much deeper northern basin. The annual water deficit of the Dead Sea, whose present surface area is around 625 km^2 (Hall, pers. com) is 625 million cubic meters.

The decline in the Dead Sea level is a manifestation of the negative water balance of the lake, whereby evaporation greatly exceeds inflows. This negative water balance is attributed primarily to diversion of freshwater for agricultural and domestic use from the two main sources of the Jordan River: since the 1960s, water from Lake Kinneret is pumped to the Israeli National Water Carrier, while most of the Yarmouk River is diverted by Syria and Jordan (to the King Abdullah canal) and the rest is taken up by Israel. Thus, the flow though the Jordan River, which was the single most important water source to the Dead Sea, has reduced from about 1500 million cubic meters (MCM)/year to less than 150 MCM/year (Salameh and El-Naser, 1999; Al Weshah, 2000). Furthermore, the quality of the water presently flowing in the Jordan has greatly deteriorated, and presently it consists mostly of irrigation return flow, saline groundwater which discharges to the river, and treated and untreated sewage (Farber et al., 2003). Nevertheless, during particularly rainy winters, when the dams on Lake Kinneret and the Jarmouk Rivers have to be opened, large volumes of water flow through the Jordan River. Such flows occurred in winter 1979/80 and 1991/92 when lake level rose by 1.5 and 2 meters, respectively (Beyth et al., 1993). A more moderate rise of 60 cm occurred following the rainy winter of 2002/2003. During such winters the frequency and magnitude of the flash-floods in the rivers draining to the Dead Sea are larger, adding their share to the rise in the water level of the Dead Sea. These dramatic lake level rises, however, are worn-out within 2-3 years and do not change the more general trend of the declining lake level.

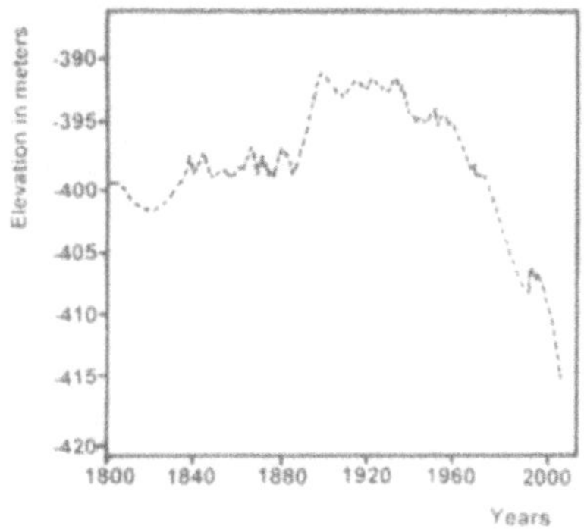

Figure 2: The water level of the Dead Sea: 1800-2000(compilation of Klein (1961) and and new data).

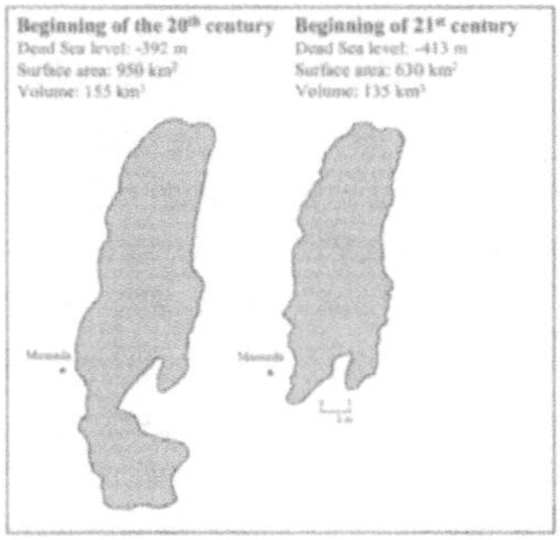

Figure 3: The Dead Sea shore line, water level, surface area, and volume in the beginning of the 20th and the 21st century.

The activities of the Israeli and the Jordanian potash industries near the Dead Sea contribute to the decline of the level by artificially increasing the evaporating surface of the remaining lake. These industries pump together 400-450 MCM of brine from the Dead Sea into the evaporation ponds, located on the otherwise dried southern basin, and return about 200 MCM of concentrated "end brine" back to the lake. Thus, about 30-40 cm/yr of the water level decline of the Dead Sea is due to evaporation in these ponds.

If the current situation will prevail, the Dead Sea level is expected to continue to decline. In fact, future inflow to the Dead Sea is only expected to decrease further, as more of the water currently flowing to the lake will be captured and diverted to meet growing needs for freshwater. More of the Wadi's on the eastern escarpment of the Dead Sea Rift, such as Wadi Mujib with its perennial flow, are planned to be dammed, while the discharge of the major spring system around the lake, En Feshcha locate on the north-western shores of the lake, will probably decrease as water is pumped upstream. These reduced flows are partially offset by some increase in the rate of groundwater discharge to the lake. This increase is due to the receding base level and the consequent increase in the hydraulic gradient and seaward migration of the brine/freshwater interface (Salameh and El-Naser, 1999, 2000a,b).

4. PHYSICAL LIMNOLOGY OF THE DEAD SEA

At the time of the first in-depth study of the properties of the Dead Sea water column, conducted in 1959-1960 by the Geological Survey of Israel (Neev and Emery, 1967), the lake was stratified (meromictic), and the southern basin was still flooded. A less saline upper water mass (around 300 g/l total dissolved salts, down to a depth of about 40 m) floated on top of a denser lower water mass (332 g/l total dissolved salts). The lower water mass was a layer of "fossil" water that had been isolated at least for several centuries (Steinhorn et al., 1979; Stiller and Chung, 1984). This lower water mass was anoxic, and contained sulfide.

With the increase in salinity of the upper water layers, the pycnocline that had existed during the meromictic state weakened. A complete overturn finally occurred in February 1979 (Beyth, 1980; Steinhorn and Gat, 1983; Steinhorn et al., 1979). The lower anaerobic and sulfide-rich "fossil" water mass ceased to exist, and the water column became homogeneous in composition.

Since 1979, periods of holomictic and meromictic regimes have alternated. In the holomictic state (1982-1991 and 1995-present), stratification develops and is destroyed annually (Anati and Stiller, 1991; Gertman and Hecht, 2002). Stratification develops in spring due to the warming of the upper water layer, or less commonly in late winter due to limited dilution of the water as a result of winter rain floods. During the summer months the stratification is maintained by a stabilizing thermocline (generally located between 25 and 30 m depth). Surface water temperatures may reach 35-36°C, while the temperature of the water mass below the thermocline remains stable in the range of 22-23°C. This stabilizing thermocline is large enough to balance the destabilizing halocline that is formed during the summer months as a result of increased evaporation. Overturn occurs following the autumn-cooling of the upper water column and the consequent increase in its density. Anati and Stiller (1991) and Anati (1997, 1998) described these cycles on a temperature-salinity diagrams, demonstrating the gradual increase in the salinity and temperature over the years (Fig. 4). For convenience the salinity in the Dead Sea is commonly expressed in the quasi-salinity units of σ_{20} whereby:

$$\sigma_{20} = (\rho_{20}-1)*1000$$

where ρ_{20} is the density at 20°C, calculated from the measured density (ρ_T) using the thermal expansion coefficient for the Dead Sea brine (Steinhorn, 1980):

$$\partial\rho/\partial T = -0.4309 \pm 0.0005\ \sigma/C:$$

The holomictic regime has been interrupted by two short meromictic episodes: from 1979-1982 and again from 1992-1995 (Anati, 1997; Anati and Stiller, 1991; Anati et al., 1987; Gavrieli et al., 1999; Stiller et al., 1984). This was the result of massive inflow of fresh water during unusually rainy winters. During the winter of 1980 the surface level rose by almost 1.5 m, and an even more dramatic rise in surface level of nearly 2 m occurred between November 1991 and May 1992, when approximately $1.5x10^9$ m^3 of fresh water entered the lake, causing a dilution of the upper 5 m of the water column to about 70% of their previous salinity (Beyth et al., 1993). During the subsequent three years, lake level dropped while the upper water became more saline and the thermocline and halocline deepened, before overturn took place in 1995 (Anati et al., 1995).

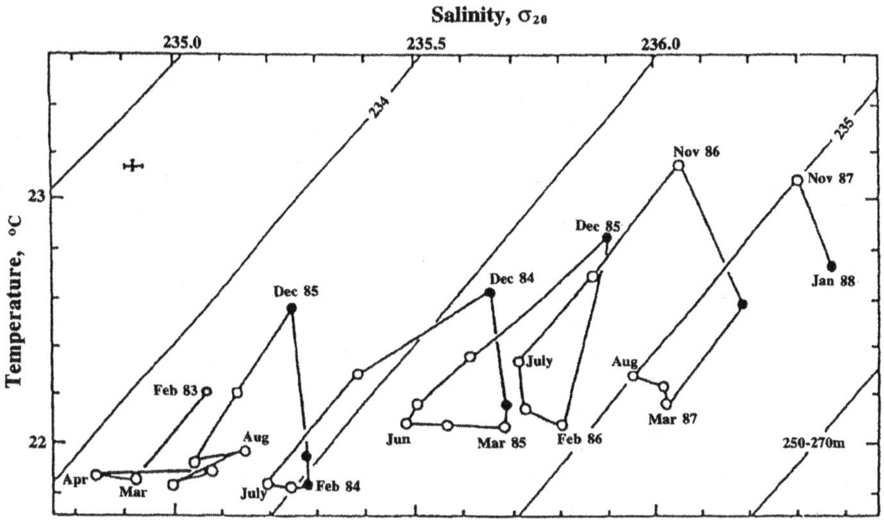

Figure 4: A T-S diagram showing the evolution of the deep the Dead Sea brine at depth of 250-270 meters during a holomictic period (1983-1988) (From Anati, 1998). The diagonal lines denote potential density in σ_{20} units. Open dots denote stratified states, and full dots denote mixed states.

5. THE CHEMISTRY OF THE DEAD SEA BRINE

Undiluted Dead Sea water contains about 342 g/l total dissolved salts and has a density of 1.237 kg/l (Table 1). Among the hypersaline lakes the Dead Sea is unique because of its peculiar Ca-chloride composition [i.e. $Ca/(SO_4+HCO_3)>1$]. Divalent cations dominate (presently $Mg^{2+} + Ca^{2+} =$ 2.33 M, as compared to $Na^+ + K^+ = 1.79$ M). Cl^- and Br^- are the dominant anions (99% and 1% of the anion sum, respectively), and concentrations of SO_4^{2-} and HCO_3^- are very low. (The density of the brine is about 1.237 g/cm^3, and its pH is about 5.9 (Ben-Yaakov and Sass, 1977).

Table 1: The composition of Dead Sea water (summer 2002) and a typical end brine composition (in g/l).

	Na	K	Ca	Mg	Cl	Br	Alkalinity (as HCO_3)	SO_4	TDS
Dead Sea	34.3	8.0	18.3	47.1	228.6	5.4	0.3	0.4	342.4
End brine	3.7	4.1	35.4	88.4	328.2	10.1	0.3	0.5	470.7

The Dead Sea brine evolved from seawater that intruded into the Rift Valley, probably during the Pliocene, and formed the Sedom lagoon (Zak, 1967, Starinsky, 1974, Stein, 2001). The shape of the lagoon, which was some 200 km long and only a few km wide, the rapid subsidence of the Rift Valley and the prevailing arid climate gave rise to deposition of thick layers of evaporitic minerals. These include mainly gypsum and halite with some carnallite and possibly more advanced evaporitic minerals (Zak, 1967). The concentrated brine later percolated into the surrounding limestones and dolomitized major part of the country rock. This process led to further gypsum precipitation and to the removal of most of the sulfate from the subsurface brine. Since then, the brines are being recycled back and forth between the surface and the subsurface, and play a major role in the geochemistry of the lakes that developed in the Rift during the Pleistocene and Holocene, namely Lake Amora, Lake Lisan (70-14 Kyr) and the Dead Sea.

The Dead Sea is presently saturated to oversaturated with respect to aragonite ($CaCO_3$), anhydrite ($CaSO_4$) and halite ($NaCl$) (Gavrieli et al., 1989). Kinetic factors which dominate over thermodynamic considerations dictate that gypsum ($CaSO_4 \cdot 2H_2O$), rather than anhydrite is the actual Ca-sulfate mineral that precipitates from the Dead Sea brine. Prior to the 1979 overturn, the lower water body was saturated with respect to these minerals, whereas the upper water body was undersaturated with respect to halite and saturated to oversaturated with respect to aragonite and anhydrite (Neev and Emery, 1967). Aragonite crystallized from the upper water body and settled to the bottom, forming the white laminae of the Dead Sea sediments, whereas gypsum crystallized on exposed and submerged surfaces along the shores. The "whitening" of the Dead Sea surface, which has been described by several observers before the overturn (Bloch et al., 1943; Neev and Emery, 1967), is attributed to spontaneous crystallization of aragonite, possibly with some gypsum, from the surface water. The Dead Sea sediments are thus characterized by alternating laminae of detrital material which precipitated following the winter flooding to the lake and laminae consisting mainly of aragonite with some gypsum which precipitated usually during summer. Similar sediments where also deposited by Lake Lisan, the late Pleistocene precursor of the Dead Sea (Begin et al., 1974), though the lake Lisan sediments have also thick gypsum layers. At present, despite the saturation to oversaturation of the Dead Sea with respect to aragonite and the Ca-sulfate phases, their precipitation is rather limited. This is due to the decreasing input of freshwater to the Dead Sea, which supplied bicarbonate and sulfate to the lake. Thus, these sediments, which until recently continued to precipitate from the Dead Sea, are currently being exposed along the shores of the lake as the water level recedes.

In 1982 halite began to precipitate from the Dead Sea (Steinhorn, 1983), and its precipitation has continued nearly uninterrupted since then (Gavrieli, 1997). A decrease in halite precipitation rate was observed in 1992-3 to 1995. This was due to the 1991-1995 stratification which diluted the upper water body and isolated the lower water body (Beyth et al., 1993; Anati et al., 1995). Massive halite precipitation was restored following the November 1995 overturn. Under the current negative water balance of the lake and increasing salinity, halite will continue to precipitate from the brine. It should be noted that since 1982, any object suspended within the deeper Dead Sea brines was immediately covered by massive halite crystals. A somewhat similar situation existed in the 1960s when gypsum, rather than halite, quickly covered exposed surfaces, although this was limited only to the upper waters. The continuous precipitation of halite from the Dead Sea brine has resulted in a change in its Na/Cl ratio which over the years has slightly decreased (Fig. 5).

Levels of biologically available nitrogen in the Dead Sea are high. The average concentration of ammonium ions in the water column was reported to be 5.9 mg/l in 1960 and 8.9 mg/l in 1991 (Nissenbaum et al., 1990; Stiller and Nissenbaum, 1999). Nitrate is present in low concentrations only (20 μg NO_3^--N/l in the 1960s, a value that had increased to 200-500 μg/l in 1981 as a result of anthropogenic pollution of the Jordan River) (Stiller and Nissenbaum, 1999). Phosphorus is not abundantly found in the Dead Sea, as its solubility in the lake's brines is limited. Stiller and Nissenbaum (1999) reported dissolved phosphorus levels of about 35 μg PO_4^{3-}-P/l. Particulate phosphorus was more variable at 30-50 μg/l. The sediments were suggested to contribute between 30 and 58% to the phosphate input in the Dead Sea water column, the remainder being derived from the Jordan River and flood waters (Nissenbaum et al., 1990; Stiller and Nissenbaum, 1999). Another source of phosphorus to the Dead Sea is dust from the atmosphere. Dust deposition over a three-year period (1997-1999) varied between 25.5-60.5 g/m^2.year. The average phosphorus content of this dust was 1.2% (calculated as P_2O_5), present mainly as apatite. The value was especially high in the winter months (average of 2.6% for December-February). Thus, between 4 and 10 mmol/m^2 of phosphorus may be estimated to enter the Dead Sea annually from atmospheric dust (Singer et al., 2003). Dissolved oxygen levels in Dead Sea water are low: concentrations measured in the water column in 1987-1989 averaged around 0.8 ml/kg (equivalent to 1 ml/l or 1.4 mg/l) at all depths (Shatkay, 1991; Shatkay et al., 1993).

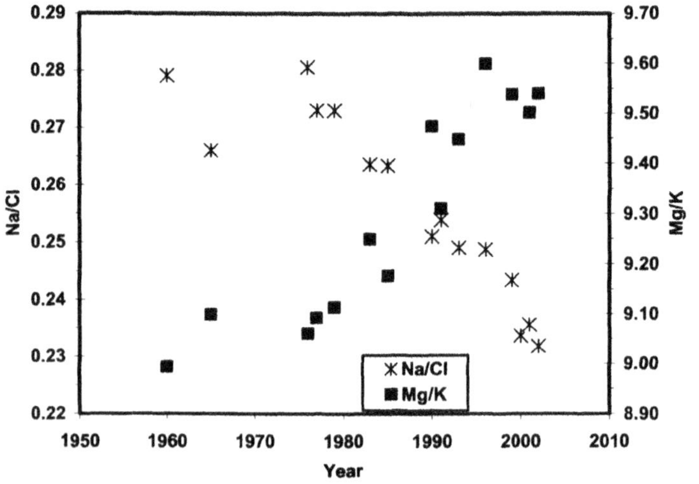

Figure 5: Changes in the molar ratios of Na/Cl and Mg/K in the Dead Sea. The former is due to halite precipitation from the Dead Sea which began in the 1980s. Halite precipitates also in the evaporation ponds of the potash industries, a process that took place long before it began precipitating in the lake. The increase in the Mg/K ratio is due to the harvesting of potassium by the potash industries.

Israel and Jordan have large chemical plants on the shores of the now dried basin of the southern Dead Sea. The primary product of these industries is potash (KCl), although magnesium and bromide products are also produced. The industries pump Dead Sea brine from the northern basin into shallow evaporation ponds constructed in the otherwise dry southern basin of the Dead Sea. As the brine evaporates halite precipitates and at a density of about 1.3 kg/l carnallite ($MgKCl_3 \cdot 6H_2O$) also begins to precipitate. In the factories the latter is dissolved and potash is produced. About 50% of the volume initially pumped from the Dead Sea is conveyed back to the lake as concentrated "end brine" (TDS: 470-500 g/l; Density: 1.33-1.35 kg/l; Table 1). These brines consist mainly of Mg-Ca-Cl with low Na/Cl and K/Cl ratios of 0.02 and 0.01, respectively.

The mixing of these brines, characterized by high Mg/K ratio has resulted in a slight increase in this ratio in the Dead Sea (Fig. 5). Concurrent with this change is a decrease in the Na/Cl ratio, due primarily to halite precipitation from the Dead Sea, and to lesser extent to precipitation in the evaporation ponds.

6. THE MICROBIOLOGY OF THE DEAD SEA - PAST AND PRESENT

In spite of its extremely high salt concentration and its unusual ionic composition, a variety of microorganisms have been shown to live in the Dead Sea. These include autotrophic unicellular green algae (*Dunaliella* sp.) and a number of aerobic heterotrophic prokaryotes. The dominant types are red halophilic Archaea belonging to the family *Halobacteriaceae*: *Haloferax volcanii*, *Halorubrum sodomense*, *Halobaculum gomorrense*, and others (Fig. 6) (for reviews see Nissenbaum, 1975; Oren, 1988, 1997, 1998, 2000).

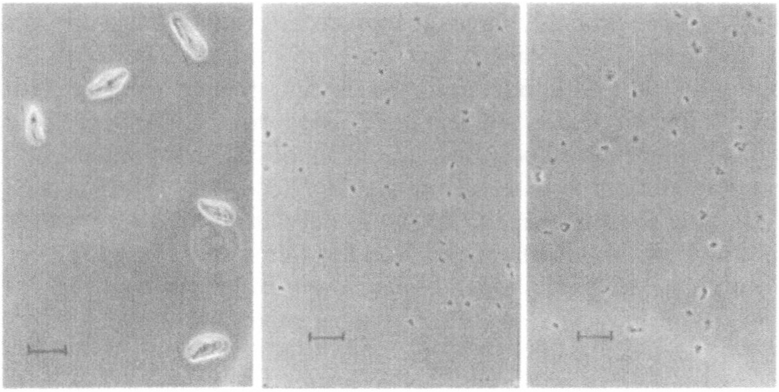

Figure 6. Dead Sea microorganisms. From left to right: *Dunaliella parva*, *Haloarcula marismortui*, and *Haloferax volcanii* (bar = 10 μm).

The Dead Sea is an exceptionally harsh environment, even for those microorganisms best adapted to life at high salt concentrations. The water activity in Dead Sea brines (0.67 calculated in 1979) (Krumgalz and Millero, 1982) is close to the lowest water activity known to support life. In addition, molar concentrations of divalent cations are poorly tolerated even by the most halophilic organisms known.

The pioneering studies of Benjamin Elazari-Volcani (1915-1999) in the late 1930s - early 1940s first showed that the Dead Sea is inhabited by an indigenous community of microorganisms (Elazari-Volcani, 1940a, 1940b, 1943, 1944; Volcani, 1944; Wilkansky, 1936). The first quantitative determinations of the sizes of the algal and archaeal/bacterial communities were performed only in 1963 (Kaplan and Friedmann, 1970). Only from 1980 has a systematic monitoring program of the microbial communities in the Dead Sea been operating. The results have shown that the Dead Sea is a highly dynamic biotope, whose biological properties vary greatly from year to year.

In the present holomictic state of the lake no growth of *Dunaliella* is possible, as the salt concentration of the brine is too high. Algal blooms only appear when the upper water layers become significantly diluted (10-20% at least) by winter rain floods. *Dunaliella* blooms have been observed in 1980 and again in 1992, in both cases triggered by massive winter floods that caused the formation of a diluted epilimnion, initiating a meromictic episode. In the summer of 1980, algal population densities of up to 8.8×10^3 cells/ml were observed (Oren and Shilo, 1982), and even higher numbers (up to 1.5×10^4 *Dunaliella* cells/ml) were counted in the spring of 1992 (Oren, 1993, 1999; Oren et al., 1995) (Fig. 7). No algae were observed the water column during the monomictic periods (1983-1991 and from 1996 onwards). The algal blooms probably develop from resting stages that survive in the bottom sediments of the lake (Oren et al., 1995). Analysis of the spatial distribution of the 1992 bloom at its onset, using remote sensing, confirmed that the bloom originated in nearshore areas around the lake, wherever shallow sediments potentially harboring *Dunaliella* cysts came in contact with the diluted surface water (Oren and Ben-Yosef, 1997). An additional factor essential for a *Dunaliella* bloom to develop in the lake is phosphate, which is the limiting nutrient in the Dead Sea (Oren and Shilo, 1985).

Concomitant with the algal blooms, red halophilic Archaea rapidly develop in high numbers at the expense of organic material produced by *Dunaliella*. Glycerol may well be the main organic compound on which the Archaea thrive as *Dunaliella* cells accumulate glycerol as osmotic stabilizer. We counted up to 1.9×10^7 bacteria/ml in the surface layers in 1980 (Oren, 1983a), and up to 3.5×10^7 cells/ml in 1992 (Oren, 1999; Oren and Gurevich, 1995) (Fig. 7). After the archaeal blooms in 1980 and 1992 had reached their peaks, the community density slowly declined. Bacteriophages may have been a cause for the declines. Electron microscopic examination of water samples collected in 1994-1995 showed large numbers of virus-like particles. Their numbers exceeded those of the prokaryotic cells ten-fold on the average (Oren et al., 1997).

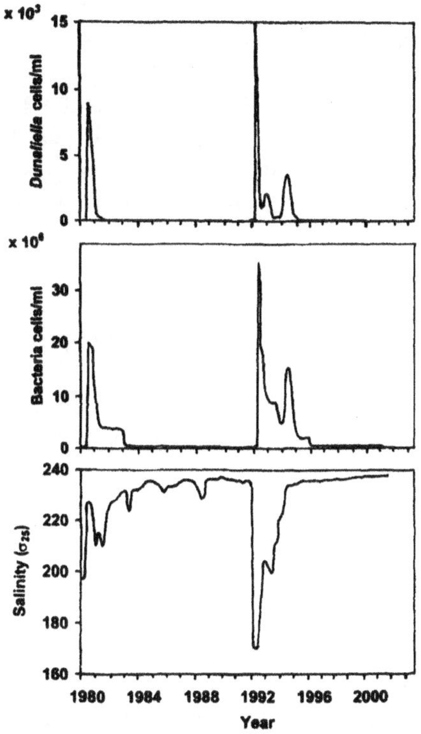

Figure 7. Population density of the unicellular green alga *Dunaliella* (upper panel) and the community of prokaryotes (mainly Archaea) (middle panel) in the upper meters of the Dead Sea water column, 1980-2002, as correlated with the salinity of the upper water layer (lower panel). Salinity is expressed in sigma units, indicating the density excess (in kg/m^3) to the standard reference density of 1000 kg/m^3. σ_{25} denotes the density in sigma units at 25 °C.

The bacterioruberin carotenoid pigments present in the cell membrane of the Archaea imparted a reddish color to the Dead Sea water during the bloom periods. Another pigment that may have contributed to the red coloration of the Dead Sea water during these blooms is bacteriorhodopsin (Oren and Shilo, 1981). Light energy absorbed by bacteriorhodopsin is converted into a pH gradient, which can be used as a source of energy. *Halorubrum sodomense*, isolated from the 1980 bloom, is one of the species able to synthesize purple membrane with bacteriorhodopsin (Oren, 1983b).

The bacterial blooms were case confined to the upper meters of the water column above the pycnocline. The vertical distribution of the bacteria in the water column could be used as a sensitive tracer of stratification: when a new holomictic episode starts, the remainder of the archaeal community that was previously confined to the upper water layers above the pycnocline and/or thermocline becomes distributed evenly over the entire water column (Anati et al., 1995; Oren, 1985, 1999; Oren and Anati, 1996). Densities of *Archaea* in the water column during the holomictic episodes have been low (below 10^6 microscopically recognizable cells/ml) (Oren, 1992).

It is unknown whether dissimilatory sulfate reduction presently occurs in the bottom sediments of the Dead Sea. Prior to the 1979 overturn, sulfide was found in the anaerobic hypolimnion (Neev and Emery, 1967). This sulfide was enriched in light sulfur isotopes, suggesting bacterial sulfate reduction as its source (Nissenbaum and Kaplan, 1976). The microorganisms responsible for the formation of this sulfide have never yet been isolated, and attempts to quantify sulfate reduction in Dead Sea sediments by following the formation of $H_2{}^{35}S$ from ${}^{35}SO_4{}^{2-}$ did not give conclusive evidence for the occurrence of the process. The anaerobic sediments have a potential for methanogenesis. ${}^{14}CH_4$ was evolved when sediment slurries were incubated with ${}^{14}C$-labeled methanol (Marvin DiPasquale et al., 1999). No methane formation was found on acetate, trimethylamine, dimethylsulfide or methionine.

7. THE FUTURE OF THE DEAD SEA

A model proposed by Yechieli et al. (1998) for the future evolution of the lake suggests that under current conditions of negative water balance the lake level will continue to decline but will approach a steady state level some 200-400 years from now, at an elevation of about –510 to 550 m, i.e. 100-150 m below the present level, depending on the volume of future inflows (Fig. 8). This steady state will be achieved when the volume of inflowing water will compensate for water evaporated from the Dead Sea surface. Such conditions will be achieved due to the diminishing surface area of the lake and the decrease in evaporation rate, the latter resulting from the increasing brine salinity. Similar steady state lake level forecasts were obtained by Krumgalz et al. (2000), who based their calculations on a thermodynamic approach. In fact, even if no water would flow to the Dead Sea it would not fully dry out; the hygroscopic nature of its solute (mainly Mg-Ca-Cl) implies that the brine would reach a steady state with the atmosphere, even at the very low humidity which characterizes the Dead Sea region.

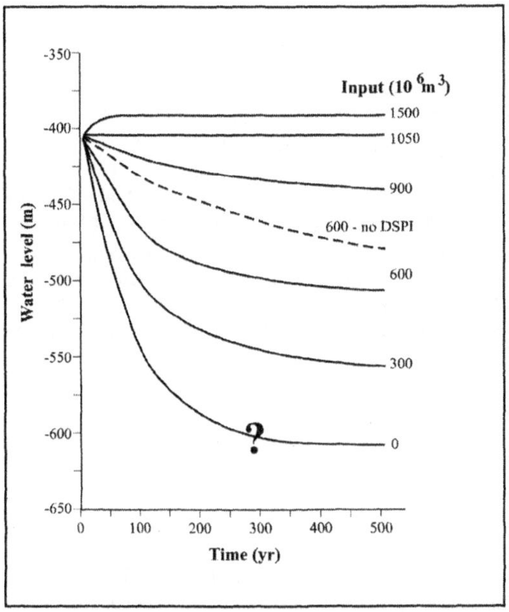

Figure 8: Calculated future Dead Sea water levels as a function of inflow volumes in million cubic meters (after Yechieli et al., 1998). The dashed line is the expected levels if the Dead Sea Potash industries where not operating and the annual inflow was 600 MCM.

There is no doubt that the best solution for the declining level of the Dead Sea is to restore its natural water inflow, mainly from the Jordan River, including its Yarmouk tributary. However, in view of the shortage of potable water and the geopolitical situation in the region, this solution does not seem to be feasible in the near future. Accordingly, the alternatives of allowing the lake level to continue to decline or stabilizing it by conveying seawater through the "Peace Conduit" must be carefully considered for the benefits of future generations.

8. REFERENCES

Abelson, M., Baer, G., Shtivelman V., Wachs, D., Raz, E., Crouvi O., Kurzon, I., and Yechieli, Y. 2003. Collapse-sinkholes and radar interferometry reveal neotectonics concealed within the Dead Sea basin. *Geoph. Res. Lett.* 30: 52 (1-4).

Al Weshah, R. 2000. The water balance of the Dead Sea. *Hydrol. Proc.* 14: 145-154.

Anati, D.A. 1997. The hydrography of a hypersaline lake, pp. 89-103. In: Niemi, T., Ben-Avraham, Z., and Gat, J.R. (Eds.), The Dead Sea - the Lake and its Setting. Oxford University Press, Oxford.

Anati, D.A. 1998. Dead Sea water trajectories in the T-S space. *Hydrobiologia* 381: 43-49.

Anati, D.A., and Shasha, S. 1989. Dead Sea surface-level changes. *Israel J. Earth Sci.* 38: 29-32.

Anati, D.A., and Stiller, M. 1991. The post-1979 thermohaline structure of the Dead Sea and the role of double-diffusivity. *Limnol. Oceanogr.* 36: 342-354.

Anati, D.A., Stiller, M., Shasha, S., and Gat, J.R. 1987. Changes in the thermo-haline structure of the Dead Sea. *Earth Planet. Sci. Lett.* 84: 109-121.

Anati, D.A., Gavrieli, I., and Oren, A., 1995. The residual effect of the 1991-1993 rainy winters on the Dead Sea stratification. *Isr. J. Earth Sci.* 44: 63-70.

Baer, G., Schattner, U., Wachs, D., Sandwell, D., Wdowinski, S., and Frydman, S. 2002. The lowest place on Earth is subsiding – an InSAR (interferometric synthetic aperture radar) perspective. *Geol. Soc. Am. Bull.* 114: 12-23.

Begin, Z.B., Ehrlich, A., and Nathan, Y. 1974. Lake Lisan- the precursor of the Dead Sea. *Geol. Surv. Isr. Bull.* 63. 30 pp.

Bloch, R., Littman, H.Z., and Elazari-Volcani, B. 1943. Occasional whiteness of the Dead Sea. *Nature* 154: 402-403.

Ben-Yaakov, S., and Sass, E. 1977. Independent estimate of the pH of Dead Sea brines. *Limnol. Oceanogr.* 22: 374-376.

Beyth, M. 1980, Recent evolution and present stage of the Dead Sea brines, pp. 155-166 In: Nissenbaum, A. (Ed.), Hypersaline Brines and Evaporitic Environments. Amsterdam, Elsevier Scientific Publushers.

Beyth, M., Gavrieli, I., Anati, D., and Katz, O. 1993. Effects of the December 1991 – May 1992 floods on the Dead Sea vertical structure. *Israel J. Earth Sci.* 42: 45-47.

Elazari-Volcani, B. 1940a. Studies on the microflora of the Dead Sea. Ph.D. thesis, The Hebrew University of Jerusalem (in Hebrew).

Elazari-Volcani, B. 1940b. Algae in the bed of the Dead Sea. *Nature* 145: 975.

Elazari-Volcani, B. 1943. A dimastigamoeba in the bed of the Dead Sea. *Nature* 152: 301-302.

Elazari-Volcani, B. 1944. A ciliate from the Dead Sea. *Nature* 154: 335-336.

Farber, E., Vengosh, A., Gavrieli, I., Amer, M., Bullen, T.D., Mayer, B., Holtzman, R., Segal-Rozenhaimer, M., and Shavit, U. 2003. The origin and mechanisms of salinization of the lower Jordan River. *Geochim. Cosmochim. Acta.* In press.

Garfunkel, Z., and Ben-Avraham, Z. 1996. The structure of the Dead Sea basin. *Tectonophysics* 266: 155-176.

Gavrieli, I. 1997. Halite deposition in the Dead Sea: 1960-1993, pp. 161-170. In: Niemi, T., Ben-Avraham, Z., and Gat, J.R. (Eds.), The Dead Sea - the Lake and its Setting. Oxford University Press, Oxford.

Gavrieli, I., Starinsky, A., and Bein, A. 1989. The solubility of halite as a function of temperature in the highly saline Dead Sea brine system. *Limnol. Oceanogr.* 34: 1224-1234.

Gavrieli, I., Beyth, M., and Yechieli, Y. 1999. The Dead Sea - A terminal lake in the Dead Sea rift: a short overview, pp. 121-127 In: Oren, A. (Ed.). Microbiology and Biogeochemistry of Hypersaline Environments. CRC Press, Boca Raton, Florida.

Gavrieli, I., Lanski, N., Yaari-Gazit, N., and Oren, A. 2002. The impact of the proposed "Peace Conduit" on the Dead Sea. Evaluation of current knowledge on Dead Sea – seawater mixing. The Geological Survey of Israel, Report GSI/23/2002, 42 pp.

Gavrieli, I., Bein, A., and Oren A. 2003. The deteriorating Dead Sea basin: limnological and environmental changes and the expected impact of the "Peace Conduit". *Mitigation and Adaptation Strategies for Global Change*, in press.

Gertman, I., and Hecht, A. 2002. The Dead Sea hydrography from 1992 to 2000. *J. Mar. Syst.* 35: 169-181.

Kaplan, I.R., and Friedmann, A. 1970. Biological productivity in the Dead Sea. Part 1. Microorganisms in the water column. *Israel J. Chem.* 8: 513-528.

Ken-Tor, R., Enzel, Y., Agnon, A., and Stein, M. 2003, Late Holocene lake-levels of the Dead Sea. *Bull. Geol. Soc. Am.*, submitted for publication.

Klein, C. 1961. On the fluctuations of the level of the Dead Sea since the beginning of the 19th century. *Hydrological paper no.* 7, Ministry of Agriculture, Hydrological Service of Israel.

Krumgalz, B.S., Hecht, A., Starinsky, A., and Katz, A. 2000. Thermodynamic constraints on Dead Sea evaporation: can the Dead Sea dry up? *Chem. Geol.* 165: 1-11.

Krumgalz, B.S., and Millero, F.J. 1982. Physico-chemical study of the Dead Sea waters. 1. Activity coefficients of major ions in Dead Sea water. *Mar. Chem.* 11: 209-222.

Marvin DiPasquale, M., Oren, A., Cohen, Y., and Oremland, R.S. 1999. Radiotracer studies of bacterial methanogenesis in sediments from the Dead Sea and Solar Lake (Sinai), pp. 149-160 In: Oren, A. (Ed.). Microbiology and Biogeochemistry of Hypersaline Environments. CRC Press, Boca Raton, Florida.

Neev, D., and Emery, K.O. 1967. The Dead Sea. Depositional Processes and Environments of Evaporites. Bulletin No. 41, State of Israel, Ministry of Development, Geological Survey, 147 pp.

Nissenbaum, A. 1975. The microbiology and biogeochemistry of the Dead Sea. *Microb. Ecol.* 2: 139-161.

Nissenbaum A. and Kaplan I., 1976. Sulfur and carbon isotopic evidence for biogeochemical processes in the Dead Sea ecosystem, pp. 309-325 In: Nriagu, J. (Ed.). Environmental Biogeochemistry, Ann Arbor Sci., Publ., Ann Arbor, Michigan.

Nissenbaum, A., Stiller, M., and Nishri, A. 1990. Nutrients in pore waters from Dead Sea sediments. *Hydrobiologia* 197: 83-90.

Oren, A. 1983a. Population dynamics of halobacteria in the Dead Sea water column. *Limnol. Oceanogr.* 28: 1094-1103.

Oren, A. 1983b. *Halobacterium sodomense* sp. nov., a Dead Sea halobacterium with extremely high magnesium requirement and tolerance. *Int. J. Syst. Bacteriol.* 33: 381-386.

Oren, A. 1985. The rise and decline of a bloom of halobacteria in the Dead Sea. *Limnol. Oceanogr.* 30: 911-915.

Oren, A. 1988. The microbial ecology of the Dead Sea, pp. 193-229 In: Marshall, K.C. (Ed.), Advances in Microbial Ecology, Vol. 10. Plenum Publishing Company, New York.

Oren, A. 1992. Bacterial activities in the Dead Sea, 1980-1991: survival at the upper limit of salinity. *Int. J. Salt Lake Res.* 1: 7-20.

Oren, A. 1993. The Dead Sea - alive again. Experientia 49: 518-522.

Oren, A. 1997. Microbiological studies in the Dead Sea: 1892-1992, pp. 205-213 In: Niemi, T., Ben-Avraham, Z., and Gat, J.R. (Eds.), The Dead Sea - the Lake and its Setting. Oxford University Press, Oxford.

Oren, A. 1998. The rise and decline of a bloom of halophilic algae and Archaea in the Dead Sea: 1992-1995, pp. 129-138 In: Oren, A. (ed.), Microbiology and Biogeochemistry of Hypersaline Environments. CRC Press, Boca Raton.

Oren, A. 1999. Microbiological studies in the Dead Sea: future challenges toward the understanding of life at the limit of salt concentrations. *Hydrobiologia* 205: 1-9.

Oren, A. 2000. Biological processes in the Dead Sea as influenced by short-term and long-term salinity changes. *Arch. Hydrobiol. Spec. Issues Advanc. Limnol.* 55: 531-542.

Oren, A., and Anati, D.A. 1996. Termination of the Dead Sea 1991-1995 stratification: biological and physical evidence. *Israel J. Earth Sci.* 45: 81-88.

Oren, A., and Ben-Yosef, N. 1997. Development and spatial distribution of an algal bloom in the Dead Sea: a remote sensing study. *Aquat. Microb. Ecol.* 13: 219-223.

Oren, A., and Gurevich, P. 1995. Dynamics of a bloom of halophilic archaea in the Dead Sea. *Hydrobiologia* 315: 149-158.
Oren, A., and Shilo, M. 1981. Bacteriorhodopsin in a bloom of halobacteria in the Dead Sea. *Arch. Microbiol.* 130: 185-187.
Oren, A., and Shilo, M. 1982. Population dynamics of *Dunaliella parva* in the Dead Sea. *Limnol. Oceanogr.* 27: 201-211.
Oren, A., and Shilo, M. 1985. Factors determining the development of algal and bacterial blooms in the Dead Sea: a study of simulation experiments in outdoor ponds. *FEMS Microbiol. Ecol.* 31: 229-237.
Oren, A., Gurevich, P., Anati, D.A., Barkan, E., and Luz, B. 1995. A bloom of *Dunaliella parva* in the Dead Sea in 1992: biological and biogeochemical aspects. *Hydrobiologia* 297: 173-185.
Oren, A., Bratbak, G., and Heldal, M. 1997. Occurrence of virus-like particles in the Dead Sea. *Extremophiles* 1: 143-149.
Quennell, A.M. 1959. Tectonics of the Dead Sea rift. *Inter. Geol. Cong. 20th, Assoc. African Geol. Surv.*: 385-405.
Salameh, E., and El-Naser, H. 1999. Does the actual drop in the Dead Sea level reflect the development of water sources within its drainage basin? *Acta Hydrochim. Hydrobiol.* 27: 5-11.
Salameh, E., and El-Naser, H. 2000a. Changes in the Dead Sea level and their impacts on the surrounding groundwater bodies. *Acta Hydrochim. Hydrobiol.* 28: 24-33.
Salameh, E., and El-Naser, H. 2000b. The interface configuration of the fresh-/Dead Sea water- theory and measurements. *Acta Hydrochim. Hydrobiol.* 28: 323-328.
Shatkay, M. 1991. Dissolved oxygen in highly saline sodium chloride solutions and in the Dead Sea – measurements of its concentration and isotopic composition. *Mar. Chem.* 32: 89-99.
Shatkay, M., Anati, D.A., and Gat, J.R. 1993. Dissolved oxygen in the Dead Sea – seasonal changes during the holomictic stage. *Int. J. Salt Lake Res.* 2: 93-110.
Singer, A., Ganor, E., Dultz, S., and Fische, W. 2003. Dust deposition over the Dead Sea. *J. Arid Environ.* 53: 41-59.
Starinsky, A. 1974. Relationship between Ca-chloride brines and sedimentary rocks in Israel. PhD dissertation, Hebrew Univ. Jerusalem, 176 pp. (in Hebrew, English Abstr.).
Stein, M. 2001. The sedimentary and geochemical record of Neogene – Quaternary water bodies in the Dead Sea Basin – inferences for the regional paleoclimatic history. *J. Paleolimnol.* 26: 271-282.
Steinhorn, I. 1980. The density of the Dead Sea water as a function of temperature and salt concentration. *Isr. J. Earth Sci.* 29: 191-196.
Steinhorn, I. 1983. In situ salt precipitation at the Dead Sea. *Limnol. Oceanogr.* 28: 580-583.
Steinhorn, I., and Gat, J.R. 1983. The Dead Sea. *Sci. Am.* 249: 102-109.
Steinhorn, I., Assaf, G., Gat, J.R., Nishri, A., Nissenbaum, A., Stiller, M., Beyth, M., Neev, D., Garber, R., Friedman, G.M., and Weiss, W. 1979. The Dead Sea: deepening of the mixolimnion signifies the overture to overturn of the water column. *Science* 206: 55-57.
Stiller, M., and Chung, Y.C. 1984. Radium in the Dead Sea: A possible tracer for the duration of meromixis. *Limnol. Oceanogr.* 29: 574-586.
Stiller, M., and Nissenbaum, A. 1999. Geochemical investigation of phosphorus and nitrogen in the hypersaline Dead Sea. *Geochim. Cosmochim. Acta* 63: 3467-3475.
Stiller, M., Gat, J.R., Boumann, N., and Shasha, S. 1984. A short meromictic episode in the Dead Sea: 1979-1982. *Verh. Int. Verein Limnol.* 22: 132-135.
Volcani, B.E., 1944. The microorganisms of the Dead Sea. pp. 71-85 In: Papers collected to commemorate the 70th anniversary of Dr. Chaim Weizmann. Collective volume. Daniel Sieff Research Institute, Rehovoth.
Wilkansky, B. 1936. Life in the Dead Sea. *Nature* 138: 467.

Yechieli, Y., Gavrieli, I., Berkowitz, B., and Ronen D. 1998. Will the Dead Sea die? *Geology* 26: 755-758.

Zak, I. 1967. The geology of Mount Sedom. PhD dissertation, Hebrew Univ. Jerusalem, 207 pp. (in Hebrew, English summary).

Chapter 12

THE CONTINUED DEGRADATION OF LAKE CORANGAMITE, AUSTRALIA

Brian V.Timms

School of Environmental and Life Sciences, University of Newcastle, Callaghan, NSW, Australia, 2308

1. INTRODUCTION

Lake Corangamite is unique. Firstly, it is Australia's largest permanent saline lake. There are only a few other permanent saline lakes and they are different to Corangamite, being in deep craters (>15m) while Corangamite is on an open plain and is shallow (~ 6m) (Williams, 1981). Of course there are many larger lakes in Australia, but they are either fresh and in Tasmania or episodic like Lake Eyre and usually do not hold water (Timms, 1992). Lake Corangamite also has been subject to the most monitoring and research of any lake in Australia. This is because, unlike other saline lakes in remote parts, it is easily accessible to a large population centre (Melbourne). Moreover, its contentious salinity and water level fluctuations have meant aspects of its limnology have been monitored from early European settlement.

J.C.J. Nihoul et al. (eds.), Dying and Dead Seas, 307–319.

While Lake Corangamite varies naturally from being hyposaline to hypersaline in a fluctuating episodic fashion, in recent decades Lake Corangamite has been steadily increasing in salinity, due largely to diversion of its main water input (Williams, 1995). Not surprisingly, this has changed the nature of the lake. It is the purpose of this paper to review the limnology of Lake Corangamite and to update the assessment by Williams (1995). The possibility of remedial action in the near future makes this paper doubly worthwhile as it focuses on the present sorry state of the lake for all to see, and to act on.

2. LAKE CORANGAMITE THROUGH TIME

2.1 Location, Geomorphology and Hydrology

Lake Corangamite lies on the volcanic plains of western Victoria, about 160 km west of Melbourne and 15 km north-west of the city of Colac (Fig. 1). In 1979 at ~116m AHD (≈ above sea level) it was 25,160 ha, shoreline length 159 km, maximum depth ~ 6m, volume 1509 x 10^6 m^2, lake area to catchment area 0.20 (Gutteridge, Haskins & Davey, 1980). As shown in Fig. 1 its shores are generally steeply shelved to a flat floor. The lake lies between lava flows and tongues of lava protrude into the lake in the southern half. Detailed mapping in this area would increase the shoreline length considerably, but the exact length would change with small changes in water level.

Lake Corangamite is a closed lake, receiving input from many streams, particularly the Woady Yaloak in the north (35% of catchment) and to a much lesser extent the Pirron Yalloak in the south (5%). Freshwater springs also discharge into the lake, of which McVean Springs in the southwest are the most significant. There is no doubt groundwater input too, including from the Warrion aquifer along the mid-eastern shore (Warrion Groundwater Supply Protection Area Consultative Committee, 2002). Dickenson (1995) believes there is groundwater outflow to the east, so the lake is probably not truly closed hydrologically. When it does overfill, water flows eastward from the north-east end of the lake to the Lough Culvert area, on to Lake Murdeduke and then spills into the Barwon River.

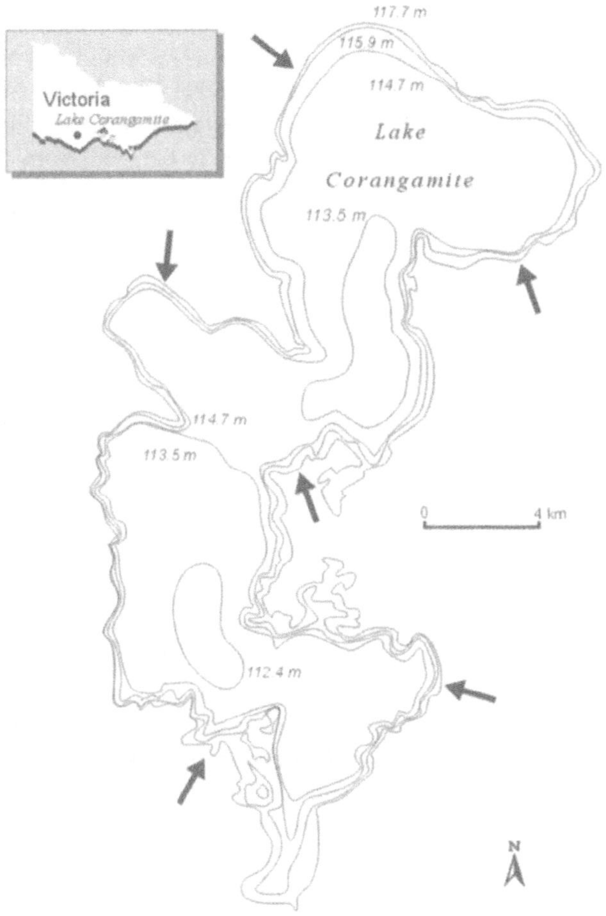

Figure 1. Map of Lake Corangamite showing collecting stations as arrows and depth contours (in m above AHD).

2.2 Pre-European conditions

Currey (1964) showed that Lake Corangamite was once much larger at 1792 km^2 as a result of the Newer Volcanics destroying previous drainage patterns and diverting the Barwon and Leigh Rivers into the Corangamite area. The larger Lake Corangamite overflowed at 120.7m AHD. Presumedly its waters were fresh at this stage. Despite the much increased catchment, a climate change invoking a greater rainfall by 330 mm is claimed to be necessary to fill this enlarged lake (Sutcliffe in Currey, 1964). With time the drainage pattern was rearranged and the lake overflow cut back to 119.2m AHD to give the present hydrological configuration.

As the ancestral lake developed it must have become saline as evidenced by the deposition of lunette dunes on drying shorelines and cut-off lakes. (It is believed these dunes build under seasonally fluctuating water levels in saline conditions – Bowler, 1983). Certainly the original inhabitants experienced the lake as saline as its name is derived from a native word 'corang' meaning 'bitter' (Hebb, 1970). Apparently the lake fluctuated in level so much that the aborigines at times walked across the lake dryshod (Hebb, 1970). Nothing is known of the biology of the lake during these times, but there are aboriginal fish traps in places, indicating the presence of fish at times.

The lake is the focus of an extensive wetland system, including many freshwater springs, creeks and wetlands. These act as important refugia for fauna during times of low levels and high salinities in the main lake. These reseed the lake when it fills again after a series of dry years.

2.3 European settlement to 1960

The area was first settled by Europeans in the 1840s (Hebb, 1970). Its general saline nature was of concern to many, and a Mr F. Acheson proposed to drain it and convert it to holding freshwater for irrigation but its costs and unbelievable explanation for the saline-freshwater conversion meant the plan was shelved. A succession of wet years in the 1870s saw the lake greatly expand in area and it was noted it teamed with waterbirds, 'chiefly black swans, ducks, pelicans, snow white cranes (i.e. egrets), etc' (Hebb, 1970). The flooding of adjacent properties, lead to a much cheaper new proposal by Mr Gordon to drain the excess water either directly or by diverting the waters of Woady Yaloak Creek (Hebb, 1970). Again no action resulted perhaps aided by the natural receding water level. A simple monitoring of water level and rainfall for the next 100 years shows how the water level of the lake is directly related to local rainfall (Fig. 2). The correlation coefficient between the two is 0.8153 and is highly significant ($p>0.001$).

As Williams (1985) shows, various spot measurements of water level, salinity and water analysis were made over the next 50 years to 1939. Salinities between 18.4 g/l and 123 g/l were recorded and a brine dominated by Na and Cl ions noted. The first biological expedition to the lake was made in 1918, when many of its characteristic animals were encountered and described (Shepard et al., 1918), including what later was described as the copepod *Calamoecia clitellata*, the isopod *Haloniscus searlei*, and also the snail *Coxiella striata* and the brine shrimp *Parartemia zietziana*. Ostracods were caught, but not identified further. Curiously no mention was made of the common amphipod *Austrochiltonia subtenuis* in the lake, possibly because the level was low at the time, indicating a high salinity beyond its tolerance.

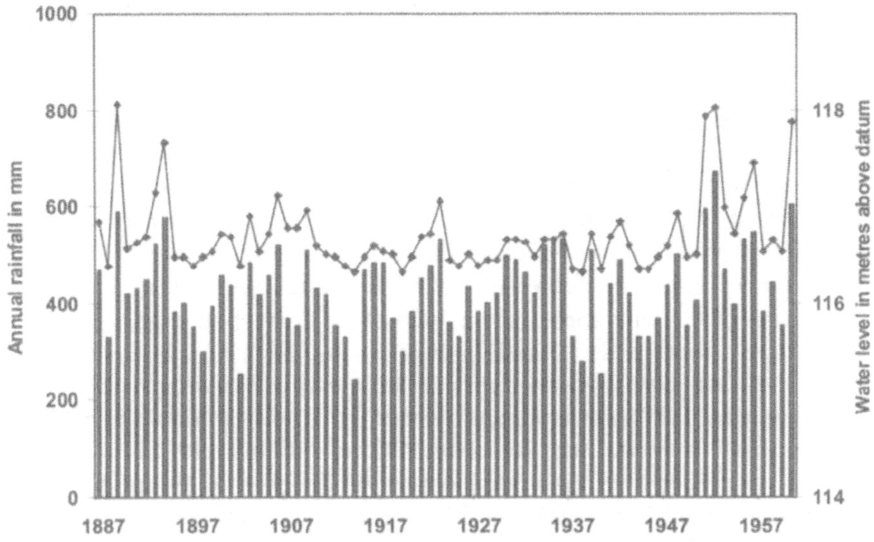

Figure 2. Annual rainfall and lake heights at Lake Corangamite for the period 1887 to 1960. These data were collected by an unknown settler near Lake Corangamite. The rainfall figures are much lower than for Colac, southeast of the lake and more akin to those at Cressey northeast of the lake, so probably they were collected from the northern end of the lake. The heights are not adjusted to AHD and lower values seem to be too high compared to known heights when this data are overlapped with published data in Williams 1995. Despite the lack of authenticity of these data, they are presented here to show the previous direct relationship between rainfall and water level in Lake Corangamite.

A series of very wet years in the 1950s resulted in the 1870s being revisited, with extensive flooding of pastures near the lake and an inferred salinity (from Fig. 2 in Williams 1995) of less than 10g/l. Unfortunately there are no recorded biological observations from this period. Agitation from landholders and others resulted in the implementation of a scheme by the then State Rivers and Water Supply Commission to divert the waters of the Woady Yaloak into the Barwon River and onto the sea.

2.4 1960 to 1992

This was a period of relatively thorough monitoring of salinity and water levels and of many biological studies of the lake. These are detailed by Williams (1995) and the following is a summary. Water level fell over the 32 year period from ~ 118m ADH to ~ 114.5m ADH and concomitantly salinity increased from ~ 25 g/l to ~ 60 g/l. Superimposed on these trends were seasonal fluctuations of up to 20 g/l.

Studies have been limited on microflora, but there was a persistent bloom of *Nodularia spumigena* through 1960s and 1970s and an abundance of littoral masses of filamentous algae (*Entomorpha* and *Cladophora*) in 1992. Submerged macrophytes , chiefly *Ruppia megacarpa* and *Lepilaena preissii*, were common until 1980, but by 1992 none were present. Invertebrate diversity decreased over this time (see Tables 3 and 4 in Williams, 1995) with some species persisting throughout (e.g. *Calamoecia clitellata, Haloniscus searlei*), many disappearing by 1980 (e.g. *Cordylophora caspia, Austrochiltonia subtenuis, Coxiella striata*) and one addition after 1980 (*Australocypris robusta*). The fish *Galaxias maculatus* disappeared from the lake by about 1980 (Williams, 1995). Its demise would have been due to a combination of the loss of its main food (*A. subtenuis*, and *C. striata*) and the salinity in the lake exceeding its tolerance. Waterbirds using Lake Corangamite fluctuate greatly in numbers and species richness, with lower values in highly saline conditions (Table 1). Before 1980 at least 20 species were regularly present with some of these exceeding 10,000 individuals on just the northern half of the lake (Corrick, 1982). Data from the Department of Sustainability and Environment for 1979 onwards show that, while species richness can be around 30 - 36 species at times, numbers rarely exceeded 10,000 in any count. (Table 1 and unpublished data from DS& E).

2.5 1996-2003

Monitoring of water level and conductivity has continued by the Corangamite Management Authority. Conductivity increased marginally from 1997 to 2000, but major increases were experienced during the dry summers of 2000-1, 2001-2 and 2002-3, so that in January 2003 conductivity had increased 171250 EC units from 56000 in July 1997 (Fig. 3). Salinity (actually TDS x 0.9 according to Williams, 1986) at various sites around the lake in February 2003 varied from 104 to 116 g/l. Lake level had decreased from 115.6 m in July 1997 to 113.8m in February 2003

(CCMA, unpublished data) These values are similar to those experienced during the dry years of the early 1930s (unpublished data, Rural Water Commission of Victoria).

Table 1. Waterbirds on Lake Corangamite 1979-2002.

Parameter	Year					
	1979	1980	1990	1993	1998	2000-2
Indicat. lake salinity (g/l)*	~ 40	39	~ 67	~ 50	~ 40	80-110
Number of bird counts	11	36	9	6	7	3
Species richness	36	30	19	16	33	18
Common species^	5	2	0	0	8	5
Dominant species#	H.H. Grebe Aust.Coot Black Swan	H.H.Grebe Aust.Coot (Musk Duck)	Black Swan A. Shellduck Silver Gull	Black Swan A. Shellduck Silver Gull	A. Shellduck H.H. Grebe Banded Stilt	A. Shellduck Banded Stilt Chestnut Teal

Based on a large and incomplete data set belonging to the Victorian Department of Sustainability and Environment, Colac. Only broad trends should be noted from this data since they are heterogeneous (e.g. counts made at different places in no set pattern) and some parameters are not standardized (e.g. different numbers of bird counts across the years).
* data from Williams (1995), Fig. 3 and text.
^ species with > 1000 individuals seen at any one time
# brackets used for species with fewer than 1000 birds seen at any one time

Little biological data are able for this period except for bird censuses in most years by the Department of Sustainability and Environment and aquatic invertebrate collections by the author in September 2002 and February 2003.

During the early 1990s when salinities reached 50-70 g/l, bird numbers and species richness declined (Table 1). With decreasing salinities by 1998, bird numbers and diversity recovered, only to be reduced markedly during the summers of 2000-2002 as salinities again soared (Table 1). This recent decade is notable for the overall reduction in large and small waders and most ducks, but the rising relative importance of Australian shelducks and more recently of banded stilts (Table 1 and unpublished data of DS& E).

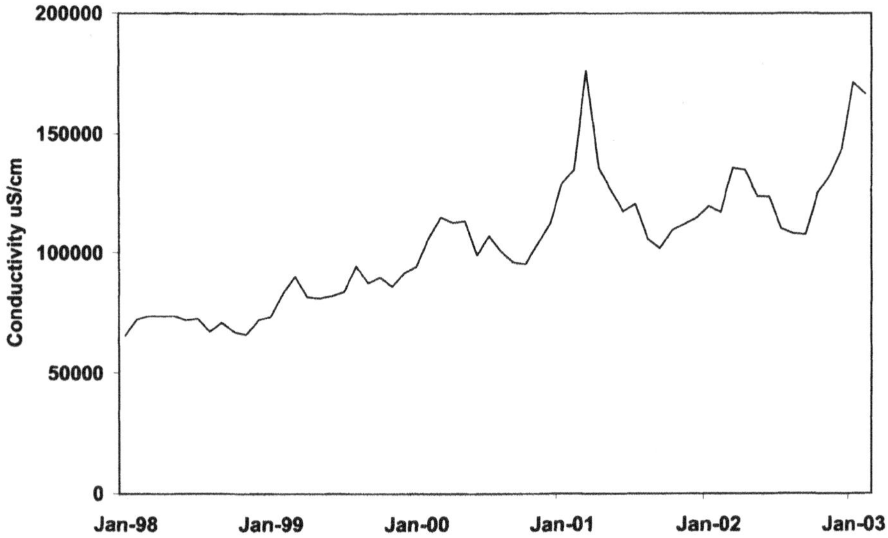

Figure 3. Conductivity of Lake Corangamite waters 1998 to 2003. Data from Corangamite Catchment Management Authority.

From six stations around the lake (see Fig 1), the following invertebrates were encountered by netting in open water and hand collecting in the littoral: *Parartemia zietziana, Calamoecia clitellata, Australocypris robusta, Haloniscus searlei* (all crustaceans) and an unidentified ceratopogonid larva. No detailed attempt was made to collect benthos, but the littoral net was deliberately skimmed through the top centimetre of mud, the only catch being ceratopogonid larvae and the ostracod. No fish or macrophytes were encountered and very few birds, mainly banded stilts and silver gulls, were seen. Macroscopic invertebrates were typically present in large numbers, so it cannot be said the lake is dead. However there is no doubt diversity is much reduced.

In a comparison of these results with those of Williams 1995, a decade earlier, there has been further decline in invertebrate diversity. No rotifers were encountered in 2002-3 though the collecting gear was suitable to catch the four species Williams recorded in 1992. Other losses have been the copepod *Mesochra baylyi* and the ostracods *Diacypris* spp. An addition to the list of Williams (1995) is the brine shrimp *Parartemia zietiana*, though it had been recorded over 90 years ago when the lake was similarly very saline.

3. CONSERVATION ISSUES

Williams (1995) details the conservation values of Lake Corangamite. Foremost of these is its importance to waterbirds. Besides supporting a diversity of species, some of which breed in the lake (e.g Australian pelican (*Pelecanus conspicillatus*), the lake is very important for the Australian shoveller (*Anas rhynchotis*) with, for instance, an average of near 20,000 birds in 1988-89. Significantly the lake also supports some endangered species such as the freckled duck (*Stictonetta naevosa*) and Cape Barren goose (*Cereopsis novaehollandiae*). Many birds use the lake as a resting site, but feed elsewhere (R. Missen, pers com). The Ramsar status of the lakes then is not surprising. This recognises its importance internationally, and since migratory waders use the lake, this international recognition is supported by the Japan/Australia Migratory Bird Agreement (JAMBA) and the China/Australia Migratory Bird Agreement (CAMBA).

Lake Corangamite is also recognised as a 'high value wetland' by the Victorian Wetlands Scientific Committee (1991) based on its ecological, scientific, educational, cultural and scenic features. Besides its importance to birds, the lake is listed among the 199 most significant geological features in Victoria, its environs has some significant plant species (*Lepidium aschersonii, L. hyssopifolium, Cuscuta victoriana* and *Leptorhynchos waitzia*), the area has a high diversity of habitats, and the lake is well researched and used for educational purposes.

The lake is a dominant feature of the volcanic plains of western Victoria. As such it is important in this context and exchanges fauna with other lakes in the area. This has been recognised as many of the nearby larger lakes such as Murdeduke, Colongulac and Terangpom are also Ramsar sites. However this fails to recognise the role of smaller wetlands adjacent to the lakes, such as those in the Pomborneit, Dreeite and Cundare areas. The Cundare Pool, an artificial body of water created by the ponding of Woady Yalock water before it is drained away to the Barwon, is also frequented by many water birds, and is inhabited by fish (*Galaxias maculatus*) and many invertebrates characteristic of hyposaline and mesosaline water (R. Missen, pers com and author, unpublished data). Besides being used extensively by waterbirds seen on Lake Corangamite, all these wetlands are potential reseeding sources of Lake Corangamite as it fluctuates episodically in water level and salinity.

The freshwater springs around the lake and particularly along the southwestern shore are also an important component in the lake's ecology. Besides contributing some water to the lakes, they act as refugia for fauna.

4. MANAGEMENT ISSUES

It must be understood that Lake Corangamite fluctuates naturally in area, salinity and biological properties. In its present simplified state it is not dead, nor is it dying; it is just different. As such the lake is unsuitable for use by many waterbirds, has no fish, no macrophytes, a very limited array of invertebrates. One interesting phenonomon not seen in recent decades is the presence of many banded stilts feeding on the brine shrimps now present in the lake. Of most concern for wetland managers is the virtual lack of waterbirds at a Ramsar site. They are absent because there are many fewer islands for safe roosting, and the lake lacks fish and macrophytes and has different invertebrates dominant. No doubt also contributing to the lack of birds is the effect of drought on surrounding wetlands, so that many have moved away from the area. However many still frequent the lower salinity waters of Cundare Pool and Lake Terangpom.

The lake has been through low water levels, high salinity and simple biodiversity before, namely the late 1930s and has recovered to a deeper, less saline state. What is different this time is that man has been largely responsible for the decline and his engineering structures will impede change back to a more usual state. The situation in Corangamite is a direct result of greatly increased salinity induced by lack of inflow of freshwater from the Woady Yaloak, which was diverted in the early 1960s. To the best of scientific knowledge, the lake seems to function best for waterbirds (and the aquatic community generally) at a salinity below 30-35 g/l which means increasing water levels to at least 116 m ADH (Williams, 1995).

The lake needs managing to achieve a muted fluctuation in water level and thus mimic natural conditions to a large degree. High levels (above about 118m ADH) are undesirable because of extensive flooding of littoral lands that occurred in the 1870s, late 1880s, and the early and late 1950s. On the other hand low levels (below about 116m ADH) and concomitant high salinity are immical to the lake's biology. The present management strategy of excluding ALL of the inflow from Woady Yaloak Creek is cumbersome and favours excessively low water levels and ultimate drying of the lake in a succession of drought years. The scheme does however create a lower salinity Cundare Pool which provides habitat for some wildlife.

Although the Ramsar declaration of the lake goes back to 1981, the authority responsible for management, Parks Victoria, has done very little in managing the lake for the benefit of waterbirds. This is largely due to lack of funds and resources. In the 1980s not a lot needed to be done, but since 1992 the changes has been critical, but still there has been no action to arrest the declining environment for waterbirds. Only recently has a 3 year agreement been struck between the Corangamite Catchment Authority, Greening Australia and Parks Victoria for more proactive management with increased funds. Among many tasks is one to investigate and manage nutrient inputs into the southern part of the lake. As worthy as this project is, the main focus should be on the Woady Yaloak Diversion Scheme.

After many calls, including those of Williams (1992, 1995), the Woady Yaloak Diversion Scheme is being reviewed by the Corangamite Catchment Management Authority during 2003. What is needed is a balance between the needs of landowners near the lake and to re-establish the normal limnological regime in the lake, expressed visibly by a usual abundance of waterbirds on the lake. With present land use around the lake, it is obvious the lake cannot be allowed to fluctuate widely in water level so that there is extensive flooding of valuable pasture. So diversion of some water entering the lake maybe necessary at times. An appropriate mean level for the lake should be at least 116m ADH and even more at 117m ADH. At 116m the lake would be in the salinity range of 30-35 g/l (Williams, 1995), and the lake would have a sufficient buffer to accept water during wet years so as not to initiate nuisance flooding of pastures. However this salinity is too high for some of its normal inhabitants, most notably the fish *Galaxias maculatus,* a key member of the aquatic food chain. Clearly a lower salinity of about 20-25 g/l is needed and based on historical records this implies a level of 116.5 to 117m ADH (Williams, 1995). This higher level allows little capacity for more water during wet years and some flooding of pastures may result. However careful modelling of lake levels, rainfall, and inputs from the Woady Yaloak could manage situations of excessive rainfall when the lake is allowed to oscillate around the higher level of 117m. Evenso the price to pay for a lake functioning to the advantage of waterbirds will be the resumption of some land around the lake. Otherwise there is little point in having the lake declared a Ramsar site if it is rarely suitable for large numbers of waterbirds.

Should the increase in salinity be reversed, there is no question that the lake cannot recover its former flora and fauna (Williams, 1995). Species known to occur in Lake Corangamite, live in other wetlands within its catchment (Williams 1981, 1992; author, unpublished data) and would be expected to colonise the lake soon (i.e within a year or two) after appropriate salinites returned. An example, at other the end of the salinity spectrum, is the return of the brine shrimp, *Parartemia zietziana* in 2002/3 (and possibly up to a few years earlier) after it had not been recorded in the lake since 1918 (though it probably was there during the high salinities of the 1930s but not noticed).

Besides the major problem of less water than is needed to maintain the lake in a 'normal' salinity regime, some waters entering the lake are highly charged with nutrients. Pirron Yalloak Creek at the southern end of the lake is a major problem, as is the increasing nitrogen content in spring water (Craig Allen, CCMA, pers com). Some effort is being made to control these inputs. These contaminated waters may limit the role of these habitats as refugia, and moreover excessive groundwater extraction may dry the springs and so destroy completely their value as refugia.
Unlike many other saline lakes under stress, the future for Lake Corangamite is not without hope. It hinges on an appropriate decision from the review of the Woady Yaloak Scheme and then its speedy implementation. This will then allow Lake Corangamite some latitude to vary as it always has done. Man will at last have learnt to live with a lake that fluctuates in extent and salinity.

5. ACKNOWLEDGEMENTS

I owe a debt of gratitude to the late Professor Williams whose studies on the lake formed a base for this review. I also extend my thanks to the Corangamite Catchment Management Authority for their assistance, to Olivier Ray-Lescure for drafting Figure 1, to Tony Mahony for information and unearthing the information for Figure 2, to Steve McDougall of the Department of Sustainability and Environment, Colac for access to data on waterbirds, to Robert and Dawn Missen for hospitality and information and to Craig Allen and Robert Missen for comments on the manuscript.

6. REFERENCES

Bowler, J. M., 1983. Lunettes as indices of hydrologic change: A review of Australian evidence. Proc. r. Soc. Vict. 95: 147-168.
Corrick, A.H., 1982. Wetlands of Victoria. III Wetlands and waterbirds between Port Phillip Bay and Mount Emu Creek. Proc. r. Soc. Vict. 94: 69-87.
Currey, D.T., 1964. The former extent of Lake Corangamite. Proc. r. Soc. Vict.77: 377-386.
Dickenson, C.G.,1995. Numerical modelling of groundwater flow systems around Lake Corangamite. B. App. Sc (Hons) Research Thesis, University of Ballarat.

Gutteridge, Haskins & Davey 1980. The waters of the Western District Lakes. Addendum to Draft Environmental Protection Policy No. W-34B with Explanatory Notes and Study Report. Environment Protection Authority of Victoria, Melbourne.

Hebb, I, 1970. The History of Colac and District. The Hawthorn Press, Melbourne. 404pp. (First published in series in the Colac Herald, 1888).

Shepard J., Searle J. & Hardy, A.D., 1918. Excursion to Lake Corangamite. Vict. Nat. 35: 22-30.

Timms, B.V., 1992. Lake Geomorphology. Gleneagles Press, Adelaide. 180pp.

Warrion Groundwater Supply Protection Area Consultative Committee 2002. Warrion Groundwater Management Supply Protection Area. Warrion Groundwater Management Plan. Southern Rural Water, Colac, Victoria. April 2002.

Wetlands Scientific Committee 1991. Evaluation of Lake Corangamite for the 'Wetlands Conservation Program for Victoria'. March 1991.

Williams, W.D. 1981. The limnology of saline lakes in western Victoria: A review of some recent studies. Hydrobiologia 82: 233-259.

Williams, W.D. 1986. Conductivity and salinity of Australian salt lakes. Aust. J. Mar. Freshwat. Res. 37: 177-182.

Williams, W.D. 1992. The biological status of Lake Corangamite and other lakes in western Victoria. Report to the Department of Conservation and Environment, Colac, Victoria, April 1992.

Williams, W.D., 1995. Lake Corangamite, Australia, a permanent saline lake: Conservation and management issues. Lakes & Res: Res & Managem. 1: 55-64.

Chapter 13

LAKE CHAD : A CHANGING ENVIRONMENT

Jacques LEMOALLE

Laboratoire Divha, IRD, BP 64501, 34394 Montpellier cedex 5, France (lemoalle@mpl.ird.fr)

1 INTRODUCTION

In closed lakes, an equilibrium water level is reached when the surface area allows for an evaporation that balances the rain and stream flow into the lake. As these inflows are closely associated with rainfall over the basin, such water bodies are highly sensitive to the climate variability, and have been described as amplifier lakes. In natural conditions, they are good indicators of climate changes, but they also react rapidly to man-made changes in their water budget. An analysis of the hydrological and ecological functioning of some closed lakes experiencing modifications to their water budget may provide some pointers to future management strategies for water resources in endoreic basins subject to global climatic change. The recent evolution of Lake Chad is presented here in this context.

J.C.J. Nihoul et al. (eds.), Dying and Dead Seas, 321–339.

Lake Chad lies in an endoreic basin in the centre of Africa on the southern margin of the Sahara (Figure 1). There is no surface outflow. River and rainfall inputs are offset by evaporation losses, some seepage and by volume and area fluctuations according both to seasonal and annual variations of the water budget.
As a result of a decrease in rainfall over the basin, the lake level and area have progressively decreased. In 1973-74, previously shallow areas of the lake bed have emerged and the lake has subsequently split into separate pools. This change in the hydrology and landscape of Lake Chad led to concerns that it might be dry up altogether (the death of the Lake), causing major ecosystem and livelihood changes in the region., with the loss of the natural resources provided by the lake.

Figure 1. The situation of Lake Chad in Africa, at the southern border of the Sahara.

The purpose of the this chapter is to describe the present functioning of the lake, and to analyse the conditions necessary to enable the Lake once again to achieve a "normal" state. The chapter draws extensively on previously published papers and data summarized in Carmouze, Durand and Lévêque (1983), Lemoalle (1991) and Olivry et al. (1996), with updates from recent satellite and ground data.

2 LAKE CHAD : VARIABILITY AND LANDSCAPES

Lake Chad has a long history of wet and dry periods which span different time scales. During the quaternary, the last lacustrine transgressions, although interspersed with dry shorter episodes, occurred between 12,000 and 6,000 BP (when the water level reached up to 300-310 m asl and the total lake extent achieved was 250,000 km^2) and from 3,200 to 1,800 BP (287-290 m asl) (Servant & Servant, 1983).
Over a shorter time scale, during the last millennium, data from sediment stratigraphy and pollen analyses and other historical analyses indicate a similar succession of high (up to 285 m asl) and low levels, with four to five drought periods between 900 and 1900 AD. A very dry period of 25- to 30- years duration at the end of the 15th century has been well documented (Maley, 1980).

Some European explorers have visited the lake in the 19th century : Denham in 1822, Barth and Overweg in 1851, Rohlfs in 1866, Nachtigal in 1870 and Monteil in 1892. They described L. Chad as a very large water body, but reported from local tradition that it was probably much smaller at some time between 1822 and 1850 (Carmouze et al., 1983). Later military missions indicated that the lake had receded between 1904 and 1915, with large areas of previously open waters emerging as marshes or dry land. Detailed descriptions of this new environment have been provided by Destenave (1903a, b), Freydenberg (1907), Tilho (1910) and Garde (1911).

These important changes initiated the first dispute between scientists on the potential desiccation of the lake. However, a quick recovery in water level led to a hypothesis that the Lake periodically dried up and re-emerged over time. A classification in three main states of the lake was proposed by Tilho (1928), who showed that the lake level variations were directly related to rainfall changes on its basin :

- The Large L. Chad, holds 25,000 km^2 of open waters with a limited coastal sand dune archipelago, a water surface altitude of 283.5 m and occasional slight overflow towards the North-East through the Bahr El Ghazal.
- The Small Lake Chad, is made up of different separated bodies with a permanent open water pool of about 1,700 km^2 at a maximum

altitude of circa 280 m and permanent or seasonal marshes ranging from 2,000 to 14,000 km^2.

- The Normal (intermediate) Lake Chad has an intermediate level of 281 to 282 m, an archipelago of some 2,000 dune islands, some marshy vegetation on the shores and a single body of water covering about 20,000 km^2 (Table 1) .

Table 1. The main characteristics of the different states of Lake Chad

Lake Chad	Small	Normal	Large
Chari discharge (km^3/y)	10 - 35	40	45
Water level (m asl)	279 - 280	281 - 282	283.5
Number of water bodies	several	one	one
Total lake area (km^2)	3,000 – 14,000	18,000-22,000	24,000
North basin area (km^2)	0 – 7,000	9,000	10,000
Landscape	marshes	dune islands	open waters
Vegetation	+++	++	+

The Normal Chad may be divided in two main basins, north and south, which are separated by a narrow and shallow belt, locally known as the Great Barrier, often covered by aquatic macrophytes (Figure 2). On the north-east side of both basins, the lake is bordered by an erg of ancient dunes, the altitude of which progressively decreased along a NE-SW axis. This creates an archipelago of sandy islands, with a decreasing altitude towards the open waters of the two basins. The seasonal variation of the lake level determined the distribution of the riparian vegetation, commonly limited to a narrow band along the steepest shores of the dune islands and close to the lake border, though more extensive towards the open waters, along the south and west shores of the south basin and over the Great Barrier. The summit of some drowned dunes would also be covered by some reeds, including Papyrus, Typha and Phragmites, constituting « bank islands », spots of helophytes apparently developing in open waters. The open water areas were quite extensive, covering about 4,000 to 6,000 km^2 in each basin.

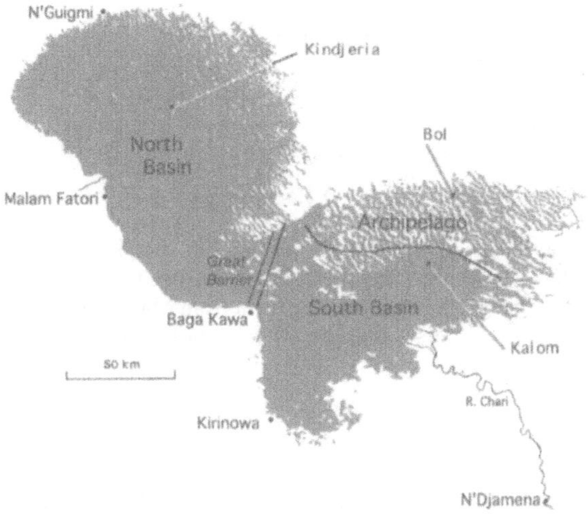

Figure 2. The main regions of the Normal Lake Chad and the situation of the main sills, between the open waters and archipelago in the southern basin and between southern and northern basin (Great Barrier).

3 THE TRANSITION TO A SMALL LAKE CHAD

After a period of Normal Lake since the beginning of water level monitoring in 1953 the lake level progressively decreased as a result of reduced rainfall in the basin. The lake level decreased from 283.27 m on January 1, 1965 to 279.70 m on January 1, 1973, a rate of 0.45 m/y (Figure 3). The aquatic vegetation along the shores of the lake was not able to withstand this decline, and by January 1973 had totally disappeared. At this time, the lake was still made of a single body of water, although very shallow in some regions, especially in the Great Barrier and at the limit between the open waters and the archipelago of the south basin.

In March 1973, the sediment of the shallow regions emerged and the lake was split into smaller pools, either connected with a water input (e.g open water facing the R. Chari delta), or only subject to evaporation losses. The level of these pools therefore behaved differently (Figure 4).

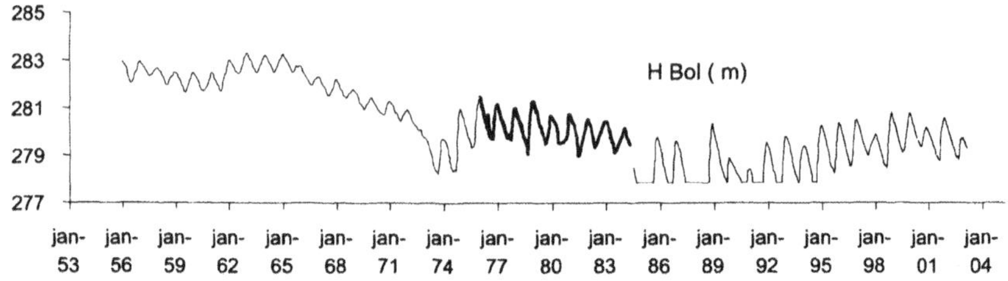

Figure 3. The water level as monitored at the Bol gauge (data from ORSTOM and DREM). Since March 1973, the Lake has split into basins of differing water levels and can not be described by a single water level. The level from March 1979 to May 1984 has been calculated from Kirinowa levels using a correlation between the two stations.

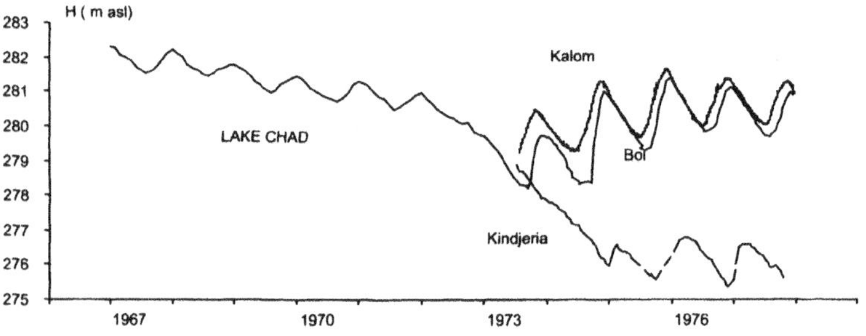

Figure 4. The water level in the three main regions of L. Chad during the transition from Normal to Small lake. The location of the gauges is given in Figure 2.

In the large pool facing the R. Chari delta, and fed by the river discharge, the level decreased only slightly. In the numerous small pools of the eastern archipelago, isolated from any input, evaporation led to marked changes, as in Bol where the decrease was 1.19 m between March 1 and September 15, 1973. Further large areas of the archipelago sediment became exposed to the air. In July of that year, the onset of the rainy season triggered the germination of seeds present in these exposed sediments. A dense vegetation grew up, comprising *Cyperus papyrus*, *Phragmites australis*, *Typha australis* and *Aeschynomemne elaphroxylon* and other smaller plants. When the R. Chari flood arrived in the south basin, these plants remained in place and constituted a large marshy area (Iltis and Lemoalle, 1983). The areas of open water in the south basin at

the end of 1973 were those which had remained inundated during the low level period. This distribution of the vegetated/open water areas has remained almost unchanged since 1974.
After March 1973, the north basin was separated from the south basin by the emergence of the Great Barrier. It received some water from the River Yobe, with only a limited impact around its estuary, but no water from the south basin passed over the Great Barrier until January 1975. During this 22 month duration, the water level in Kindjeria, the centre of the basin, decreased by 3.82 m. After a small spill over from the south basin in January 1975, the level decreased again until June 1975 when the basin completely dried up (Figure 4). At that time, although some rain fell around August 1973 and 1974, vegetation did not develop, probably because the soil salinity, higher than in the south basin, was detrimental to *C. papyrus* and *A. elaphroxylon*. With the exception of the Great Barrier, where the marsh vegetation developed in 1973 as in the south basin, vegetation in the north basin appeared only at the end of 1976, and was a mixture of reeds and shrubs including *Ziziphus* and *Aeschynomene spp.* More permanent humidity in the estuary of the K. Yobe River promoted some marsh vegetation in this area.

Since 1976, the inundation of the north basin has varied from year to year and season to season, depending on the level in the south basin and spillage over the Great Barrier. The maximum yearly flood extent varied from zero (in 1985, 1987 and 1988) to more than 7,000 km2 (January and February 1976, 1977, 1979, 1989, and from 1999 to 2003). From 1975 to 2001, the basin was almost completely dry for a number of months each year between October and December (Lemoalle, 1991). From 1999 to 2002, however, some water remained all the year round (Leblanc et al. 2003, and Landsat data,). Local populations used the former lake bed as arable land and range land during the draw-down periods, and fisheries were extremely active when the flood occurred.

In both basins, the new hydrological regime and the development of the fixed and floating vegetation created a new environment with changes in water chemistry as well as in the lacustrine communities ranging from phytoplankton to fish. These have been described for the first years of the Small Lake Chad (see Carmouze et al. 1983).

4 THE WATER FLUXES AND BALANCE

The water level of the lake may be described and analysed only during periods of Large or Normal Lake. In periods of Small Lake, the different level recorders and gauges are only representative of the pool in which they stand; extrapolation of data to the whole lake has sometimes led to misinterpretations of the lake regime.
The Normal Lake level varies with time, seasonally and from year to year, and is mainly a function of the discharge of the River Chari. The river has a tropical regime, with very low discharge during the dry season and a high flood: the mean ratio between the highest and the lowest monthly discharge at N'Djamena was 18.5 during the Normal Lake Chad period. This seasonality of inflows was reflected in seasonal level variations of about 0.7 m in the lake, with peak levels in December or January, and low levels in August. The variation between years (January to January) were generally lower, with a maximum range of 0.6 m.
The water level may also differ at any given time, from place to place. This partly results from the flooding pattern from the R. Chari delta to the extremities of the south and north basins, and because of the wind pattern, with a diurnal and seasonal period which creates wind set-up. The seasonal change in wind direction from northeast (November-May) to southwest (June-October), associated with the migration of the Inter Tropical Convergence Zone (ITCZ), creates a tilting of the water surface of 0.4 m between Bol (northeast) and Kirinowa (southwest) in the south basin (Talling & Lemoalle, 1998).

4.1 Normal Lake Chad

The water budget of Lake Chad has been studied by a number of authors, from Touchebeuf de Lussigny (1968) to Olivry et al. (1996). Mean values for the period 1954-1969 of Normal Lake Chad have been calculated by Vuillaume (1981) (Table 2). During this period, the lake level has varied between a minimum of 281.02 m in July 1954 and a maximum of 283.32 m in February 1963, but the initial and final levels, on May 1, 1954 and May 1, 1969 differ by only 0.03 m.

Table 2. Mean values for 1954-69 of the components of the water budget of Lake Chad. Values expressed as water height (mm/y) over the lake surface area (from Vuillaume, 1981)

	Units	Normal L. Chad
River inputs	mm/y	+1946
Direct rainfall	mm/y	+329
Evaporation	mm/y	- 2170
Net seepage	mm/y	- 102

The major part of the input to the lake is the discharge of the R Chari (82.3 % of total inputs), supplemented by direct rainfall (14 %) and inflow from the other small tributaries, El Beid and K. Yobe (3.6 %). It has been estimated that the losses in the R. Chari discharge between N'Djamena, where the discharge was measured, and the lake, are compensated for by inflows from the other tributaries. The respective mean figures were :

- R. Chari at N'Djamena : 41.7 km^3/y
- El Beid 1.43 km^3/y
- K. Yobe 0.49 km^3/y.

The losses result mainly from evaporation (95.5 %) and seepage out of the lake (4.5 %). The seepage may change in direction, seasonally or from year to year, according to the trend in the change of the water level. This may explain some differences between mean annual seepage rates and instantaneous local observations (Isiorho & Matisoff, 1990).

4.2 Small Lake Chad

The hydrology of the Small Lake results from two main sills separating three pool systems (Figure 2). One sill divides the south basin into the south pool, directly fed by R. Chari inflows, and the Bol archipelago, which is in fact made of a great number of small pools. The other sill is the Great Barrier. The altitude of the sills is close to 279.3 m. But they are covered with a dense vegetation acting as a break to the circulation of water (the Great Barrier extends more than 40 km between the two basins). It has been observed that significant transfer of water over the ridges occurs only when the water level in the south pool reaches 280.0 m. The flow is also dependent on the physiological state of the macrophytes and on the occurrence of channels cut through the vegetation to allow for navigation.

During the Small Chad period, the annual cycle of the water level in the southern pool has been relatively stable compared with that of the north basin, as shown by the Bol and Kalom gauge readings (DREM) complemented by accurate Topex/Poseidon satellite altimetric data (Birkett, 2000 ; Mercier et al. 2002). Starting from the seasonal low level in July or August in the southern pool, an increase in level by about 1.5 m, larger than during the Normal Chad period, occurs with the input of the R. Chari flood, until a level of 279.5 to 281 m is reached in December. When this level is reached, overflow occurs toward the northern basin of the lake through the Great Barrier and to the eastern archipelago (Figure 5).

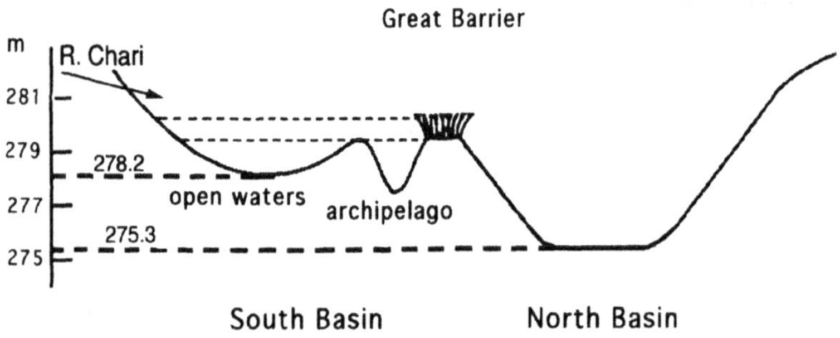

Figure 5. A schematic cross section of lake Chad showing the two main basins, with indications of the altitude of the sills and of the basin floors.

When compared with a Normal Lake Chad period, the Chari discharge maintaining a Small Lake is much lower (Table 3). The rainfall over the lake, calculated from five or more stations around the lake has also decreased since 1972 (Olivry et al. 1996).

Density and health of the vegetation over the Great Barrier, limits the possibility of high seasonal water levels in the southern pool. The annual cycle in the northern pool during a Small Chad period is, on the contrary, highly variable, as described above.

With these lower inputs an equilibrium level may only be achieved through lower losses, especially in their main component, evaporation, caused by a reduction in the lake surface area.

Table 3. The Chari discharge at N'Djamena during Normal and Small Lake Chad periods (data from Orstom and DREM-Tchad), and calculated rainfall over the lake.

	Period	Value	Period	Value
Chari discharge (N'Djamena) (km^3/y)	1950 - 71	39.1	1972 -2000	21.8
Direct rainfall (mm/y)	1950 - 71	320	1972 - 89	207

5 THE CAUSES OF THE PRESENT STATE AND CONDITIONS FOR RECOVERY

Historical descriptions of Lake Chad indicate clearly that its level can vary according to natural causes such as rainfall changes over the basin. Human activity in the basin may be considered as having had no effect on its hydrology at least until 1960. The impact of global climatic change on the basin climate, or the influence of the land use changes within the basin on water runoff, which may be attributed respectively to global or local human activities, remain uncertain and will not be discussed here. The more recent implementation of irrigated schemes and their possible impact on the lake water budget is, however, discussed below.

5.1 Rainfall over the basin and river discharge

A long period of above average rainfall from 1950 to 1967 over the West African Sahel was followed by below average rainy seasons after 1970, with harmful droughts in 1972, 1973 and 1984 (L'Hôte et al. 2002) (Figure 6). Over the Chari-Logone basin, annual rainfall has decreased by about 150 mm, with a North-South displacement of the rainfall gradient of 150 km (L'Hôte & Mahé, 1996 ; Climatic Research Unit, 2003). As a result, the annual discharge of the main tributary to the lake has also been lowered (Figure 7). This is the main reason for the decrease in Lake Chad level and the shrinking of surface area from a Normal to a Small Lake.

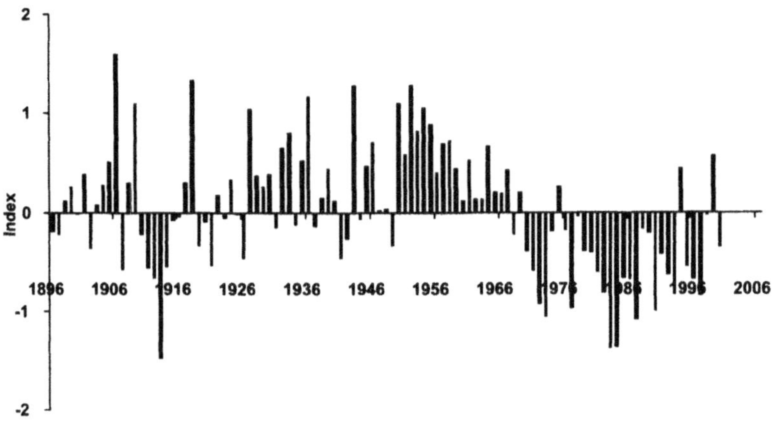

Figure 6. Time series from 1896 to 2000 of a normalized rainfall anomaly index over the West African Sahel (mean value computed for 1921-2000). From L'Hôte et al., 2002.

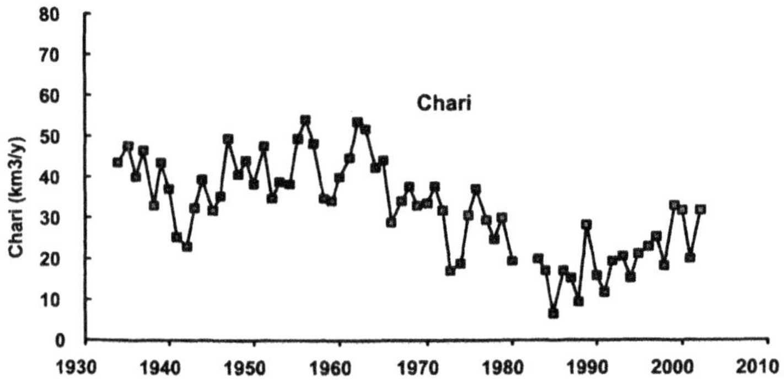

Figure 7. The River Chari annual discharge at N'Djamena. Data from ORSTOM-IRD and DREM (Chad).

If the values presented in Table 3 are applied to a totally closed lake (assumed to have no seepage) and with an annual evaporation E of 2.17 m as has been calculated for Lake Chad, the water surface area S_{eq} that would allow for a balance between the losses and inputs Q_r and direct rain P is

$$S_{eq} = Q_r/(P-E)$$

which gives an area of 21,1 km^2 in the wet period 1950-71 and of 10,8 km^2 during the dry period 1972-89. With this first approximation, it may be considered that the observed level and area of the lake are in accordance with the inflows measured at N'Djamena.

5.2 Water abstraction for irrigation

As the rain-fed agriculture in the basin does not provide sufficient output for local subsistence, a number of irrigation schemes have been implemented along the rivers valleys flowing into the lake and around lake Chad itself. The water abstraction resulting from these irrigation schemes is still poorly documented, but an order of magnitude can be estimated. We shall here consider successively the rivers, the lake surroundings and the lake bed.

Recent estimations of water abstraction in Chad along the Chari and Logone rivers amount to 100 Mm^3 (Rép. du Tchad-PNUD, 2003). Part of this volume is used for traditional irrigation, i.e. water overflowing onto the floodplains which would in any case be lost to the lake. An other portion (40 Mm^3) is used for sugar cane near Sahr, in Southern Chad. The Maga Reservoir, in Cameroon upstream of the Logone floodplain, has a capacity of 400 Mm^3 with a surface area of 90 to 360 km^2. It was constructed in 1979, together with a system of dams along the River Logone, in order to retain part of the natural input to the floodplain for the development of paddy. However, it contributed to the destruction of the floodplain ecosystem by severely decreasing the water input. After a preliminary test in 1994, the whole irrigation system has subsequently been modified to allow seasonal inundation of the floodplain and to provide the necessary water resources for fisheries, farmland and rangeland. Its impact on the water budget is low since most of the river input to the floodplain is naturally lost through evaporation (Wesseling et al., 1996 ; Olivry et al., 1996).

Two large reservoirs have been constructed on the upper basin of the River Yobe basin in Nigeria : Tiga Dam (1492 Mm^3, in 1974) and Challawa Dam (972 Mm^3, in 1992) while a third one (Kafin Zaki) is currently under study. They are used for drinking water supplies and for irrigation, but it appears that their effect on the discharge in the R. Yobe

lower basin and to Lake Chad is more important on the seasonal flow pattern than on annual discharge which remains close to 650 Mm^3/year (Goes, 2001).
After successful tests at a pilot scale south of Lake Chad in Nigeria, the South Chad Irrigation Project (67000 ha) and the Baga Polder Project (20000 ha) were inaugurated respectively in 1979 and 1982. Their full operation would have needed 2.5 km^3 of water per year. They both had long intake canals to pump water from the lake, but the lake recession did not allow for the irrigation of more than 7000 ha in the first years of operation before a complete halt in operations (Thambyahpillay, 1983 ; Hollis, 1990).
A number of polders have been constructed in the eastern archipelago of Lake Chad. According to the dam construction and irrigation system, their water needs have been estimated at between 6,000 to 16,000 m^3/ha/year. When these irrigation needs are compared with 21,000 m^3/ha/year of lake surface evaporation, it appears that the creation of polders does not contribute to a net water loss from the initial whole lake extent. At present there is total area of 5000 ha of polders, with plans to extend this to 12000 ha in the near future.

Currently, therefore, and with the available data, it is estimated that less than 200 Mm^3 are abstracted from the Chari-Logone system upstream of N'Djamena where discharge is measured. The gross uptake for the polders is 75 Mm^3, but this is less than would have evaporated from the lake surface had they not been implemented. The net uptake for irrigation in the Lake Chad basin probably therefore amounts to some 200 Mm^3/year. This value is much lower than estimates presented elsewhere, and would not contribute substantially to the present low level of the lake (Coe and Foley, 2001).

5.3 The conditions for a recovery to a Normal Lake

A simple hydrological model has been set up in order to evaluate some local and more general environmental issues. It has been used to estimate the necessary inflow to restore "Normal" Lake Chad levels with a single water body covering all basins. It has also been used to investigate how the sills between the sub-basins in a lake contribute to the maintenance of a lacustrine environment during low level periods.
The model uses the stage-area relationship partially derived from satellite data (Lemoalle, 1978), observed monthly levels or areas in the lake and

measured inflow from the River Chari for 1990-94 (data from DREM, Chad). The monthly rainfall and evaporation have been set as constants (Vuillaume, 1981) and it is assumed that seepage is a constant fraction of evaporation (i.e. a function of lake area). The daily transfer of water over the sills through the overlying vegetation is a fraction of the volume of water situated in the southern pool above the sill altitude (2.5 % for the northern basin and 1 % for the archipelago).

Once calibrated for the 1990-94 period, the model indicates that one annual inflow of 38 to 40 km^3/year would immediately re-establish a Normal Lake Chad comprising a single body of water. The same annual inflow is further needed to maintain this status of the lake in the ensuing years.

It has been observed that a permanent lacustrine water body is maintained in the southern pool of the lake, even for an annual Chari discharge as low as 10 km^3/year. The delimitation of a rather small pool around the R. Chari delta by the two main sills allows for this situation as the depth of this pool is sufficient to avoid desiccation during the low inflow period. Such a division into smaller basins when the water level falls significantly is a feature common to other closed lakes (e.g. the Aral Sea, this volume) which is useful in lowering salinity and maintaining lacustrine communities when evaporation over the whole lake would exceed inflows.

6 Discussion

Since 1973-76, when the lake split into different water bodies and the first total desiccation of the northern basin took place, Lake Chad has been in a of Small Chad. equilibrium state. It decreased in level and area between 1965 and 1973 but has not subsequently. Contrary to news reports in the Press or on the web (e.g. NASA, 2003), Lake Chad is, at the time of writing (2003) neither shrinking, disappearing nor dying. It is just a Small Lake Chad, as it has been several times in the past prior to recovering Normal or Large Lake Chad levels.

The present status of the Small Lake Chad is highly dependent on the annual Chari discharge, which itself can vary according to natural (or global) as well as man-made local causes. A significant decrease in the discharge, down to about 12 km^3/year, would lead to a permanent

desiccation of the northern basin of the lake but would maintain the present functioning of the southern basin.
A recovery of the inflow to its 1950-70 values would allow for full recovery of the Normal Lake Chad.
The models presently available on global climate change do not allow for reliable estimates of trends in the rainfall over the Lake Chad basin over the coming decades. Although the models do not forecast significant changes in annual rainfall over the Chari-Logone basin, some modifications in the intensity and distribution over the rainy season may occur, in association with a net increase in mean air temperature (Dokken et al. , 2001; Climatic Research Unit, 2003). It is thus not possible to predict whether the natural Chari discharge will decrease or increase as a result of global change. Future human development in the basin may increase water abstraction through new or rehabilitated irrigated schemes along R. Chari and R. Logone. The possibility of water supply augmentation is also under study through water transfers from the upper Ubangui-Zaïre basin to the Chad basin (LCBC, 2000). At least part of the projected 40 km^3/year transfer would be discharged into Lake Chad, with the objective of maintaining a Normal or Large Lake.

With the decreased rainfall in the Sahel since the end of the 1960s, and the severe droughts of 1972, 1973 and 1984, many people have lost access to their livelihood resources (either pastures or rain-fed agriculture). Some communities have moved toward places where water would be less limited, especially along the shores of Lake Chad where intensive fisheries, cattle grazing and cultivation in the draw down zone have developed. The increased seasonal level variation of the Small Lake Chad in the southern basin provides a larger area for the growth of aquatic macrophytes which are grazed by the cattle before the cultivation of cereals or legumes (Kolawole, 1988; Sarch and Birkett, 2000). The fisheries have adapted to the new conditions and to the new dominant fish species of the marshy areas (Jolley and Neiland, 1998; Oualbadet et al. 1996). In the northern basin, the variability from year to year is particularly high, but the use of the natural resources of the lake bed has also increased. The high-density population in close proximity of the lake has been exploiting the natural resources of a Small Lake Chad since the beginning of the 1970s, and has been able to cope with its natural variability.

These services should be taken into account in the planning of future water resources management in the basin, either in the case of increased inputs by water transfer, or in the case of decreased water inflows to the lake through the development of irrigation along the rivers and around the lake itself. Superimposed onto these man-made modifications, the natural variability in Lake Chad levels will most probably remain an important driving force.

Acknowledgements

The Meteosat data used in this study have been provided by CMS/ORSTOM in Lannion, France (1973-88), by UTIS (1988-2001) and Laboratoire d'Enseignement et de Recherche en Géomatique (LERG) (2001-2003), in Dakar, Senegal. Hydrological data are from ORSTOM-IRD Hydrology (Montpellier, France) and Direction des Ressources en Eau et de la Météorologie (DREM), (N'Djamena, Chad).

A. Nicol (ODI, UK) has been very helpful in improving the English.

References

Birkett C.M., 2000.- Synergistic remote sensing of Lake Chad: variability of basin inundation. Remote Sens. Environ. 72 : 218-236.

Carmouze J.P., Dejoux C., Durand J.R., Gras R., Iltis A., Lauzanne L., Lemoalle J., Lévêque C., Loubens G, & Saint-Jean L., 1972 .- Grandes zones écologiques du lac Tchad. Cah. ORSTOM, sér. Hydrobiol. 6, 2: 103-169.

Carmouze J.P., Durand J.R. & Lévêque C., 1983.- Lake Chad, Monog. Biologicae vol. 53, Junk, The Hague, 575 p.

Coe M. T. & Foley J., A., 2001.- Human and natural impacts on the water resources of Lake Chad basin.. J. Geophysical Research (D) 106 : 3349-3356.

Climatic Research Unit, 2003.- http://www.cru.uea.ac.uk/~mikeh/datasets/

Destenave, 1903a.- Le Lac Tchad. Première partie: le lac, les affluents, les archipels. Rev. gén. Sci. , Paris 1903 : 649-662.

Destenave, 1903b.- Le Lac Tchad. deuxième partie : les habitants, la faune, la flore. Rev. gén.Sci. , Paris 1903 : 717-727.

Dokken D., Noguer M., Van Der Linden P., Johnson C., Pan J. et al. (eds.), 2001.- Climate Change 2001: Synthesis Report. Cambridge University Press, Cambridge, UK and New York, USA, 397p.

Freydenberg, 1907.- Explorations dans le bassin du Tchad. La Géographie 15 : 161-170.

Garde G., 1911.- Description géologique des régions situées entre le Niger et le Tchad, et à l'Est et au Nord-Est du Tchad. Paris, Lib. A. Hermann et Fils.

Goes B.J.M , 2001.- Effects of damming the Hadejia River in semiarid northern Nigeria : lessons learnt for future management. IAHS Publ. 268 : 73-80.

Hollis G.E., 1990.- Environmental impacts of development on wetlands in arid and semi-arid lands.. Hydrol. Sci. J. 35 : 411-428.

Iltis A., Lemoalle J., 1983 .- The aquatic vegetation of Lake Chad, pp. 125-143 *In* Lake Chad, Carmouze J.P., Durand J.R. et Lévêque C. (eds), Monog. Biologicae vol. 53, Junk, The Hague, 575 p.

Isiorho S.A. & Matisoff G., 1990.- Groundwater recharge from Lake Chad. Limnol. Oceanogr. 35 : 931-938.

Jolley T.H. & Neiland A.E., 1998.- Changes in the fish stock composition of the wetlands of the Lake Chad Basin 1960-1996: a multi-disciplinary analysis. CEMARE-Misc-Publ Portsmouth-UK Univ-of-Portsmouth 42 : 1-13.

Kolawole A., 1988.- Cultivation of the floor of Lake Chad : A response to environmental hazard in eastern Borno, Nigeria. Geogr.-J. 154:243-250.

LCBC (Lake Chad Basin Commission), 2000.- http://www.ramsar.org/w.n.chad_summit_e.htm

Leblanc M., Leduc C., Razack M., Lemoalle J., Dagorne D. & Mofor L., 2003.- Applications of remote sensing and GIS for groundwater modelling of large semiarid areas : example of the Lake Chad basin in Africa. IAHS Publ. 278 : 186-192.

Lemoalle J., 1978.- Application des images Landsat à la courbe bathymétrique du lac Tchad. Cah. ORSTOM, sér. Hydrobiol. 12 : 83-87.

Lemoalle J., 1991.- The hydrology of Lake Chad during a drought period (1973-1989). FAO Fisheries Reports 445 : 54-61.

Lemoalle J., in press.- African wetlands in a semi arid climate : the Chad basin as an example pp.XXX *In* World's largest wetlands, P. Keddy & L. Fraser (eds). Cambridge University Press, XXp.

L'Hôte Y. & Mahé G.,1996.- Afrique de l'Ouest et centrale, précipitations moyennes annuelles (période 1951-1989). Carte à l'échelle 1/6 000 000. Paris, ORSTOM.

L'Hôte Y., Mahé G., Somé B., Triboulet J.-P., 2002.- Analysis of a Sahelian annual rainfall index from 1896 to 2000 : the drought continues. Hydrological Sciences Journal 47 : 563-572.

Maley J., 1980.- Les changements climatiques de la fin du Tertiaire en Afrique : leur conséquence sur l'apparition du Sahara et de sa végétation pp. 63-86 *In* The Sahara and the Nile, Williams, M.A.J. & Faure H. (eds), Rotterdam, A.A.Balkema Publ.

Mercier F., Cazenave A. & Maheu C., 2002.- Interannual lake level fluctuations in Africa (1993-1999) from Topex-Poseidon: connections with ocean-atmosphere interactions over the Indian Ocean. Global and Planetary Change 32 : 141-163.

NASA,2003;- http://earthobservatory.nasa.gov/Newsroom/NewImages

Olivry J.C., Chouret A., Vuillaume G., Lemoalle J., Bricquet J.P., 1996.- Hydrologie du Lac Tchad. ORSTOM, Paris, 259 p.

Oualbadet M., Dara L., Hochet A.M., 1996.- Les "gens du poisson" au Tchad. Bull.-CE-Coop.-Pêche 9 : 20-25.

République du Tchad/PNUD-ONU-DAES, 2003.- Schéma directeur de l'Eau et de l'assainissement. Projet PNUD CHD/98/004.

Sarch M.T. & Birkett C., 2000.- Fishing and farming at Lake Chad: Responses to lake-level fluctuations. Geographic Journal 166: 156-172.

Servant M. & Servant S., 1983.- Paleolimnogy of an upper quaternary endhoreic lake in the Chad basin, pp. 10- 26 *In* Lake Chad, J-P. Carmouze, J.R. Durand & C. Lévêque (eds). Monogr. Biol. 53, Junk, 575 p.

Talling J.F. & Lemoalle J., 1998.- Ecological dynamics of tropical inland waters. Cambridge University Press, 441 p.

Thambyahpillay G.G.R., 1983.- Hydrogeography of Lake Chad and environs: contemporary, historical and palaeoclimatic. Annals of Borno, 1:105-145.
Tilho J., 1910.- Documents scientifiques de la mission Tilho (1906-1909). Imprimerie Nationale, Paris, Tome 1 : 412 p., Tome Il 598 p.
Tilho J., 1928.- Variations et disparition possible du lac Tchad. Ann. Géogr. Fr. 37 238-260.
Touchebeuf de Lussigny P., 1969.- Monographie hydrologique du lac Tchad. ORSTOM, Paris, 169 p.
Vuillaume G., 1981.- Monthly hydrological balance and simple modelisation of Lake Chad hydrological regime. Cah. ORSTOM, sér. Hydrol. 18 : 23-72.
Wesseling J.W., Naah E., Drijver C.A. & Ngantou D., 1996.- Rehabilitation of the Logone floodplain, Cameroon, through hydrological management pp. 185-198 *In* Water management and wetlands in sub-saharan, Acreman,-M.C. & Hollis,-G.E (eds) Africa. Gland, Switzerland, IUCN.

Chapter 14

GENETIC TRACES OF ENVIRONMENTAL VARIATIONS IN ANCIENT LAKES

Sergei V.Semovski and Dmitry Yu.Sherbakov

Limnological Institute Siberian Branch Russian Academy of Sciences, Ulanbatorskaya, 3, Irkutsk, 664033, Russia

Erik Verheyen

Royal Belgian Institute of Natural Sciences, Rue Vautier, 29, Brussels, B-1000, Belgium

1. INTRODUCTION

Ecosystems of deep ancient intercontinental lakes during their long history went through different stages of development. Due to geological transformations and global climatic variability environmental conditions became more or less favourable for water organisms. Some lakes changed their state from being dead into being alive and back. Modern data on DNA structure can serve as a source of information on the evolutionary history of species beyond the sequence of speciation events. Correlation of genetic dating and environmental history in ancient lakes will provide new information on ecological consequences of variations in lakes regime and can be used as a source of knowledge of possible future changes in lacustrine ecosystem.

J.C.J. Nihoul et al. (eds.), Dying and Dead Seas, 341–357.

In the past decade an increasing amount of data on genetic diversity of populations and species has allowed the further examination of the mechanisms that generate and support current species richness. Ancient deep lakes like Lake Baikal, lakes Tanganyika, Victoria, Malawi and others are often called "natural laboratories" for evolutionary studies. The reason for this view is the great peculiarity of the hydrophysical conditions in these lakes and, due to this, quite specific environmental conditions for aquatic organisms. In these isolated lakes many lineage evolved endemic species, that developed in isolation from related taxa in other places. Species flocks are among the most interesting objects for evolutionary studies, they also reflect a whole history of speciation and divergence in the same lacustrine environment. Modern methods used to analyse genetic data are able to produce estimations of the moment, when significant changes occurred in the intensity of evolutionary branching, such as explosive speciation. The correlation of speciation events with paleoclimatic evidence can be used as a tool to infer the importance of environmental factors on ecosystem functioning. Spatial distributions of stenotopic species can be determined by geological processes, that control the shape of lake shore and zones of genetic contacts among populations. Due to that combination of phylogenetic and phylogeographic data can produce new evidence on the presumed connection between geological, climatic and evolutionary events.
We present several examples of genetic data applications to reveal correlation between geological and climatic history of some great ancient lakes and the evolution of their faunas. We will not discuss the simplification of the choice of neutral part of DNA or analysis of synonymic substitutions, calibration of molecular clock and other pure genetic aspects (see, for instance, [1] and numerous other studies). It is important to note possibility to use neutral nucleotides sequences for phylogeneic trees producing, which reflect a divergence history of genetic structure of organisms and correlated to the history of speciation. Various parts of DNA can be used for phylogenetic studies, some are transported on maternal line (mitohondrial DNA or mtDNA), some on paternal line (Y-chromosome). Recombination of genes can produce some additional effects, which can be also studied using phylogenetic approach.
The first section contains a review of the statistical methods we use to detect the fingerprints of the significant changes in evolutionary history of lacustrine faunas, which can be used to correlate with paleoclimatic and geological history of ancient lakes. We use an individual-based numerical model of adaptive population dynamics that include neutral dynamics of nucleotide sequences. Such model [2, 3, 4] is a flexible tool to describe the influence of environmental and ecological factors on variability of DNA structure.

Possibly, this approach in the future may lead to methods that allow incorporating different mechanisms of speciation into a model. This would enable one to infer the importance of ecological conditions for speciation events from real phylogenetic and phylogeographic data based on mitochondrial DNA sequences. The comparison of data sets for simulations involving neutrally mutating sequences and diverse speciation mechanisms to the real molecular data available on species flocks in great ancient lakes would allow us to evaluate the putative importance of environmental changes on speciation processes.

The fact that the neutrally evolving DNA sequences required in our simulation models behave exactly like mitochondrial sequences, the most commonly used molecular data used in evolutionary studies, allows one to estimate the limitations of the data sets to which this approach may apply. The examples of simulations given here acquire maximum 5% of divergence during the simulation. This implies that, if we assume that average rate of substitution accumulation in mitochondrial DNA is about 2% of substitution per million years, that the evolutionary time-span covered should not exceed approximately 2.5 million years (as discussed in [5]). In addition, for a data set consisting of 100 sequences with a length of 1000 base pairs, the width of the time interval which one may use to observe significant changes in the pair-wise distances distribution plot is 50 thousand years, a time span approximately corresponds to the estimated time that is required for speciation processes to complete (Chapter 13 of [6] at least for sympatric and parapatric processes).

Therefore, the presented approach offers interesting possibilities for the study of the evolutionary history and speciation processes of various components of faunas occur in a closed ecosystem. Particularly interesting examples that would benefit from this approach are the famous amphipod species flock from Lake Baikal in Siberia and the cichlid species flocks of the Great African Rift Lakes. For these species flocks, that evolved in the confines of the same lacustrine ecosystem, sufficiently important mitochondrial DNA data sets are available to investigate the probably importance of environmentally induced ecological changes for the patterns of evolutionary diversification of these highly diverse faunas. It would be particularly interesting to compare patterns obtained for cichlid genera with important ecological differences with regards to habitat and food choice. In view of the fact that the water level in lakes Tanganyika and Malawi have changed dramatically during the Quaternary (lake Victoria even assumed to be dried out completely), it is realistic to expect that the proposed approach may enable us to evaluate whether or not rock-dwelling cichlids may be differently affected by environmental disturbances than benthic sand-dwellers.

2. STATISTICAL MEASURES TO DETECT CHANGES IN POPULATION SIZE AND EXTENT NUMBER OF CLADOGENETIC EVENTS, EXAMPLES OF MODELLING

Since the inference of demographic changes in populations during their evolutionary histories is important, considerable efforts have been made towards the development of a procedure, which would allow one to detect moments of significant changes in population size. The most common approach to this is to count the number of surviving ancestor lineages from a phylogenetic tree inferred from molecular data with the assumption of a molecular clock ([7, 8]) or to produce a "skyline", which is the estimation of the harmonic mean of the effective population size based on the first attempt ([9]). This type of analysis is designed to be applied on phylogenetically closely related taxa, at or above the species-level.

Alternative methods of demographic analyses are based on the analysis of frequency distributions of pair-wise genetic distances ([10, 11]). These methods attempt to reveal dramatic changes of population size such as near extinctions (bottle-neck events) or the explosive increase of populations of a single species. Although the utility of pair-wise difference distributions for revealing population growth has been demonstrated, note that the method fails to characterize clearly a history of constant population size [10]. More exact estimations based on big samples are made in assumption of increasing of population size [11].

The division of a genetically isolated populations, which is the definition of speciation, is not necessary correlated with population growth. It means that statistical procedures designed to detect demographical changes will not always be suitable for the investigations of the history of cladogenesis.

In the first section we present two examples that are based on the analysis of frequency distributions of pair-wise distances to the speciation process of two sub-groups of one initial population, which, due to different environmental challenge, will also differ during their simulated evolutionary histories. The described procedure is compared to the existing approaches that analyse evolutionary histories from tree-based estimations of the number of ancestral lineages.

In [12] it was proposed to apply a modified logistic equation at the limited capacity of environment, the Gaussian intensity of competition between individuals and random mutation to describe branching of population due to disruptive selection. Each individual is described by a quantitative trait (polygene) $\boldsymbol{x}$, which determines the efficiency of resource use, such as for example body size or beak size on which the range of accessible food items

depends. We propose that the intensity of competition between individuals depends on their genetic distance according to the function ***C(t)***, where ***t*** is time. In [12] this arrangement has been used to build an individual-oriented model for asexual populations in which individuals were described with a parameter ***x***, in which individuals are born at a constant rate. In case of asexual population and at certain values of the parameters, the model displays the branching of lineages into genetically isolated populations. This is considered to be the simulation of the speciation (cladogenesis). Here we only consider the case for asexual populations, since in [12] and other studies it was shown that bisexual populations start to branch in the similar conditions (parameter values) if additional assumption of assortative mating [12] or general genetic isolation expression [2, 3, 4] are introduced.
The observed evolutionary dynamics results from continuous mutation pressure and disruptive selection. Therefore the dynamics of a population with phenotype ***x*** can be described by the following equation:

$$\frac{dN(x,t)}{dt} = r \cdot N(x,t) \cdot \left[1 - \frac{N(x,t)}{K(x)} \int C(x-y) N(y,t) dy\right]$$

$$C(z) = \exp\left[-\frac{z^2}{2\sigma_C^2}\right] , \qquad (1)$$

where ***N(x,t)*** is the size of the population with phenotype ***x*** at time ***t***. The capacity of the environment (distribution of resources) is the stable solution of the population dynamics. Individuals that differ by the value of the ecological parameter (phenotype) interact according to their density, where ***z*** is the difference between ecological parameters and describes the intensity of competition between populations. A mutation that appears in a parent with phenotype ***y*** produces offspring with phenotype ***x*** according to probability distribution with possible values ***1*** and ***0***.

In order to bring a neutrally evolving marker into the individually oriented model - comparable with a situation where a given specimen carries a neutrally evolving nucleotide sequence such as a mitochondrial gene — we use an integer vector consisting of elements with possible values of ***1*** to ***4***, which corresponds to the four possible character states at each position of a DNA sequence. For simplicity, we consider that all mutations in this sequence are neutral i.e. they have no impact on the individual fitness. Each specimen inherits a parental sequence that mutates with a probability of ***0.01*** per generation per nucleotide. If applied to populations of sexually reproducing organisms, this sequence is considered to be inherited from one of the parents only (e.g. maternal inheritance of mitochondrial genome). A randomly chosen subset from the set of neutral sequences, used in the described simulation was used for phylogenetic inferences which were performed with programs *kitsch* and *neighbour* from the package PHYLIP [13].

Simulations.

Two groups of parameters were chosen for simulations so that they cause different branching pattern if the fitness function changes. In the first case the branching was more likely after the fitness change due to higher environmental capacity.

$$K(x,t) = K_0(t)\exp\left[-\frac{(x-x_0)^2}{2\sigma_K^2}\right], \quad K_0(t) = \begin{cases} K_1, \, t < 600, \\ 2K_1, \, t \geq 600. \end{cases}$$

In this case evolutionary branching (formation of genetically isolated subpopulations, Fig.1a) correlates with growth of population size (Fig.1b). In the second case, the branching occurs due to fitness changing after intensive disruptive selection

$$K(x,t) = K_0\left[\alpha F_1(x) + (1-\alpha)F_2(x)\right], \quad \alpha = t/1500.,$$

$$F_1(x) = \exp\left(-\frac{x^2}{2\sigma_K^2}\right), F_2(x) = \exp\left[-\frac{(x-0.05)^2}{0.2\sigma_K^2}\right] + \exp\left[-\frac{(x+0.05)^2}{0.2\sigma_K^2}\right]$$

Two "narrow" ecological niches are formed after several generations and we have situation that resembled "microaalopatric" speciation rather then sympatric one, because environment is not more homogeneous in this case. The population size declines during the splitting of population into two parts (Fig.2b).We performed individually oriented simulation of genetic diversity accumulation in changing environment for these two groups of parameters.

At the first stage we assigned each of 100 initial individuals with a neutrally evolving nucleotide sequence of length considered to be 1000 "base pairs" long. After the completion of the simulation the obtained "nucleotide sequences" were used to infer phylogenies for both cases (see Figs.1c, 2c). For simplicity, 10% of the generated sequences were randomly sampled for phylogenetic analysis. Reassuringly, the obtained trees closely resemble the evolutionary patterns obtained in course of simulation (Figs.1a, 2a). It is important to note, that for the second case, the branching in the phylogenetic tree occurs much earlier then the actual formation of subpopulations in the model (this feature is known from the literature, see, for instance, [8]).

Since the inference of phylogenetic trees is prone for artefacts due to various reasons, we used the matrix of pair-wise mutational distances to characterize molecular evolution of neutrally mutating DNA sequences representing the "individuals" in our model. The range between minimal mutational distance (always equal to zero under the conditions of our model) and maximal distance was split into an arbitrary number of intervals; and the number of values for each of these intervals were counted. The thus obtained frequency distributions are shown in Figs. 1d, 2d. In both cases these distributions appear to be polymodal, whereas one single maximum was obtained in simulations where the populations evolved under constant environmental conditions only. The first peak on the beginning on the time scale corresponds to the growth of genetic diversity from the initial constant value for all individuals and is typical for all calculations. The second peak corresponds to theoretical curve for the population in stable conditions [10]. However, the peaks in the middle part of evolutionary history (marked by arrow) in both cases can be used to detect significant changes in the genetic structure of the initial population, that becomes divided into genetically isolated subpopulations. The peaks on these frequency distribution curves represent the time intervals where the lineage splitting is highest, i.e. the time intervals when the highest number of sequence pairs with given genetic distance originated. Accordingly, we suggest that the observed minima (i.e. lowest values on the frequency curves) correspond to time intervals when diversification rate is minimal.

Estimations of population size by "skyline" (Figs.1e, 2e) fails in both cases. If in the first case (Fig.1e) growth of population is detected but it is complicated by the moment of change in environment conditions (begin of population growth). For the second case (Fig.2e) skyline estimation is far from reality.

Note that application of "skyline" plot to the present situation is not reliable due to the initial assumptions. As it was noted in [9] that results of "skyline" estimation of effective population size are valid for population without subdivisions and migrations, which is not valid in present case.

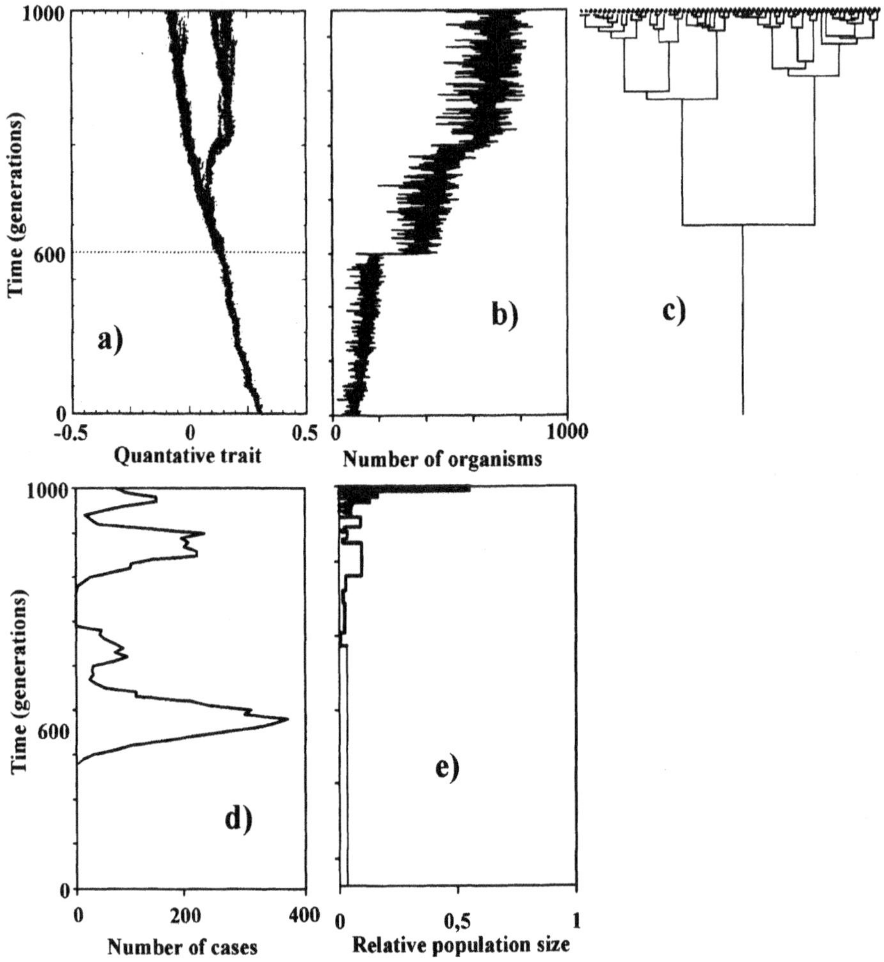

Figure 1. Results of evolution process simulation according to equation (1). After 600 steps (generations) fitness is increasing twice. (a) Each dot represents at least one individual. Dots on a horizontal line represent individuals belonging to the same generation. (b) Numbers of individuals; (c) phylogenetic tree produced by resulting neutral genetic markers (in the same time scale); (d) histogram of pair-wise genetic distributions; (d) "skyline" estimation of population size.

The simplicity of the model of pair-wise frequency distribution, and the method we used to reveal discontinuities in the accumulation of sequence diversity allowed us to apply parametric bootstrapping to estimate the statistic significance of the observed distribution features (see [3]). We argue that the approach described there may be used to access the confidence of inferences of ecological discontinuities in the evolutionary histories for real data sets (mitochondrial DNA sequences).

The developed model describes deviations of nucleotide substitution in a neutrally evolving gene in specimens from a population living in changing environmental conditions. We have shown that environmental changes impacted the population by changing the competition level among individual specimens. The decrease of the competition level, or the width of niche, results in an increasing number of nucleotide sequences, mirrored in the increased number of individuals. The main result obtained with our model shows that the signatures of such periods in the pattern of genetic diversity persist in the later generations even when the environmental changes were relatively small.

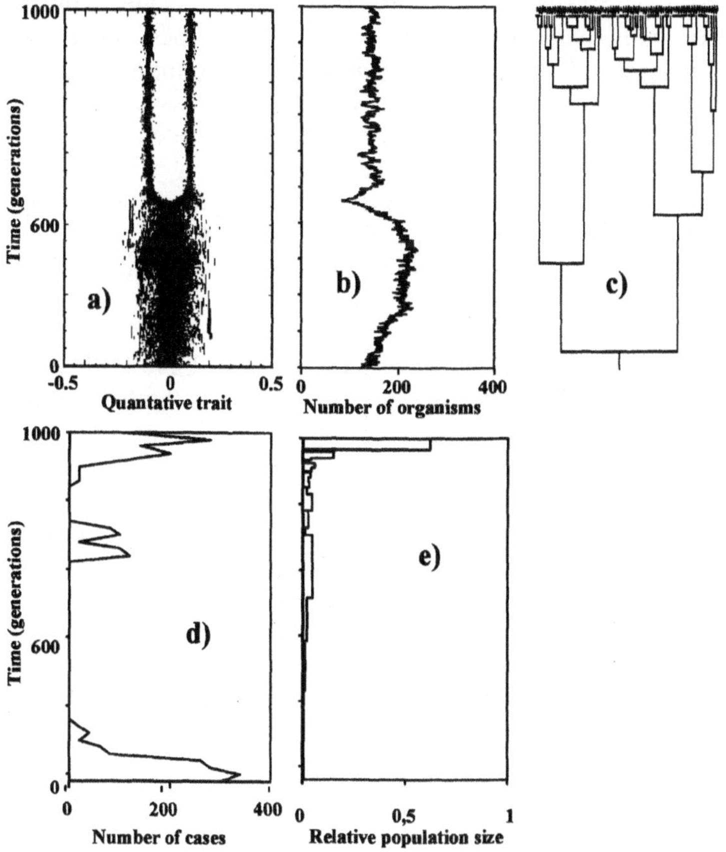

Figure 2. Results of evolution process simulation according to equation (1). Intensive disruptive selection produce splitting of population on two parts. (a) Realisation of the model (see Fig.1a). (b) Numbers of individual; (c) phylogenetic tree produced by resulting neutral genetic markers (in the same time scale); (d) histogram of pair-wise genetic distributions; (d) "skyline" estimation of population size.

3. LAKE BAIKAL, EXPLOSIVE SPECIATION FOR BOTTOM SPECIES IN LATE PLIOCENE, SECONDARY CONTACT ZONES

Lake Baikal is one of the oldest and the most voluminous freshwater lake in the World presents as well reach water ecosystem. According to some sources number of known animal species if estimated from 2500 (8% is endemic, [14]) to 1825 (54% is endemic, [15]).
During long geological history [16], dramatic changes occurred in the lake regime, climate, sedimentation and other dynamic factors, which essentially have drive the evolutionary processes in the water ecosystem.
Lake Baikal has oxygenated waters down to the deepest point (1642 m), whereas Tanganyika is effectively anoxic below 100 (North basin) to 200 m (Southern basin) – one would expect the Baikal abyss to be much more diverse then has been established to date. The oxygenated abyss is a result of deep-water renewal in the present Baikal, but in Miocene (6 Mya), Siberia had a warm subtropical climate and the paleothermal regime of Baikal probably was very much like that of modern Tanganyika [17]. The Baikal abyss as a biosphere is thus recent, relative to the age of the lake as a whole.

There are several diverse mollusc species flocks in Lake Baikal, but only three have been investigated using molecular methods (both using the same fragment of mtDNA): the endemic prosobrachian family *Baicaliidae* and the subendemic pilmonate genus *Choanomphallus* (Planorbidae) (Fig.3, [18]). Shells of *Baicaiidae* species that are more then 20 million years and have been found in the Tankhoi swift (southern shore of Lake Baikal). If published estimates for the rates of substitutions in molluscan mitohondrial protein-codes genes are applied, the existence of common ancestor for all contemporary baicaliids cannot be earlier then 3 Mya. Interestingly, this date approximately coincides with the onset of general cooling of the climate, when shallow-water-dwelling baicaliids would have been affected most, first by irregular freezing of the lake and then by glaciers themselves.

Therefore, the relatively recent origin of contemporary baicaliids and their apparent monophyly could be explained by the hypothesis that certain deep-water dwellers of the earlier flocks survived in the most stable (abyssal) environment and subsequently radiated back into shallower habitats, as those again became available.

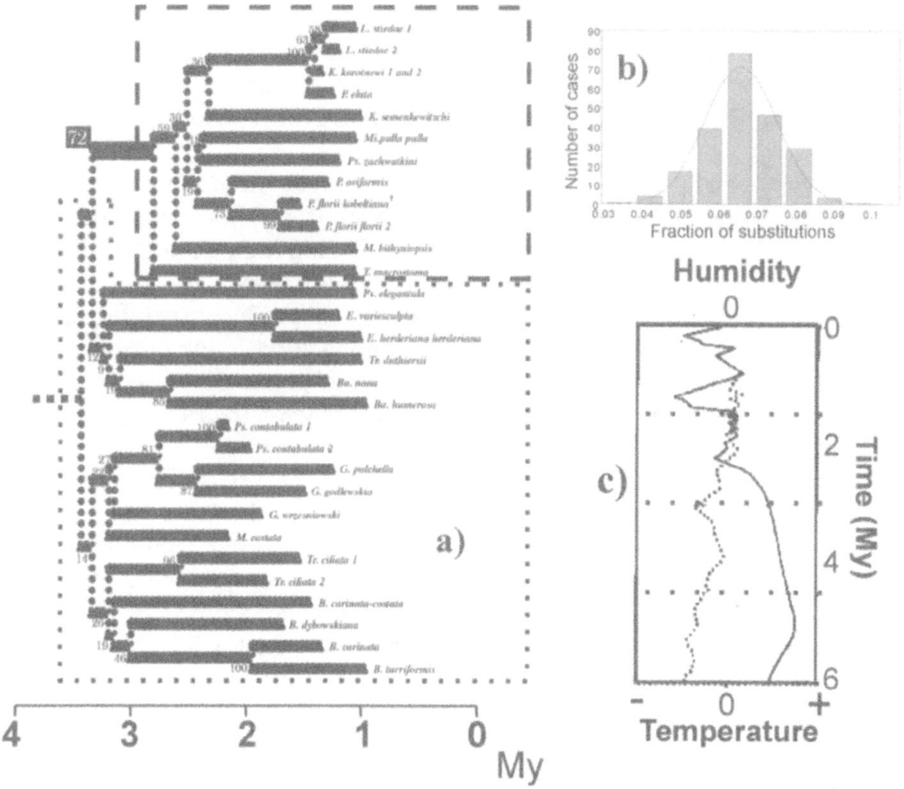

Figure 3. (a) Phylogenetic tree (distant) of *Baicaliidae* [19], (b) Distribution of distances using all types of substitutions between species for *Baicaliidae* family [19] (c) Trends in annual temperature (solid line) and in humidity (broken line) for the past 5 million years [20].

Estimative time of the youngest radiation resulting in formation of species flock for Lamprodrilus (*Oligohaeta*) also is estimated as 3.8-2.5 Mya [18]. Divergence of species inside this group is correlated in time with thebeginning of neobaikalian stage of Baikal Rift formation, rich in different geological, climatic and ecological variations.

From other hand, according to modern data, pelagic organisms (fish, amphipods) did not demonstrate such intensive evolutionary transformations during this period of time [21, 22, 23]. It's reasonable to assume that the factors influence with evolutionary "explosion" of Baikalian benthic organisms (in particular, oligohets and molluscs) were connected with sharp transformations of bottom dynamics. Probably, the most influenced were changes in bottom relief and great depths formation and variations in sedimentation (food flows) and, probably, higher deep ventilation due to global cooling.

Note that this example of explosive speciation due to changes in the environment can be described by the first model of section 1. Below we discuss several alternative examples to use genetic data to correlate evolutionary history and distribution of species with variability of lakes in the past.

Great number of gammarid species and subspecies, more then 300, occurs in Lake Baikal, 89% of them are endemic. It is shown that they are originated from few ancestral species [24]. During long geological history of Lake Baikal processes of sympatric speciation as well as allopatric speciation in temporary separated lake basins produce specific spatial distribution of genetically separated subpopulations. Joint analysis of spatial distribution and genetic diversification of gammarides made in [25] using allozyme analysis demonstrate some genetic traces of geological processes in the lake.

Endemic gammarids *Eulimnogammarus cyaneus* occurs exclusively in the narrow littoral zone of the stony and rocky shore of the lake. The samples taken all around the lake (Fig.4) have been analysed using allozimic analysis. In Fig.4b two genetically distinct groups are presented. The two groups are contacted in narrow zone in Olkhon Gates strait. The formation of this contact zone between groups separated about 180000 years ago according to molecular clock [25]. One of hypothesis of this separation is correlated with probable drop of the lake level during Quaternaly glaciations. Due to level change North Baikal population of gammarides exists in isolation during some period. After level rise and formation of the modern water body two population form hybrid zone. Another version is to correlate formation of contact zone with separation of Olkhon island from land.

The southern subgroup of the population is divided into further two groups,spatially divided by Angara river, which is only outlet of the Baikal. The genetic age if divergence of these subgroups in 60000 years is corresponds to formation of Angara during second half of the Late Pleistocene (80000-120000 years ago). The Angara is the greatest outlet in the World, its intensive current (1 m/s in average) effects as a barrier for genetic flow between subpopulations of gammarides on both sides of the river.

4. GREAT LAKES OF AFRICA, GEOLOGICAL HISTORY, DATING OF BIOLOGICAL EVOLUTION.

The Great East African Rift Lakes have been established as suitable model systems for the study of adaptive radiation. One or more flocks of cichlid fishes have been found in each lake including hundreds of endemic species.

Despite several stunning similarities, the species flocks for three lakes differ from each other in age, species number, complexity, and overall degree of morphological diversity.

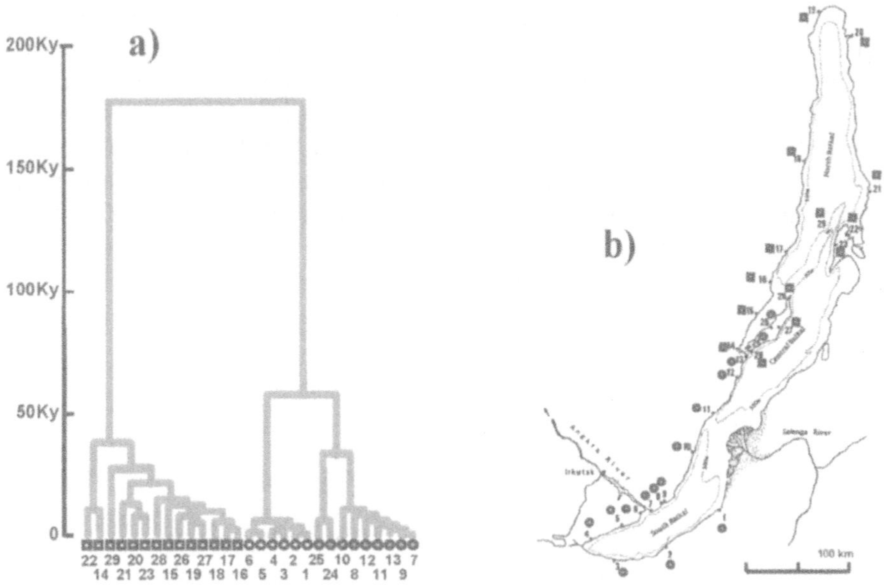

Figure 4. (a) UPGMA dendrogram obtained by allozyme analysis of 21 gene loci in the 29 populations of *Eulimnogammarus cyaneus* (from [25]). (b) Locations, from which specimens were collected.

Lake Tanganyika, the oldest of the three lakes (basin age estimated to be between 9 and 12 million years old) [26] supports at least 197 endemic cichlids in 49 endemic genera. These genera can be grouped into 12 separate tribes that are thought to have diverged from seven distinct ancestral lineages (reviewed in [27]). These tribes are relatively old compared to other East African cichlid lineages; phylogenetic evidence suggests that some of the tribes originated at least five million years ago making them older than both the lakes Victoria and Malawi. More than 300 endemic species of cichlids occur in Lake Victoria, all of which were thought to be derived from a single common ancestor [28]. The cichlid flock of Lake Malawi is intermediate to those in Lakes Tanganyika and Victoria in almost every respect. Cichlids are thought to have invaded Lake Malawi approximately 700 000 years ago, and their morphological diversity is considerably greater than the much younger species flock of Lake Victoria [27].

All three lakes have complex geological histories, characterized by dynamic basin morphology. Lake level fluctuations are affected by variations in amount of rainfall, temperature, evaporation, and, for some lakes, tectonic activity [26].

The age of a species flock may not correspond to the geological age of a lake, since climatic or geological events may have caused a temporary dry-up, such that pre-existing species flocks may have gone extinct. Moreover, it was also shown by recent molecular studies that the dynamics of diversification events in African cichlid fishes are likely to be connected to fluctuations in the lake level (see, for instance, [29]). The evolutionary consequences of such severe environmental changes are well known for European terrestrial faunas, but are much less understood for tropical freshwater ecosystems.

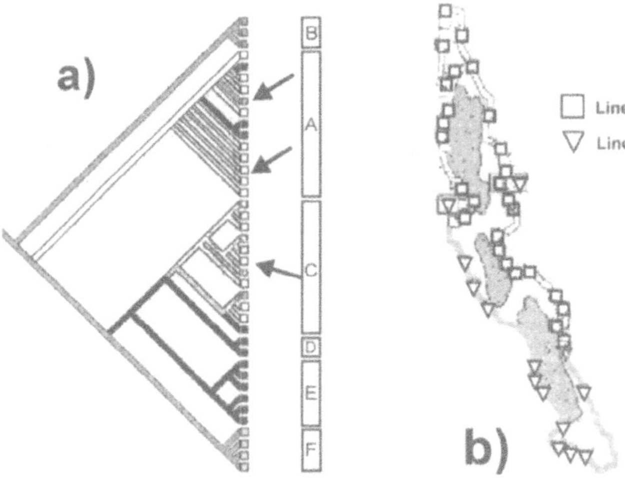

Figure 5. (a) Reconstruction of the evolution of tooth shape within the cichlid fishes of the tribe *Eretmodini* based on one of the 12 most parsimonious trees by using the combined mtDNA data set and no weight, arrows are for species, showed on Fig.6b ([29]). (b) Maps of Lake Tanganyika showing the haplotype distribution of the *Eretmodini* studied according to their tooth shape. The symbols follow the assignments to lineages A and C as in Fig. 6a and are summarized in the box. The three isolated Pleistocene sub-basins are indicated in gray (see [29] for the complete picture).

Sharp climate changes during Pleistocene affect African lakes level variability through changes in precipitation in the area. Several studies agree that the lake levels of all three lakes were substantially lower during the late ice ages, when the climate in much of north and equatorial Africa

became progressively more arid. Lake level fluctuations temporarily form or break down barriers among habitats and thus either promote or prevent gene flow among adjacent populations and/or incipient species [30].
The degree of habitat change enforced by water level fluctuations may range from small-scale effects to major vicariant events that affect species communities in most habitats. Any drop in the lake level will establish secondary contact and admixes among previously isolated populations in shallow regions of a lake, leading to an increase in genetic diversity in admixed populations. A rise in the lake level may promote population subdivision due to the colonization of new habitats. Newly formed ecological barriers interrupt gene flow, such that genetic differences can accumulate independently and lineage sorting can proceed.
The age estimates for various endemic Tanganyika lineages suggest that during the geological history of the lake, when the actual rift formation occurred, cichlids rapidly filled the available niches over the entire lake (review in [31]). When, more recently in the Pleistocene, climatic changes resulted in lowered water level, Lake Tanganyika basin became divided into three paleo-lakes, probably for many thousand years. These fluctuations in level effectively isolated populations of cichlids into three basin populations. If the period of isolation existed long enough, genetic differences between populations would have been likely to arise by genetic drift, and could result in formation of new species (Fig.5).
Sturmbauer et al. [30] founds further genetic traces in cichlids for a much more recent climate change (about 17 Ka) resulting in a dramatic drop of the lake level in Lakes Tanganyika and Malawi, and in the presumed desiccation of Lake Victoria. This event was suggested to have synchronized the most recent diversification events of cichlid fish in all three lakes.
Eretmodine cichlids are restricted along shallow rocky and pebble shores and are unable to disperse across open water. Each of the six eretmodine lineages shows a limited distribution within the lake (Fig.5). The high degree of intralacustrine endemism and the pronounced phylogeographic structuring of eretmodines can be partly explained by the influence of major lake level fluctuations in the Pleistocene that are generally assumed to have had a strong influence on phylogeographic patterns and speciation of rock-dwelling cichlids. During this time, the single lake basin of Lake Tanganyika split up into three isolated sub-basins (shown in gray in Fig.5b); this event is still reflected in the distribution of mtDNA lineages.
Analysis of mtDNA data for cichlid fish from all three Great African lakes in [32] demonstrate that the strikingly similar average numbers of base substitutions in all phylogeographically informative haplotype clusters suggest that the latest periods of low lake level in Lakes Tanganyika,

Malawi, and Victoria happened, or at least ended, roughly at the same time. As a result, rising level influenced in explosive formation of new groups of subspecies. The finding of identical genotypes of *Tropheus* at opposite shores in the central region of Lake Tanganyika suggests a retreat of the lake level by a minimum of 550 m, which would be sufficient to shift a continuous band of rock bottom into the depth limit of *Tropheus* (about 50 m). The distribution of identical or very closely related genotypes in Lake Malawi, and particularly their occurrence at the edge of the deep basin, suggests a retreat of about 500 m, but certainly more than 400 m, below its present level.

5. REFERENCES

1. Kimura, M. *The neutral theory of molecular evolution*, Cambridge University Press, 1983.
2. Semovski, S.V., Bukin, Yu.S. and Sherbakov, D.Yu. Speciation in ancient lakes - modelling attempt, Hydrobiologia, in press.
3. Semovski, S.V., Verheyen E. and Sherbakov, D.Yu. Simulating the evolution of neutrally evolving sequences in a population under environmental changes, accepted to Ecological modelling.
4. Semovski, S.V., Bukin, Yu.S. and Sherbakov, D.Yu. Speciation and neutral evolution in one-dimensional closed population, Int.J.Mod.Phys., 2003; 14(8).
5. Knowlton, N. and Weigt, L.A. New dates and new rates for divergence across the Isthmus of Panama, Proc.R.Soc.Lond.B, 1998; 265:2257-2263.
6. McCune, A.R. and Lovejoy, N.R. The relative rate of sympatric and allopatric speciation in fishes: tests using DNA sequence divergence between sister species and among clades. in: Howard, D.J. , Berlocher, S.H., eds. Endless Forms: Species and Speciation. Oxford University Press, New York, Oxford, 1998.
7. Nee, S. Inferring speciation rates from phylogenies. Int.J.Org. Evolution. 2001; 55(4): 661-668.
8. Barraclough, T.G. and Nee, S. Phylogenetics and speciation, TREE, 2001: 16(7):391-399.
9. Pybus, O. G., Rambaut, A. and Harvey, P. H. An integrated framework for the inference of demographic history from reconstructed genealogies. Genetics, 2000: 155:1429-1437.
10. Rogers, A.R. and Harpending, H. Population growth makes waves in the distribution of pairwise genetic differences, Mol.Biol.Evol., 1992; 9(3):552-569.
11. Polanski A., Kimmel M., and Chakraborty R. Application of a time-dependent coalescence process for inferring the history of population size changes from DNA sequence data, PNAS USA, 1998; 95:5456–5461,
12. Dieckmann, U., Doebeli, M. On the origin of species by sympatric speciation, Nature, 1999; 400:354-357.
13. Felsenstein, J. Phylogeny Inference Package. Retrieved March 2003 from: http://evolution.genetics.washington.edu/phylip.html.
14. Timoshkin, O.A. Biodiversity of Lake Baikal fauna: present state of the art and perspectives of the investigations. in: Guide and Atlas of pelagic animals of the Lake Baikal with ecological notes. O.A. Timoshkin, ed. Novosibirsk: Nauka., 1995: 25-51.
15. Martens, K. Speciation in ancient lakes. TREE, 1997; 12(5):177-182.

16. Mats, V.D. The structure and development of the Baikal rift depression. Earth-Sci.Rev. 1993; 34:81-118.
17. Kuimova, L.N., and Sherstyankin, P.P. Development of the Mechanisms of Paleo-Hydrophysical Processes on Lake Baikal During Last 5 M.Y. in: Third International Symposium' Ancient Lakes: Speciation, Development in Time and Space, Natural History; 2002:93.
18. Sherbakov, D.Yu. Molecular phylogenetic studies on the origin of biodiversity in Lake Baikal, TREE, 1999;14(3): 92-95.
19. Zubakov, D.Yu., Sherbakov, D.Yu., and Sitnikova, T.Ya. Analysis of phylogenetic relationships of Baikalian endemic molluscs form Baicaliidae, Klessin 1878 (Gastropoda, Pectinibranchia) based upon partial nucleotide sequences of mitochondrial gene CO I, Mol.Biol. 1997; 31:1097-1102.
20. Vorobjova, G.A., Mats, V.D. and Shimaraeva, M.K. Paleoclimates of late Kainosoi in Pribaikalje Region, Geol.Geophys. 1995; No.8:82-96.
21. Sherbakov, D.Yu., Kamaltynov, R.M., Ogarkov, J.B. and Verheyen, E. Patterns of evolutionary change in Baikalian Gammaridae inferred from DNA sequences (Crustacea, Amphipoda). Mol..Phylogeny and Evol. 1998; 10(2):160-167.
22. Sherbakov, D. Y., Ogarkov, O. B., Kamaltynov, R. M., Vainio, J. K., Verheyen, E., and Vainola R. On the phylogeny of Lake Baikal amphipods in the light of mitochondrial and nuclear DNA sequence data. in: Proceedings and abstracts of the 4th international Crustacean congress, 20-24 July 1998. Amsterdam:59.
23. Slobodyanyuk, S. Ya., Pavlova, M. E., Kirilchik, S. V., and Novitskii, A. V. The evolutionary relationships of two families of cottoid fishes of Lake Baikal (East Siberia) as suggested by analysis of mtDNA. J. Mol. Evol. 1994; 40:392-399.
24. *Lake Baikal, Evolution and Biodiversity*, O.M.Kozhova and L.R.Izmest'eva, eds., Backhuys Publ., Leiden, 1998.
25. Mashiko, K., Kamaltynov, R. M., Sherbakov, D. Yu., and Morino H. Genetic separation of gammarid (*Eulimnogammarus cyaneus*) population by localized topographic changes in ancient Lake Baikal. Arch. Hydrobiol. 1997; 139:379-387.
26. Cohen, A.S, Lezzar, K.E, Tiercelin, J.J, and Soreghan M. New palaeogeographic and lake-level reconstructions of Lake Tanganyika: Implications for tectonic climatic and biological evolution in a rift lake. Basin Res. 1997; 9:107–132
27. Meyer A. Phylogenetic relationships and evolutionary processes in East African cichlid fishes. TREE, 1993; 8:279–284
28. Verheyen E., Salzburger W., Snoeks J., and Meyer A. Origin of the Superflock of Cichlid Fishes from Lake Victoria, East Africa. Science. 2003; 299:325-329.
29. Rueber, L., Verheyen, E., Sturmbauer, C., and Meyer A. Lake level fluctuations and speciation in rock-dwelling cichlids in Lake Tanganyika, East Africa. In *Evolution on Islands*. Oxford Grant P., ed. University Press, Oxford, 1998.
30. Sturmbauer, C, Baric, S, Salzburger, W, Riiber, L, Verheyen, E. Lake level fluctuations synchronize genetic divergences of cichlid fishes in African lakes, Mol.Biol.Evol. 2001; 18:144–154.
31. Coulter, G.W. Zoogeography, affinities and evolution with special regard to the fishes. In Lake Tanganyika and its life, G.W.Coulter, ed. Oxford University Press, 1991.
32. Baric, S., Salzburger, W., and Sturmbauer, C. Phylogeography and Evolution of the Tanganyikan Cichlid Genus Tropheus Based upon Mitochondrial DNA Sequences, J Mol Evol. 2003; 56:54–68.

Chapter 15

HYPOXIA AND THE PHYSICS OF THE LOUISIANA COASTAL CURRENT

Wm. J. Wiseman, Jr.

Coastal Studies Institute, and Department of Oceanography and Coastal Sciences, Louisiana State University, Baton Rouge, LA 70803

N. N. Rabalais

Louisiana Universities Marine Consortium, 8124 Highway 56, Chauvin, LA, 70344

R. E. Turner, and D. Justic

Coastal Ecology Institute, and Department of Oceanography and Coastal Sciences, Louisiana State University, Baton Rouge, LA 70803

1. INTRODUCTION

The Mississippi River is the largest river in North America. It drains 2.979 10^6 km^2, including 41% of the lower 48 United States and parts of two Canadian provinces. The effluent from this system is debouched into the Gulf of Mexico through the modern birdfoot delta of the Mississippi River and through a secondary outlet, the Atchafalaya River delta. Under the influence of the Coriolis force, a significant portion of these waters flows westward along the Louisiana coast as the highly stratified Louisiana Coastal Current (Wiseman and Kelly, 1994). Each summer, hypoxic or, occasionally, anoxic conditions develop beneath the associated halocline.

J.C.J. Nihoul et al. (eds.), Dying and Dead Seas, 359–372.

(Hypoxia is defined operationally as dissolved oxygen concentrations less than 2 mg/l (Rabalais, et al. 1991).)

The simplest form of the paradigm describing the development of hypoxia in the region is the following. The spring flood of the Mississippi River system delivers a massive buoyancy flux and nutrient flux to the coastal waters of the northwestern Gulf of Mexico. At the same time, light conditions conducive to primary production are improving. Finally, the winter storms that can mix surface waters across the halocline are diminishing in intensity and frequency. Phytoplankton, held in the near-surface waters by the strong pycnocline associated with haline stratification, bloom. These then sink to the bottom as both dead and live cells, or repackaged as zooplankton fecal pellets. This carbon flux to the bottom produces an organic substrate that, as it oxidizes, consumes the dissolved oxygen in the lower layers more rapidly than it can be replaced by advective-diffusive processes. (Wiseman, et al., 1997). This process commences in the spring, is modulated by storms, and persists until the first strong cold air outbreaks of autumn.

This process has been ongoing for many years (Nelsen, et al. 1994; Rabalais, et al. 2002a), but there is clear evidence that the environmental response is intensifying (Rabalais, et al. 2000b). Indeed, the area of mid-summer hypoxic bottom waters has doubled since 1985 to more than 20,000 km^2. While the records are inconclusive, there is evidence that phytoplankton assemblages have altered (Dortch, et al. 2001) in response changes in the nutrient ratios of the waters delivered by the river system to the coast (Justic, et al., 1995). Prudence suggests that a pro-active stance be taken to avoid catastrophic impacts of seasonal or persistent hypoxia (Caddy, 1993).

2. HISTORICAL UNDERSTANDING OF THE PHYSICAL OCEANOGRAPHY

Cochrane and Kelly (1986), present the first careful synthesis of understanding of the circulation over the northwestern shelves of the Gulf of Mexico. They describe a seasonal cycle dominated by a cyclonic gyre consisting of westward nearshore flow of low salinity waters derived from Mississippi River system runoff, under the influence of prevailing southeasterly winds, and a return flow over the outer shelf. During the mid-summer months, the winds become southerly along the south Texas coast. This forces a nearshore return flow towards the north and east. The westward flow of low salinity waters separates from the coast somewhere near the Texas-Louisiana border. The waters from the Mississippi River system must leave the shelf, either flowing downcoast into Mexican waters or dispersing offshore across the shelf break. The mechanisms responsible and the importance of these two fates are poorly understood, at present (Dinnel and Wiseman, 1986). More recent work (Cho, et al., 1998) generally supports this picture of the long term mean seasonal circulation, but stresses the interannual variability.

The strength of the halocline of the Louisiana Coastal Current is a dominant characteristic of this region (Wiseman, et al., 1982; Murray, 1997). It has been associated with the isolation of the sub-pycnocline waters from direct air-sea interaction (Crout, 1983; Dagg, 1988; Rabalais, et al. 2002b). Thus, the present understanding of the physics of these waters is that they respond largely as a two-layered system similar to the coastal boundary layers described in the Laurentian Great Lakes by Csanady (1981).

3. MODELING EFFORTS

A number of models of both the physics and the nutrient and oxygen dynamics of the Louisiana Coastal Current have been developed over the years. They are of varying degrees of complexity and success. Early statistical relationships between alongshore currents and winds (Crout et al. 1984) indicate the strong coherence suggestive of a successful linear model, as applied by Murray (1997). Chuang and Wiseman (1983) characterized the wind-driven flow offshore of the Atchafalaya Delta using a spectral model. Their success, though, depended upon specification of a bottom friction coefficient that was significantly larger than is typically applied to such environments. Current (1996) used a barotropic spectral shelf wave model to describe the measurements obtained during a three-year observational program (Nowlin, et al. 1998). Finally, Lewis and Reid (1985) modeled the vertical variability of a two-layered, wind-driven flow regime offshore of Texas and compared it favorably with observations. They required, though, a significantly larger interfacial drag coefficient than is considered realistic. Thus, like the study of Chuang and Wiseman, they may be characterizing missing physics by anomalous engineering coefficients. Numerous authors are attempting to apply three-dimensional models to the region. That effort which has seen the most scrutiny is an application of the Princeton-Dynalysis Ocean Model (Herring, et al. 1999).

The model was forced with observed winds and runoff for a three-year period. When compared with isolated tide gauges or current meters, the model output was skillful (Wiseman, et al., 2000), but not significantly more so than the barotropic spectral model of Current (C. Current, pers. comm.). The model was also compared with quantities that responded to the integrated forcing of advection, stirring, and mixing, such as the salinity distribution. The general character of these fields were reproduced by the model. Certain issues, though, remained. Aspects of the comparison in which the model did not fare well were attributed to lack of resolution of processes within the coastal estuaries, particularly Atchafalaya Bay, and inadequacies of the mixing parameterizations, both lateral and vertical.

Efforts to incorporate both the physics and biology have proven less successful. Scavia, et al. (2003) applied a simplified engineering oxygen sag model to the hypoxia region with some success at hindcasting hypoxic conditions. Justic (1996, 2003a, b), assuming a two-box vertical model, was able to hindcast the seasonal cycle of dissolved oxygen at a station with long-term monitoring data. He later applied the same model, in a statistical fashion, to predict the response of the system to climatological changes expected from increased CO_2 scenarios. Chen, et al. (1997) applied a two-dimensional version of the Princeton Ocean Model to the region. While their nutrient distributions reproduced the gross features of the observations, their corresponding velocity structure was significantly stronger than observations. Bierman, et al., (1994) applied the WASP4 methodology to the region in a steady-state, screening application. Their results were then modified to consider various management scenarios concerning nutrient delivery reduction from the river system. They were unable to adapt this model successfully to the flow fields from the Princeton-Dynalysis Ocean Model output. Thus, there continues to be a need for an ecosystem model that successfully represents dissolved oxygen, nutrient, and phytoplankton dynamics on a seasonal basis.

4. NEW OBSERVATIONS

The west Louisiana inner continental shelf is oriented approximately east-west. It first trends northwest from the birdfoot delta for a short distance to the longitude of the Mississippi Canyon, approximately 89^0 50'W. It then turns southwest for a brief run and finally is oriented west-northwest until the Texas border. The shelf width broadens westward from the delta except for the region immediately west of the delta, which is cut by the Mississippi

Canyon. The coast is interrupted by numerous tidal passes, which connect estuaries associated with the growth and deterioration of the deltaic plain to the open Gulf of Mexico. Winds are predominantly from the southeast. During the fall, winter, and spring, cold air outbreaks cross the coast with time scales between 3 and 10 days. During the summer, weak sea breeze systems are occasionally interrupted by tropical storms and hurricanes. The main sources of freshwater flow are the major outlets of the Mississippi-Atchafalaya River system, which discharges 850 km^3yr^{-1} to the shelf through the Atchafalaya River delta and, subsequently, Atchafalaya Bay (30%) and through the various passes of the Mississippi birdfoot delta (70%).

From 14 March through 12 November 2002, we deployed an acoustic Doppler current profiler (ADCP) at a station in 20 meters of water midway between the Mississippi River birdfoot delta and the Atchafalaya River delta (29.5 N 91.25 W) (Fig. 1). Ancillary data included bottom mounted pressure cells, roughly at the same isobath, east and west of the mooring and temperature-salinity recorders at the site of the ADCP mooring. Weather data were obtained from a station at Grand Isle (29.27 N 89.96 W).

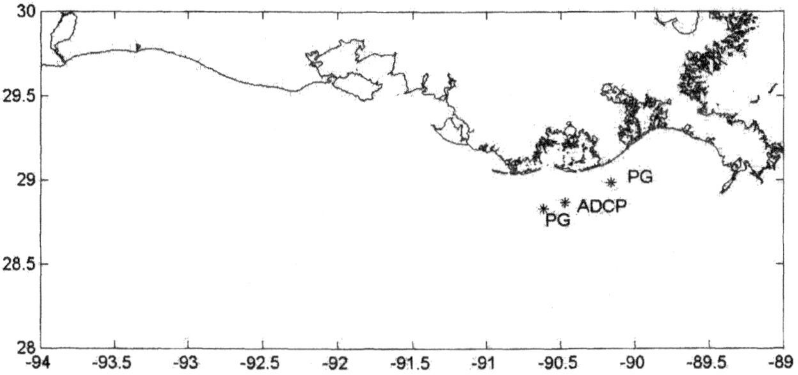

Figure -1. Location map indicating the ADCP and pressure gauge locations. All were deployed at approximately 20 m water depth.

All data were processed in a similar fashion. To eliminate the semi-diurnal and diurnal tides, as well as the strong inertial oscillations present near 30 N, a 101-point, zero-phase filter approximating a flat response out to 40 hours and a linear decay to 27 hour periods (Ormsby, 1961) was applied to the data. The low-passed currents and pseudostress were rotated into a coordinate system that is alongshore/cross-shore. This entailed a rotation of 12° counterclockwise.

The anticipated strong dominance of the alongshore currents is clear (Fig. 2). More surprising is the highly variable structure of the flow. Periods of strong vertical shear e.g. hours 3200-3500, are interspersed with long periods of very weakly sheared flow, e.g. hours 4800-5000. Long periods of flow contrary to the expected westward flow regime were also observed, e.g. hours 2700-3100. It is noteworthy that the vertical density difference between sensors at approximately 7 and 20 m depth showed reduced density difference associated with strong upper layer flow and, consequently, strong shear, even when the river was in flood and the flow was responsible for a strong lateral buoyancy flux (Fig. 3). The significant importance of shear-induced mixing is implied.

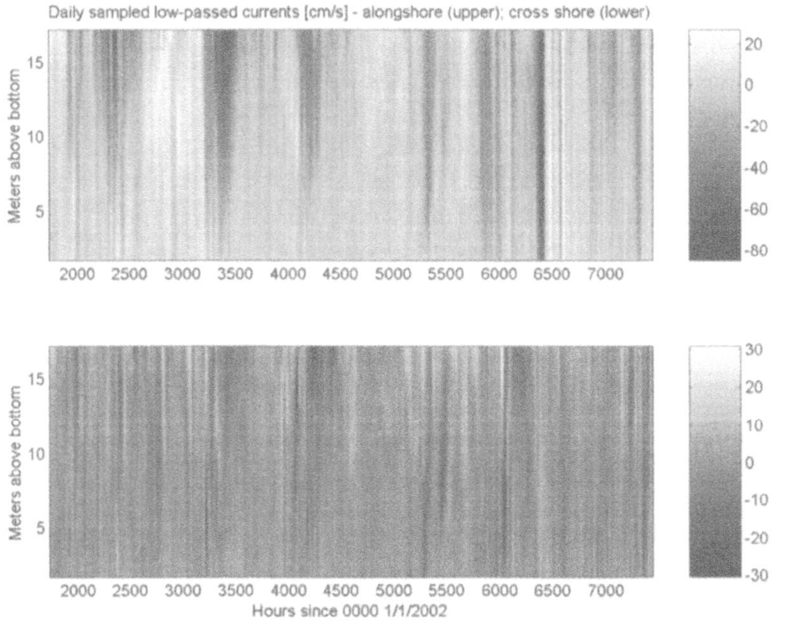

Figure -2. Contoured daily samples of low-pass filtered currents (sm/s) after rotation into alongshore (top) and cross-shore (bottom) directions. Bins are 0.5 m thick and begin 1.71 m above bottom.

The mean flow is consistent with expectations. The flow is strongly sheared in the alongshore direction with weak lower layer flow and a weak indication of reversal (Fig. 4). The cross shore flow is consistent with a wind-driven downwelling regime as would be expected under the predominantly southeasterly winds affecting the region.

Two-layered flows or flow confined to the upper layer were anticipated from both theory and prior isolated observations, particularly during the highly stratified summer season. Yet, the complex EOF analysis (Kundu and Allen, 1976) indicates that 82% of the variance in the low-passed currents is accounted for by the first mode, which describes nearly unidirectional, vertically-sheared flow (Fig. 5). The second mode accounts for only 13% of the variance and describes a two-layered flow field. Yet, the vertical sigma-t gradient during the deployment was strong, generally greater than .075 m^{-1} and often 2 or 3 times this (Fig. 3).

Depth-frequency spectra of the alongshore flow indicate the importance of the diurnal and semi-diurnal tidal-inertial band currents. At the lowest frequencies, the flows penetrate to the bottom with significant intensity, particularly the alongshore component, while the higher frequency signals suggest surface trapping of the energy (Fig. 6). A similar structure is evident in the depth-frequency spectra of the cross shore flows.

The alongshore wind pseudostress is generally coherent with the alongshore currents at the 95% significance level below 0.025 cph, except in a narrow band around 0.015 cph. This coherence is slightly stronger at depth than in the upper layers, but it is significant throughout the entire water column. The same component of pseudostress is coherent with the lower layer cross shore currents (1.7 – 7.2 m height) below 0.014 cph, except for a band between 0.01 and 0.0075 cph. During the first deployment, the upper layer alongshore currents were coherent with the alongshore pressure gradient in bands near 0.01 and 0.02 cph. Similar coherence was observed during the second deployment, but with a more barotropic resonse, and in a third band of significant coherence near 0.005 cph. It should be noted, though, that two hurricanes passed through the region, approximately a week apart, during the second deployment. The associated signals were very strong in all records and may dominate the analyses.

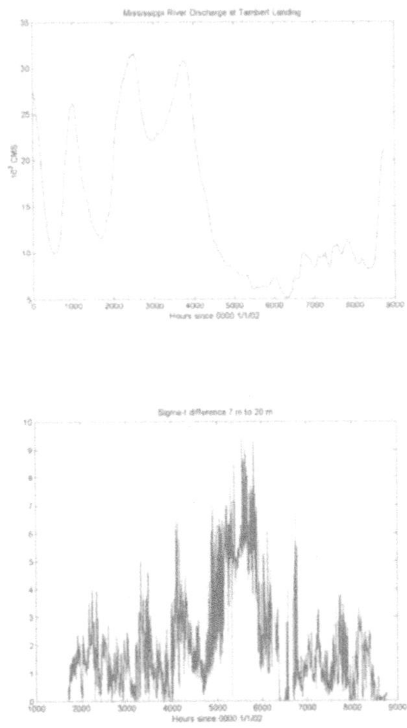

Figure -3. Top: discharge of the Mississippi River during 2002. Bottom: vertical difference between 7 and 20 m depth at the ADCP mooring.

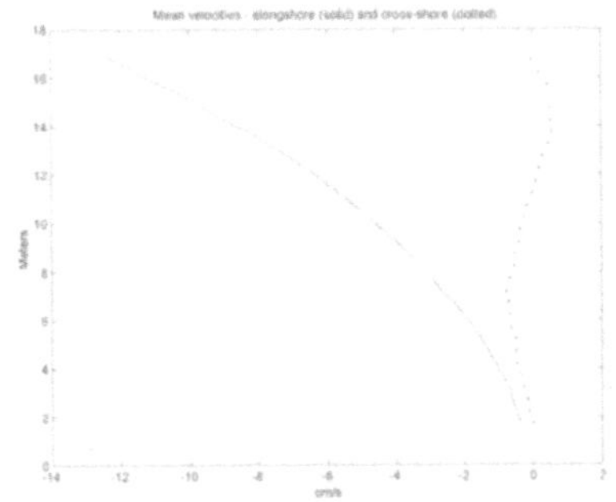

Figure -4. Mean alongshore and cross shore flows during the ADCP deployment.

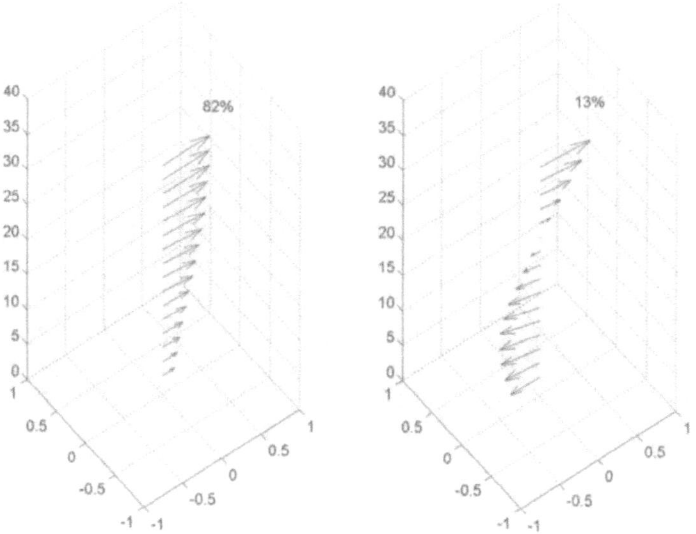

Figure -5. The first (left) and second (right) empirical orthogonal functions from the ADCP data. Values associated with every second ADCP bin are plotted.

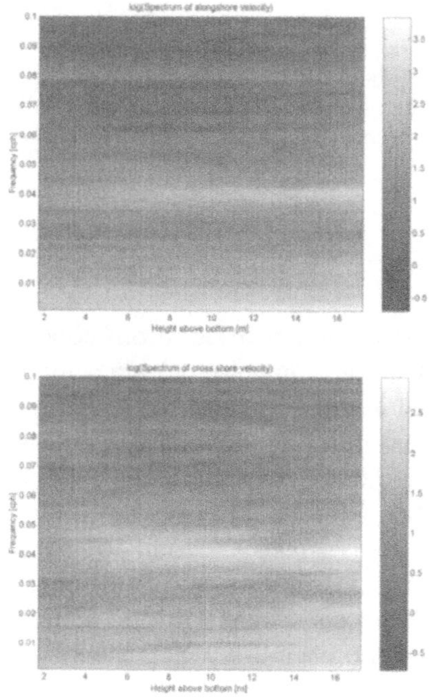

Figure -6. Current spectra: left (alongshore flow), right (cross shore flow).

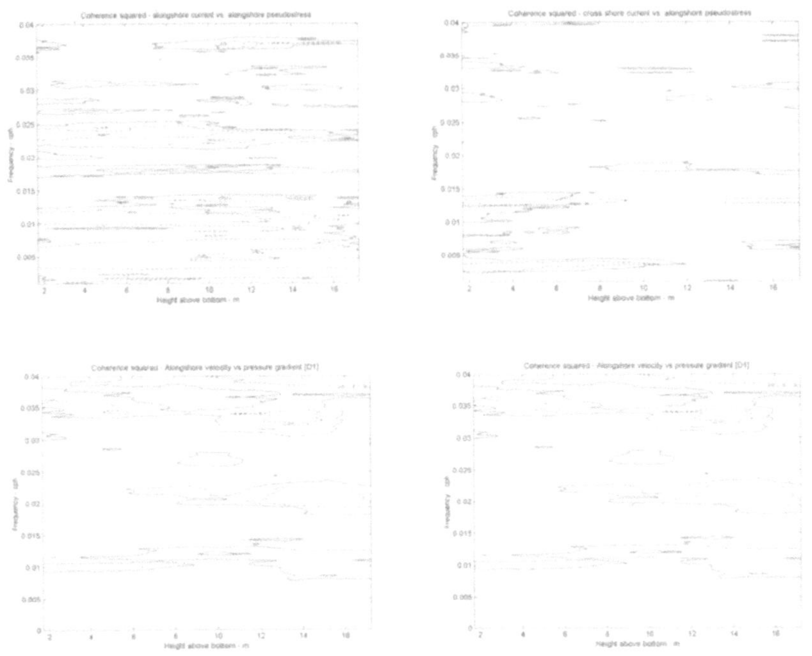

Figure -7. Coherence squared estimates. Upper left: alongshore current and alongshore pseudostress. Upper right: cross shore current and alongshore pseudostress. Bottom left: alongshore current and alongshore bottom pressure gradient during deployment 1. Bottom right: alongshore current and alongshore bottom pressure gradient during deployment 2. Solid contour is 95% significance level (0.2831), dashed contour is 0.5, dotted contour is 0.75.

5. DISCUSSION

These observations raise a number of questions concerning the dynamics of the local flow. What is the momentum source for the lower layer flow? Our normal concepts of highly stratified coastal flow as two-layered would preclude local wind forcing of the lower layer. Csanady (1981) suggests that, if the lower layer moves cross-shore as a column in response to the downwelling winds, then Coriolis force acting on this motion will cause an alongshore motion in the correct direction to match observations. The conceptual model, though, assumes two-dimensional and frictionless flow. These assumptions are not consistent with the real environment. Is lateral advection of alongshore, wind-driven waters associated with the downwelling system sufficiently intense to account for the observations or are they associated with alongshore pressure gradients? Allen, et al. (1995)

and Allen and Newberger (1996) have described the flows arising within a two-dimensional model of upwelling and downwelling dynamics and the multiple circulation cells that may result, even without the addition of a coastal buoyancy source. Our data cannot resolve the spatial complexity of their model predictions, but the observed cross-shore flows are often multi-layered. Also, it may be that the wind field, interacting with the strong stratification and the crenulated shoreline, produces spatially and temporally varying alongshore pressure gradients that drive the flow in the lower layer. Some evidence for this may be seen in the coherence analyses discussed above. If the lateral flows are important mechanisms for the advection of momentum, how does this affect the oxygen dynamics? Our near-bottom oxygen records corresponding to the current measurements presented above are still being processed, but historical records (Rabalais, et al. 2002b) suggest that near-bottom dissolved oxygen concentrations are modulated at a variety of advective scales, including those associated with upwelling/downwelling cycles.

Another observation that begs explanation is why the upper layer cross-shore flow is incoherent with the alongshore wind stress while the lower layer flow is significantly coherent with it? Fine scale models of the region suggest a strong eddy field in the upper layers. Thermal infra-red and visible satellite images of the region indicate some eddying motion in the surface fields of the coastal current, but there are no water column observations for comparison.

6. CONCLUSIONS

The lower layer waters of the inner Louisiana shelf suffer from recurrent seasonal hypoxia and anoxia. This phenomenon is related to the buoyancy and nutrient fluxes from the Mississippi River system. It has been ongoing for many years, but the duration, spatial coverage, and intensity of the phenomenon seem to be increasing. While fisheries records do not yet allow a clear determination of detrimental effects, there is anecdotal evidence of such. Caution suggests that efforts be initiated to ameliorate the situation. Because of the size of the drainage basin involved, national action is required. Such action has been undertaken in response to a series of white papers submitted to the White House Committee on Natural Resources and Environment.

As part of this effort, models are needed for hindcasting the history of the phenomenon and predicting its future development and its potential response to various management scenarios. Existing models of various types have exhibited moderate skill in hindcasting, although each has its weak points. In particular, the physical models have not proven capable of reproducing the observed variability of the important hydrographic structure in the region.

Recent observations in the region indicate numerous aspects of the flow regime that a successful model must reproduce and suggest that the dynamics are neither as local nor as simple as has been assumed in the past. Important issues remain concerning the processes responsible for both vertical and lateral dispersion within this highly stratified system. The role of short-scale pressure gradients also remains to be determined.

ACKNOWLEDGEMENTS

This work was supported by the NOAA Coastal Ocean Program, Contract No. NA060P0529 and by the U.S. Minerals Management Service, Contract No. 1425-00-01-CA-30660/19936.

REFERENCES

Allen, JS and Newberger PA (1996) Downwelling circulation on the Oregon continental shelf. Part I:response to idealized forcing. J. Phys. Oceanogr., 26(10):2011-2035.

Allen, JS, Newberger PA and Federiuk J (1995) Upwelling circulation on the Oregon continental shelf. Part I: response to idealized forcing. J. Phys. Oceanogr., 25(8):1843-1866.

Bierman, VJ, Jr., Hinz SC, Zhu D, Wiseman WJ, Jr., Rabalais NN and Turner RE (1994) A preliminary mass balance model of primary productivity and dissolved oxygen in the Mississippi River plume/inner Gulf shelf. Estuaries, 17(4):886-899.

Caddy, JF (1993) Toward a comparative evaluation of human impacts onfisheries ecosystems of enclosed and semi-enclosed seas. Rev. Fish. Sci.. 1:57-95.

Chen, C., Wiesenburg D and Xie L (1997) Influences of river discharges on biological production over the inner shelf: a coupled biological and physical model of the Louisiana-Texas shelf. J. Mar. Res., 55, 293-320.

Cho, K, Reid RO and Nowlin WD Jr. (1998) Objectively mapped stream function fields on the Texas-Louisiana shelf based on 32 months of moored current meter data. J. Geophys. Res., 103(C5): 10377-10390.

Chuang, W-S and Wiseman WJ Jr., (1983) Coastal sea level response to frontal passages on the Louisiana-Texas shelf. J. Geophys. Res., 88(C4):2615-2620.

Cochrane, JD and Kelly FJ Jr., (1986) Low-frequency circulation on the Texas-Louisiana shelf. J. Geophys. Res., 91(C9):10645-10659.

Crout, RL (1983) Wind-driven, near-bottom currents over the west Louisiana inner continental shelf. Unpublished M. S. thesis. Dept. of Marine Sciences, Louisiana State University, Baton Rouge, LA. pp. 125.

Crout, RL, Wiseman WJ Jr. and Chuang W-S (1984) Variability of wind-driven currents, west Louisiana inner continental shelf: 1978-1979. Contributions in Marine Science, 27:1-11.

Csanady, GT (1981) Circulation in the coastal ocean. Advances in Geophysics, 23:101-1183.

Current, C. L. 1996. Spectral model simulation of wind driven subinertial circulation on the inner Texas-Louisiana shelf. Unpublished Ph. D. dissertation. Dept. of Oceanography. Texas A&M University. College Station, TX. pp. 159.

Dagg, MJ (1988) Physical and biological responses to the passage of a winter storm in the coastal and inner shelf waters of the northern Gulf of Mexico. Cont. Shelf Res., 8:167-178.

Dinnel, SP and Wiseman WJ Jr. (1986) Fresh water on the Louisiana and Texas shelves. Cont. Shelf Res., 6(6):765-784.

Dortch, Q, Rabalais NN, Turner RE, and Qureshi NA (2001) Impacts of changing Si/N ratios and phytoplankton species composition. In (N. N. Rabalais and R. E. Turner, eds.) Coastal Hypoxia: Consequences for Living Resources and Ecosystems. Coastal and Estuarine Studies 58. Washington, DC. Am. Geophys. Union. pp.37-48.

Herring, JH, Inoue M, Mellor GL, Mooers CNK, Niiler PP, Oey L-Y, Patchen RC, Vukovitch FM, and Wiseman WJ, Jr. (1999) Coastal Ocean Modeling Program for the Gulf of Mexico. Final Report to the Minerals Management Service, Environmental Division, Technical Analysis Group. (3 volumes and 17 Appendices).

Justic, D, Rabalais NN, and Turner RE (1996) Effects of climate change on hypoxia in coastal waters: a doubled CO_2 scenario for the northern Gulf of Mexico. Limnology and Oceanography 41(5):992-1003.

Justic, D, Rabalais NN, and Turner RE (2003a) Climatic influences on riverine nitrate flux: Implications for coastal marine eutrophication and hypoxia. Estuaries, 26:1-11.

Justic, D, Rabalais NN, and Turner RE (2003b) Simulated responses of the Gulf of Mexico hypoxia to variations in climate and anthropogenic nutrient loading. J. Mar. Syst., in press.

Justic, D, Rabalais NN, Turner RE, and Dortch Q (1995) Changes in nutrient structure of river-dominated coastal waters: Stoichiometric nutrient balance and its consequences. Estuarine, Coastal and Shelf Science, 40:339-356.

Kundu, PK and Allen JS (1976) Some three-dimensional characteristics of low-frequency current fluctuations near the Oregon coast. J. Phys. Oceanogr., 6(2):181-199.

Lewis, JK and Reid RO (1985) Local wind forcing of a coastal sea at subinertial frequencies. J. Geophys. Res., 90(C1):935-944.

Murray, SP (1997) An observational study of the Mississippi-Atchafalaya coastal plume: Final report. OCS Study MMS 98-0040. U.S. Dept. of the Interior, Minerals Management Service, Gulf of Mexico OCS Region, New Orleans, LA. 513 pp.

Nelsen, TA, Blackwelder P, Hood T, McKee B, Romer N, Alvarez-Zarikian C, and Metz S (1994) Time-based correlation of biogenic, lithogenic, and authigenic sediment components with anthropogenic inputs in the Gulf of Mexico NECOP study area. Estuaries,17(4):873-885.

Nowlin, WD, Jr., Jochens AE, Reid RO, and DiMarco SF (1998) Texas-Louisiana shelf circulation and transport processes study: Synthesis Report, Vol. I: Technical report. OCS Study MMS 98-0035. U.S. Dept. of the Interior, Minerals Management Service, Gulf of Mexico Region, New Orleans, LA. 502 pp.

Ormsby, JFA (1961) Design of numerical filters with applications to missile data processing. Journal of the ACM, 8(3):440-466.

Rabalais, NN, Turner RE, and Wiseman WJ Jr. (2002a) Gulf of Mexico Hypoxia, a.k.a. "The Dead Zone". Annual Reviews of Ecology and Systematics, 33:235-263.

Rabalais, NN, Turner RE, and Scavia D (2002b) Beyond science into policy: Gulf of Mexico hypoxia and the Mississippi River. BioScience, 52:129-142.

Scavia, D, Rabalais NN, Turner RE, Justic D, and Wiseman WJ Jr. Predicting the Response of Gulf of Mexico Hypoxia to Variations in Mississippi River Nitrogen Load. Limnol. Oceanogr., 48(3):.951-956.

Wiseman, WJ Jr. and Kelly FJ (1994) Salinity variability within the Louisiana Coastal Current during the 1982 flood season. Estuaries, 17(4):732-739.

Wiseman, WJ Jr., Murray SP, Bane JM and Tubman MW (1982) Physical environment of the Louisiana Bight. Contributions in Marine Science, 25:109-120.

Wiseman, WJ Jr., Rabalais NN, Turner RE, Dinnel SP and MacNaughton A (1997) Seasonal and interannual variability within the Louisiana Coastal Current: stratification and hypoxia. J. Mar. Sys.. 12:237-248.

Wiseman, WJ, Jr., Inoue M, Patchen R, Ransibrahmanakul V, Murray SP, and DiMarco S (2000) Hydrography of the Louisiana Coastal Current: Model-Data Comparison. In (T. Yanagi, ed.) Interactions between Estuaries, Coastal Seas and Shelf Seas, Terra Scientific Publishing Company (TERRAPUB), Tokyo. pp. 287-302.

LIST OF PARTICIPANTS

Dr. Abdrakhmatov K.
Institute of Seismology
National Academy of Science
Asanbay 52/1
7020060 Bishkek, Kyrgyz Republic
kanab@elcat.kg
Dr. Aladin N.V.
Russian Academy of Sciences
Zoological Institute, Brackish Water Laboratory
Brackish Water Laboratory
119034 St. Petersburg, Russia
aral@zin.ru
Mrs. Alekseeva I.
Hamburg University
Institute of Oceanology
Troplowitstrasse 7
D-22529 Hamburg, Germany
A.Alveraazcarate@ulg.ac.be
Miss Alvera A.
University of Liège
GHER
B5 Sart Tilman
B-4000 Liège, Belgium
A.Alveraazcarate@ulg.ac.be
Dr. Beckers J.-M.
University of Liège
GHER.
B5 Sart Tilman
B-4000 Liège, Belgium
JM.Beckers@ulg.ac.be
Dr. Benduhn F.
Université de Neuchâtel
Centre d'Hydrogéologie de Neuchâtel (CHyN)
Rue Emile-Argant 11
CH-2007 Neuchatel, Suisse
benef5@yahoo.com

Dr. Boroffka N.
GeoForschungsZentrum Potsdam
Blissestr. 46
D-10713 Berlin, Germany
nboroffka@t-online.de
Dr. Destouni G.
KTH
Dept. of Land and Water Resources Engineering
Brinellv. 32
S-100 44 Stockholm, Sweden
gia@kth.se
Dr. Dianova T.
Lomonosov Moscow State University,
Department of Oceanology
Leninskie Gory
119992 Moscoow, Russia
akosarev@mail.ru
Dr. Djenidi S.
University of Liège
GHER
B5 Sart Tilman
B-4000 Liège, Belgium
sdjenidi@ulg.ac.be
Dr. Dvorkin Y.
The Geological Survey of Israel
30 Malkhe Israel St.
95501 Jerusalem, Israel
yona@geos.gsi.gov.il
Dr. Elmoussaoui A.
IFREMER
Laboratoire de Physique des Océans
Centre de Brest
Brest, France
abdelali.elmoussaoui@ifremer.fr
Dr. El-Rayis O. A.
University of Alexandria
Department of Oceanography
Moharrem Bek
21511 Alexandria, Egypt
elrayis@yahoo.com

Dr. Friedrich J.
Alfred-Wegener-Institute for Polar and Marine Research
Geosystem
Am Handelshafen 12
D-27570 Bremerhaven, Germany
jfriedrich@awi-bremerhaven.de
Dr. Froebrich J.
University of Hannover
Water Quality Protection and Management Section
Am Kleinen Felde 30
D-30167 Hannover, Germany
jofr@fggm.uni-hannover.de
Dr. Gavrieli I.
Geological Survey of Israel
30 Malkhei Israel St.
95501 Jerusalem, Israel
ittai.gavrieli@mail.gsi.gov.il
Dr. Gertman I.
Head of Israel Marine Data Center
Israel Oceanographic & Limnological Research
Tel-Shikmona, P.O.B. 8030
31080 Haifa, Israel
isaac@ocean.org.il
Dr. Ginzburg A.I.
P.P. Shirshov Institute of Oceanology
RAS
Moscow, Russia
afedorov@online.ru
Dr. Giralt S.
Institute of Earth Sciences "Jaume Almera"
CSIC
Lluis Sole i Sabaris, s/n
E-08028 Barcelona, Spain
sgiralt@ija.csic.es
Dr. Grégoire M.
University of Liège
Department of Biology
B6 Sart Tilman
B-4000 Liège, Belgium
mgregoire@ulg.ac.be

Dr. Jarsjö J.
Stockholm University
Department of Physical Geography and Quaternary Geology
SE-106 91 Stockholm, Sweden
jerker.jarsjo@natgeo.su.se
Dr. Karafistan A.
Canakkale Onsekiz Mart University
SUF, Department of Fisheries
P.K. 56
17100 Canakkale, Turkey
akarafistan@yahoo.com
Dr. Karimov B.
Uzbekistan Academy of Sciences
The Institute of Water Problems
F. Khodjaev st., 25 a
700000 GSP Tashkent, Uzbekistan
Karimov@dostlink.net
Dr. Kazbekov J.
Director of Hydroinformatics and
Water Resources Training Center
TIIAME
39, Kary Niyazov Street
700000 Tashkent, Uzbekistan
hydrotech@uzsci.net
Dr. Khakimov Z.
Uzbek Academy of Sciences
Institute of Nuclear Physics
702132 Tashkent, Uzbekistan
khakimov@suninp.tashkent.su
Dr. Khamidov M. K.
Uzbek Academy of Sciences
702132 Tashkent, Uzbekistan
Dr. Khan V.M.
Hydrometeorological Research Center
of the Russian Federation
Department of Long-rame Forecasting
9-11 Bolshoy Predtechensky Pereulok
123242 Moscow, Russia
odpp@hydromet.ru
han@rhmc.mecom.ru

Dr. Kosarev A.
Lomonosov Moscow State University
Department of Oceanology
Leninskie Gory
119992 Moscow, Russia
kosarev@ocean.geogr.msu.su
Dr. Kostianoy A.
Russian Academy of Sciences
P.P. Shirshov Institute of Oceanology
36 Nakhimovsky Prosp.
117997 Moscow, Russia
kostianoy@online.ru
Dr. Kouraev A. V.
Laboratoire d'Etudes en Géophysique et
Océanographie Spatiales
LEGOS
F-31400 Toulouse, Fance
kouraev@notos.cst.cnes.fr
Dr. Lebedev S. A.
Russian Academy of Sciences
Geophysical Center
3, st. Molodezhnaya
117296 Moscow, Russia
lebedev@wdcb.ru
Dr. Lemoalle J.
Institut de Recherche pour le Développement
IRD - MSE
B.P. 5045
F-34032 Montpellier Cedex, France
lemoalle@mpl.ird.fr
Dr. Leroy S.
Brunel University
Department of Geography and Earth Sciences
Uxbridge, Middlesex, UB8 3PH, U.K.
suzanne.leroy@brunel.ac.uk

Dr. Letolle R.
Université Pierre et Marie Curie
Département de Géologie appliquée
Box 203
F-75252 Paris Cedex 05, France
rene.letolle@wanadoo.fr
Dr. Melentyev V.
Nansen International Environmental
and Remote Sensing Center
NIERC
26/28 B. Monetnaya str.
197101 St. Petersburg, Russia
vladimir.melentyev@niersc.spb.ru
Dr. Micklin Ph.
Western Michigan University
Department of Geography
Kalamazoo,MI 49006, U.S.A.
micklin@wmich.edu
Dr. Mikhaylichenko Y. G.
Science and Technology of the Russian Federation
Department of Life and Earth Science
11, Tverskaya Str.
103905 Moscow, Russia
mikhaylichenko@minstp.ru
Dr. Mirabdullayev I. M.
The Uzbekistan Academy of Sciences
Institute of Zoology
700095 Tashkent, Uzbekistan
iskandar@mail.tps.uz
Dr. Naessens E.
INVE Technologies
Hoogveld 93
B-9200 Dendermonde, Belgium
e.naessens@inve.be

Dr. Nezlin N. P.
University of California Los Angeles - UCLA
Marine Science Center
Box 951606
621 Charles E. Young Dr. South
Los Angeles, CA 90095-1606, U.S.A.
nezlin@biology.ucla.edu
Dr. Nihoul J.C.J.
Liège University
GHER
B5 Sart Tilman
B-4000 Liège, Belgium
J.Nihoul@ulg.ac.be
Dr. Nurtaev B.S.
Academy of Sciences Republic Uzbekistan
Institute of Geology and Geophysics
N. Khodjibaev str. 49
700041 Tashkent, Uzbekistan
nurtaev@ingeo.uz
Dr. Oren A.
The Hebrew University of Jerusalem
The Institute of Life Sciences
Division of Microbial and Molecular Ecology
91904 Jerusalem, Israel
orena@shum.cc.huji.ac.il
Dr. Peneva E.
University of Sofia
Dept. of Meteorological and Geophysics
5 J. Bourchier Str.
1126 Sofia, Bulgaria
elisaveta.peneva@jrc.it
Dr. Ratner Yu. B.
National Academy of Sciences of Ukrainia
Marine Hydrophysical Institute
Kapitanskaya st. 2
99000 Sevastopol, Crimea, Ukrainia
yurat@sat.mhi.iuf.net

Dr. Reder Ch.
University of Hannover
Water Quality Protection and Management Section
Am Kleinen Felde 30
D-30167 Hannover, Germany
reder@fggm.uni-hannover.de
cr@fggm.uni-hannover.de
Dr. Renard Ph.
Université de Neuchâtel
Centre d'Hydrogéologie
Rue Emile-Argant 11
CH-2007 Neuchâtel, Suisse
Philippe.Renard@unine.ch
Dr. Ressl R.
German Aerospace Center
German Remote Sensing Data Center
Kalkharsmeg 53
17235 Nenshelitz, Germany
Rainer.Ressl@dlr.de
Dr. Rixen M.
University of Liège
GHER
B5 Sart Tilman
B-4000 Liège, Belgium
m.rixen@ulg.ac.be
Dr. Ronday F.
University of Liège
Meteorology
B5 Sart Tilman
B-4000 Liège, Belgium
fronday@ulg.ac.be
Dr. Salokhiddinov A.T.
Tashkent Institute of Engineers for Irrigation
and Agricultural Mechanization
Department of Water Resources
and Environmental Engineering
Kary-Niazov street, 39
Tashkent, Uzbekistan
pepiwm@globalnet.uz

Dr. Semovski S. V.
Limnological Institute
SB / RAS
P.O. Box 4199
664033 Irkutsk, Russia
semovski@sherb.lin.irk.ru
semovski@lin.irk.ru
Dr. Sempéré A.
INTAS
58/8 Av. des Arts
B-1000 Bruxelles, Belgium
antoine.sempere@intas.be
Dr. Shabanova L.
Ministry of Environmental Protection
Regional Environmental Centre
81, K. Marx Steet
475000 Kokshetau, Republic of Kazakhstan
Lshabanova@neapsd.kz
Dr. Shalovenkov N.
Institute of Biology of the Southern Seas
Department of Shelf Ecosystems
2 Nakhimov av.
99011 Sevastopol, Ukraine
shaloven@ibss.iuf.net
Dr. Sheikholeslami M.R.
Theme for Effective Regional Assessment
of Contaminant Levels "ERACL"
Caspian Environment Programme "CEP"
P.O. Box 15875/5181
Thran IR, Iran
mrnsheikh@yahoo.com
Dr. Sheremet N.
P.P. Shirshov Institute of Oceanology
RAS
36 Nakhimovsky Pr.
117997 Moscow, Russia
sheremet@sio.rssi.ru

Dr. Sheremet V.
Woods Hole Oceanographic Institution
Department of Physical Oceanography
MS#21
Woods Hole MA 02543, U.S.A.
vsheremet@whoi.edu
Dr. Shermatov E.
Institute of Water Problems
Uzbek Academy of Science
25a F. Khodjaev Steet
700000 Tashkent, Uzbekistan
Dr. Sirjacobs D.
University of Liège
Ecohydrodynamics
Sart Tilman
B-4000 Liège, Belgium
D.Sirjacobs@student.ulg.ac.be
Dr. Stanev E.
University of Oldenburg - ICBM
Physical Oceanography
Postfach 2503
D-26111 Oldenburg, Germany
e.stanev@icbm.de
Dr. Staneva J.
Alfred-Wegener-Institute for Polar
and Marine Research
Bussestr.24, postfach 120161
D-27570 Bremerhaven, Germany
jstaneva@awi-bremerhaven.de
Dr. Stanichny S.
Marine Hydrophysical Institute
National Academy of Sciences of Ukraine
2, Kapitanskaya St.
99000 Sevastpol, Crimea, Ukraine
sstanichny@hotmail.com
stas@alpha.mhi.iuf.net

Dr. Stulina G.
Central Asian Scientific Research Institute for Irrigation
B.11, Karasu-4
Tashkent 700187, Uzbekistan
galina_stulina@mail.ru
Dr. Timms B.
University of Newcastle
School of Environmental and Life Sciences
2308 Callaghan, NSW, Australia
ggbvt@cc.newcastle.edu.au
Dr. Tuchin A.
Central Asian Scientific Research Institute
for Irrigation
B.11, Karasu-4
Tashkent 700187, Uzbekistan
Dr. Vilfand R.M.
Moscow Region
Lubertsy
Kirova st. 39, # 49
Moscow
rvilfand@rhmc.mecom.ru
Dr. Vladymyrov V.
National Academy of Sciences of Ukrainia
Marine Hydrophysical Institute
Kapitanskaya st. 2
99000 Sevastopol, Crimea, Ukrainia
vlvlad@yahoo.com
Dr. Wegerich K.
Centre for Development Research (ZEF)
Walter-Flex-Strasse 3
D-53113 Bonn, Germany
wegerich@yahoo.com
kai.wegerich@uni-bonn.de
Dr. Wiseman W. J.
Louisiana State University
Department of Oceanography and Coastal Sciences
and Coastal Studies Institute
Baton Rouge
LA 70803, U.S.A.
wwiseman@lsu.edu

Dr. Wolanski E.
Australian Institute of Marine Science
PMB No.3
Townsville MC
4810 Queensland, Australie
e.wolanski@aims.gov.au
Dr. Yeskendirova G.
Ministry of Environmental Protection
Kazakhstan
Dpyankov@climate.kz
Dr. Zavialov P.
Russian Academy of Sciences
P.P. Shirshov Institute of Oceanology
Nakhimovsky Prospect 36
Moscow 117851, Russia
peter@sio.rssi.ru
peter@lpi.ru

GPSR Compliance

The European Union's (EU) General Product Safety Regulation (GPSR) is a set of rules that requires consumer products to be safe and our obligations to ensure this.

If you have any concerns about our products, you can contact us on ProductSafety@springernature.com

In case Publisher is established outside the EU, the EU authorized representative is:

Springer Nature Customer Service Center GmbH
Europaplatz 3
69115 Heidelberg, Germany

Batch number: 10809439

Printed by Printforce, the Netherlands